Microelectronic Interconnections and Assembly

NATO ASI Series

Advanced Science Institute Series

A Series presenting the results of activities sponsored by the NATO Science Committee, which aims at the dissemination of advanced scientific and technological knowledge, with a view to strengthening links between scientific communities.

The Series is published by an international board of publishers in conjunction with the NATO Scientific Affairs Division

A Life Sciences	Plenum Publishing Corporation
B Physics	London and New York
C Mathematical and Physical Sciences	Kluwer Academic Publishers
D Behavioural and Social Sciences	Dordrecht, Boston and London
E Applied Sciences	
F Computer and Systems Sciences	Springer-Verlag
G Ecological Sciences	Berlin, Heidelberg, New York, London,
H Cell Biology	Paris and Tokyo
I Global Environment Change	

PARTNERSHIP SUB-SERIES

1. Disarmament Technologies	Kluwer Academic Publishers
2. Environment	Springer-Verlag / Kluwer Academic Publishers
3. High Technology	Kluwer Academic Publishers
4. Science and Technology Policy	Kluwer Academic Publishers
5. Computer Networking	Kluwer Academic Publishers

The Partnership Sub-Series incorporates activities undertaken in collaboration with NATO's Cooperation Partners, the countries of the CIS and Central and Eastern Europe, in Priority Areas of concern to those countries.

NATO-PCO-DATA BASE

The electronic index to the NATO ASI Series provides full bibliographical references (with keywords and/or abstracts) to about 50,000 contributions from international scientists published in all sections of the NATO ASI Series. Access to the NATO-PCO-DATA BASE is possible via a CD-ROM "NATO Science and Technology Disk" with user-friendly retrieval software in English, French, and German (©WTV GmbH and DATAWARE Technologies, Inc. 1989). The CD-ROM contains the AGARD Aerospace Database.

The CD-ROM can be ordered through any member of the Board of Publishers or through NATO-PCO, Overijse, Belgium.

3. High Technology – Vol. 54

Microelectronic Interconnections and Assembly

edited by

George Harman

National Institute of Standards and Technology,
Semiconductor Electronics Division,
Gaithersburg, Maryland, U.S.A.

and

Pavel Mach

Department of Electrotechnology,
Czech Technical University in Prague,
Prague, Czech Republic

Springer Science+Business Media, B.V.

Proceedings of the NATO Advanced Research Workshop on
Microelectronic Interconnections and Assembly
Prague, Czech Republic
18-21 May, 1996

ISBN 978-0-7923-5139-9   ISBN 978-94-011-5135-1 (eBook)
DOI 10.1007/978-94-011-5135-1

Printed on acid-free paper

TABLE OF CONTENTS

Session 1.
Packaging and Interconnection Trends - Present and Future

Session 2.
Solder and Flip Chip Interconnections and Assembly

Preface

MICROELECTRONIC INTERCONNECTIONS AND MICROASSEMBLY WORKSHOP

18-21 May 1996, Prague, Czech Republic

Conference Organizers: George Harman, NIST (USA) and Pavel Mach (Czech Republic)

Summary of the Technical Program

Thirty two presentations were given in eight technical sessions at the Workshop. A list of these sessions and their chairpersons is attached below. The Workshop was devoted to the technical aspects of advanced interconnections and microassembly, but also included papers on the education issues required to prepare students to work in these areas.

In addition to new technical developments, several papers presented overviews predicting the future directions of these technologies. The basic issue is that electronic systems will continue to be miniaturized and at the same time performance must continue to improve. Various industry roadmaps were discussed as well as new smaller packaging and interconnection concepts. The newest chip packages are often based on the selection of an appropriate interconnection method. An example is the chip-scale package, which has horizontal (x-y) dimensions $\leq$ 20% larger than the actual silicon chip itself. The chip is often flip-chip connected to a micro ball-grid-array, but direct chip attach was described also. Several papers described advances in the manufacture of such packages.

Wire bonding currently dominates chip to package interconnections, with about 95% of devices so interconnected. However, there was general agreement that some form of flip chip interconnection (currently used on ~1% of semiconductor devices) will rapidly increase its market share in the future, and this was reflected in the number of papers on the subject. Tape Automated Bonding (TAB) was also discussed, but its market share is not expected to grow significantly, being used primarily for special applications. However, such a statement might be challenged by some TAB advocates. The solder metallurgy as well as the production technology of flip chip interconnections were described in several papers. Solderability, reliability, and fatigue of the joints were modeled, and mechanical stresses resulting from temperature cycling were reported.

Multichip Module and thick-film hybrid substrate interconnections were reported in several papers. These included thick and thin film metals, as well as low temperature polymer inks. Several papers described diffusion and other metallurgical interactions in thick film interfaces.

The program was very diverse, covering materials, manufacturing, and reliability issues of chip and substrate interconnections, so that a single, simple conclusion is not possible. However, all attendees left with a broad understanding of the possible interconnection choices and technologies to implement (manufacture) them, as well as with the knowledge of who could help in such activities in the future. In addition to the technical program, a fruitful discussion with the Dean of the Faculty of Electrical Engineering and the Vice Chancellor of the Technical University was included as part of the conference (in Session 5).

THE TECHNICAL PROGRAM

Session 1   *Packaging and interconnection Trends -- Present and Future*
Session Chairs: Karel Kurzweil, Bull S.A. (France), and Soeren Noerlyng, Micronsult (Denmark).

Session 2   *Solder and Flip Chip Interconnections and Assembly*
Session Chair: Pavel Mach, Czech Technical Univ. (Czech Republic), and W. Kinzy Jones, Florida Intl. Univ. (USA).

Session 3   *Single Chip Interconnection and Qualification: Wire Bonding and TAB*
Session Chair: Carlo Cognetti, SGS Thompson (Italy).

Session 4   *Multichip Module Interconnections and Assembly I*
Session Chair: George Harman, NIST (USA).

Session 5   *Discussion with the Dean and Vice Chancellor of the Czech Technical Univ., and People from the Czech and the Slovak ISHM Chapters*
The Lead-off Talk was: New Trends in the Integration and Education in Microsystems Technology: Prof. Juraj Bansky, Technical University Kosice (Slovak Republic).

Session 6   *Multichip Module Interconnections and Assembly II*
Session Chair: Zsolt Illyefalvi-Vitez, Technical Univ. Budapest (Hungary)

Session 7   *Thick Film Interconnections and Metallurgical Interactions I*
Session Chair: Milos Somora (Slovak Republic).

Session 8   *Thick Film Interconnections and Metallurgical Interactions II*
Gabor Harsanyi, Technical Univ. of Budapest (Hungary).

In all, 32 papers were presented from the following 19 countries: Belgium, Bulgaria, Czech Republic, Denmark, Finland, France, Hungary, Ireland, Italy, the Netherlands, Norway, Poland, Slovak Republic, Slovenia, Spain, Sweden, United Kingdom, Ukraine, and the USA. It was felt that the diverse national representation served to produce a balanced program that should be useful to the entire electronics packaging community.

Organizing Committee:

Dr. W. Kinzy Jones, Florida International University, USA
Dr. Karel Kurzweil, Bull SA, France
Dr. Soren Norlyng, Microsult, Denmark
Dr. Sidney Stein, Electro-Science Laboratories, USA

Publication:

Title: Microelectronic Interconnections and Microassembly
Publishers: Kluwer Academic Publishers, Dordrecht, Netherlands.

Session 1
Packaging and Interconnection Trends – Present and Future
Session Chairs: Karel Kurzweil, *Bull S. A., France*, and Soeren Noerlyng, *Micronsult, Denmark*

1.1 The Packaging & Interconnection Trends in the Nordic Microelectronics Industry
Soeren Norlyng, *Micronsult, Denmark*

1.2 **[Paper Not Given]**

1.3 IC Packaging for Miniaturized Consumer Electronics
Co van Veen, *Philips CFT, The Netherlands*

1.4 The Move towards Further Miniaturization
Maurice Sage, *BPA, United Kingdom*

1.5 Plastic Packaged Devices survive where Hermetic Packaged Devices Fail in Severe
Climates
Nihal Sinnadurai, *TWI, United Kingdom*

1.6 Thermal Simulation and Characterization of Single Chip Packages
Orla Slattery, Ciaran Cahill, John Barrett, Martin O'Flaherty and Kenneth Rodgers
National Microelectronics, Ireland

1.7 Current Trends and Future Issues in Solderability
Gary J. Ewell, *The Aerospace Corporation*

Session 2
Solder and Flip Chip Interconnections and Assembly
Session Chair: Pavel Mach, *Czech Technical University, Czech Republic* and Kinzy Jones,
Florida International University, USA

2.1 The At-Temperature Mechanical Properties of Lead-Tin-Based Alloys
W. Kinzy Jones, *Florida International University, USA*

2.2 Flip Chip Technology: Is It Time for Mass Production?
Carlo Cognetti, *SGS Thomson Microelectronics, Italy*

2.3 **[No Paper - Use Abstract]**
Flip-Chip – The Ultimate Solution Comparing Flip-Chip and Chip Scale Packaging
Bill Brox and Katarina Boustedt, *IVF, Sweden*

2.4 **[No Paper - Use Abstract]**
Solder and Solderless Flip Chip Assembly on an MCM-D in a BGA Package
Peter Bodo, *Industrial Microelectronics Center, Sweden*

The Packaging & Interconnection Trends
in
the Nordic Microelectronics Industry

Søren Nørlyng
MICRONSULT
Tulipanhaven 82, DK-2765 Smørum, Denmark
Tel: +45 4465 1457
Fax: +45 4465 1458
e-mail: noerlyng@micronsult.dk

ABSTRACT: The use and production of microelectronics in the Nordic countries has been investigated in a recent study. The survey should clarify which technologies there were preferred and which trends there were seen in Packaging & Interconnection.

The focus on microelectronics activities varies from country to country, therefore the paper will include a listing of some of the national programmes, the national research institutes and the national microelectronics activities in a few selected companies.

1. INTRODUCTION

A recent study concerning microelectronics activities in the four Nordic countries, Denmark, Norway, Sweden and Finland has taken place. The survey was sponsored by ISHM-NORDIC and MICRONSULT and should clarify which services and sub-contractors there were at hand, and a picture should be outlined of which miniaturisation technologies there seemed to be favoured and for which kind of jobs or market sectors. The survey should disclose the present use and the trends in the newest packaging and interconnection techniques.

The ways the various countries have approached technologies and techniques as MCM (Multi Chip Modules), flip chips, TAB (Tape Automated Bonding), Advanced PCB's (Printed Circuit Boards), COB (Chip On Board), PTF (Polymer Thick Film) besides chip design and chip processing, micromachining and micromechanics - could be concluded from the study.

The background for this paper is the very interesting results from the study, where not only "traditional" users of microelectronics and users with miniaturisation on Silicon and SMT (Surface Mount Technology) were visited.

Besides these "users" it was decided to visit the very important National Laboratories to hear about the National Programmes. Finally but equally important was to know, how the educational systems in the four countries have been geared to supply the necessary well-educated technicians and engineers to sustain a continuous development in this field.

2. THE NORDIC COUNTRIES

2.1 General
The northern part of Europe is often referred to as Scandinavia. Correctly speaking, Scandinavia is the peninsula made up of Norway and Sweden, therefore a more covering designation of the countries Denmark, Sweden, Norway and Finland is the Nordic countries. Hence ISHM-NORDIC.

All the Nordic countries are so-called welfare states with high standards of living, a high level of social security - and similar a high level of taxes. The environment is very clean and well protected by many regulations. Denmark, Sweden and Norway are monarchies. Only Finland is a republic.

1

G. Harman and P. Mach (eds.), Microelectronic Interconnections and Assembly, 1-7.
© 1998 *Kluwer Academic Publishers.*

The people move freely between the nations. No passports are needed and people understand each others language - except the Finnish language being completely different from all the others. For that reason the official language of ISHM-NORDIC is English.

The industry is modern and for decades the electronics industry has had a key position from the production of components and consumer goods to the production of specialised niche products in small volumes.

A shift in the direction against a high technology industry based upon a well-educated work force is clear. Some of the Nordic countries have per tradition had a strong focus on this industry, others have invested heavily to make a fast transition into the high technology sectors possible.

The very high salaries have demanded high level of skills, high innovation and high automation to make it possible to survive in a highly competitive world.

As can be seen in the table, the Nordic countries are not equally large. They do not have the same natural resources. This is originally reflected in the size of their various trades and business sectors and the relative importance of these. Even if the four countries in many ways are regarded as equal in social, educational and political systems, within the scene of the electronics industry there are many differences.

Country	Size	Inhabitants	Population density
Denmark	43,069 km^2	5,1 million	119 per km^2
Norway	324,219 km^2	4,3 million	13 per km^2
Sweden	449,964 km^2	8,6 million	19 per km^2
Finland	337,030 km^2	5,0 million	15 per km^2

Table 1: Some statistical data for the Nordic countries.

3. DENMARK

3.1 General

The Danish microelectronics industry is mainly serving the industrial, medical, telecommunications and instrumentation markets and business sectors.

Denmark is characterised with a high degree of openness between the companies working within microelectronics. Very early on Experience Exchange groups were established within the thickfilm and thinfilm microelectronics area. Technical matters of general interest could be discussed in detail. This had saved time consuming experiments by hearing about others experience in materials, processing, equipment and suppliers.

3.2 National programmes

Denmark was remarkable in the lack of national multidisciplinary microelectronics programmes for product development including design, processing, interconnection, packaging and encapsulation to production technology development. No project was covering the whole palette from silicon semiconductors, hybrid circuits to printed circuit boards in the making of functional modules and subsystems for applications with a promising market potential and for the enhancement of the competitiveness of the Danish electronics industry. This situation was changed slightly by the end of 1995 by the start-up of a new packaging project for the interconnection and packaging of silicon micro machined components to ASIC's. The actual application is a transducer system for hearing aids.

3.3 Examples of microelectronics activities

DELTA Danish Electronics, Light & Acoustics is a research institute providing services for private enterprises and public authorities. Besides design expertise of ASIC's, the main microelectronics activities today has been in conductive adhesives for the replacement of solder. But new projects in Flip Chips and Chip Size Packaging will broaden the focus.

For the manufacturing of components as potentiometers and resistive trimmers, chip capacitors and chip inductors and pi-filters, Diplohmatic and Ferroperm Components respectively have found nice niches -

especially for pi-filters, which now have found their way into base stations for mobile phones. Thick film processing is used in both companies. For the manufacture of inductors a special ferrite paste is used.

Another but quite different component manufacturer is Microtronic. Microtronic is a leading supplier of electromechanical components (50% of the world market) and transducers to the hearing aid industry. A thick film like process is used for the potentiometers, but for high precision switches, thin film technique is used.

Thick film is also the technology used at the both Danish sub-contractors or merchant suppliers, Hybrico and Danhybrid. Danhybrid is the outsourced production department from Oticon (hearing aids).

Two of Denmark's largest companies for industrial products as compressors, hydraulics, pumps and motors - Danfoss and Grundfos - are both finding thickfilm technology as the answer to packaging and miniaturisation of their power applications working in harsh environments,. Therefore nice growth rates are seen in ceramic substrates with thick film circuitry and DBC (Direct Bonded Copper). At the opposite extreme of power hybrids, growth is seen in miniaturisation for hearing aids and transducer preamplifiers.

Companies like Widex, GN Danavox and Oticon constitutes the group of Danish manufacturers having 20-25% share of the world-market for hearing aids. Technologies used are thick film, thin film, TAB, COB and component assembly on film or flexible PC Boards. For preamplifiers and signal conditioning for measuring microphones and accelerometers Brüel & Kjær finds that thick film is an excellent solution by solving the needs for miniaturisation and for functional trimming and by offering the required insulation properties through the ceramic substrates.

Big companies as Bang & Olufsen producing audio systems, TV sets and videos, and Dancall Telecom producing cordless phones and mobile phones, are still relying on SMT combined with ASIC's as the answer to their needs regarding performance, price and volume production. But both are carefully following new possibilities in thick film for miniaturisation and for power circuits, in wirebonding and COB (Chip on Board), in PTF (Polymer Thick Film), in flip chips and CSP (Chip Size Packages).

Thick film is the dominating technology. But thin film has a few niches for miniaturisation in hearing aids and in components for these.

PTF has already found many applications in multilayer boards with printed resistors, in potentiometers and in TSO's (Touch Screen Overlays) and keyboards.

A new National Programme will start activities within flip chips and CSP's to solve future requirements for miniaturisation.

4. NORWAY

4.1 General

In contrast to Denmark where the microelectronics activities are scattered all over the country, the activities in Norway seems to be concentrated in three areas: *Horten*, 100 km south of the Capital, Oslo, *Oslo* and *Trondheim* 450 km north of Oslo.

Horten is also called "The Electronic Coast of Norway" with its high concentration of high tech companies. Norway has previously not been able to boast of a huge electronics industry. However new focus combined with a steady need for the technologies developed in the sixties are slowly changing the situation. The driving force has been a combination of National programmes at the research institute, SINTEF, the local universities and the industry.

The newest volume manufacturer is Sensonor producing micromachined sensing elements and sensors for airbags. The production volume and the reliability requirements coupled with the race for reducing costs in the automotive market are calling for intelligent packaging solutions. The latest is a two chip solution without a substrate for interconnection. Adjustment of sensitivity is done by opening electrical links in the chip.

4.2 National Programmes

The microelectronics activities in Norway have all originated from the University of Oslo and its co-operational organ called SI (the Center for Industrial Research). The work was later carried on at SINTEF now one of the very big institutes serving the industry. A three year project "Industrial Microelectronics" was from 1990-1992 half sponsored by the NTNF (The Norwegian Scientific and Industrial Research Council). The total cost was 30 M NOK (4M$). The motivation for the programme was the recognition of microelectronics as a key technology for the performance and market success for the Norwegian electronics industry. The objective was to stimulate the industrial innovation by developing, transferring and exploiting knowledge and methods based upon advanced microelectronics. The focused activities were silicon sensor technology, (silicon micromechanics for sensors and actuators, and silicon radiation sensors) analog/digital design of ASIC's, sensor packaging and thermal management.

4.3 Examples of microelectronics activities

SINTEF has a wafer fab and ongoing work in micro machining. Today this is commercialised for the above mentioned automotive safety product. SINTEF has also a nice industry in the production of silicon photo- and radiation detectors mainly for military applications.

Today two subcontractors are offering their services to the local industry. Amitech with a volume experience by producing large numbers of smoke and fire detectors and MicroComponent serves both the local needs.

AME has typically high-rel customers in military, off-shore and oil exploration using their thin film expertise. For more industrial applications thick film is used and for the highest packaging density (fine lines) combined with multilayer boards, "Combifilm" (thin film sputtered on thick film), is offered. AME Space, a spin-off company, has also in-house thin film deposition for their SAW (Surface Acoustic Wave) filters used in 15 satellites in orbit. The important fishing industry, oil exploration and military industry have expert suppliers in Simrad Subsea making echo sounders and sonars based upon PZT's (Lead Zirconium Titanates) and in Getech producing seismic streamers. Both companies buy their thick and thin film circuits or blanks from other sub-contractors.

Another military supplier is NFT-Ericsson, a joint venture between Norwegian Defence Technology and Ericsson. The focus is here communication equipment. With requirements for mobile phone-like equipment in the military field, new projects are looking for ways to solve the miniaturisation needs either by flip chips, BGA's or MCM's.

Tandberg Data is a market leader in tape streamers. Today SMT can solve the packaging needs but thick film circuits have also been used.

The strong National Programmes have early on started research in Flip Chips and MCM's. Important markets as the military and oil exploration have called upon high-rel solutions which were found in thin film technology. Today thick film has taken over some of these applications and is also used in new industrial products with larger volumes and associated requirements.

PTF is used in the manufacture of keyboards.

5. SWEDEN

5.1 General

Sweden has Ericsson, Volvo and Saab. These giants have had an enormous influence on the microelectronics environment in the country serving the telecom and the military business.

5.2 National Programmes

Sweden has Ericsson to drive and show the microelectronics needs. As Ericsson is not supposed basically to develop and to manufacture everything on their own, their enormous appetite for microelectronics has a spin-off effect in the rest of the industry willing to be at a competitive and competent level serving eventual needs for Ericsson. This has called upon national projects. Of these projects still of importance to the industry is a project from the mid 80'es NMP, (The National Microelectronics Programme) where a complete 286-computer was made in a MCM package. Another sub-project was MLC (Multi Layer Ceramics) where HTCC (High Temperature Co-fired Ceramics) was used.

5.3 Examples of microelectronics activities
The research institute, IVF, is very project oriented, initialising research projects with the industry. As examples of on-going projects can be mentioned Flip Chips, Chip Size Packaging, SLC (Surface Laminated Circuit)-Boards and conductive adhesives.

Wafer fab takes place at three locations serving different needs. IMC is a university-owned research company carrying out market-oriented applied research and small scale production. As an example of their activities can be mentioned MCM-D silicon substrates. The second wafer fab is ABB HAFO - not only a component manufacturer of SOS (Silicon-On-Silicon), silicon gate CMOS and radiation hardened SOS plus a range of optoelectronics components - but ABB HAFO is also offering packaging options and OEM products. Fully tested chips can be bumped for TAB and Flip Chip applications. One of their own products is a thick film thermal printhead with flip chip drivers. The third wafer fab is the Ericsson BiCMOS submicron facility allowing 0,35µm structures. This 100 M$ investment is for prototypes and small scale production for Ericsson products. Volume production can take place at process compatible Texas Instruments wafer fabs.

A special component manufacturer is Astra Tech producing disposable thermometers for hospitals and clinics. Also here thick film technology is used in the printing of temperature sensing resistors with an in-house developed NTC paste.

Of important contract manufacturers can be mentioned Maxitech, specialised in providing just thick film screen printing and laser trimming services. They serve the needs for the telecom market, e.g. Ericsson, and have specialised in copper thickfilm and the printing of resistor networks. Xicon has expertise using organic boards with COB, 0402 passive components and conductive adhesives. Their boards can be extremely small and light. They serve markets as industrial (ignition system for chain saws, electronic price signs for shops), telecom (fiber optic modules), automotive (intelligent Hall effect sensors). The last one to be mentioned here is Combitech Electronics from the SAAB group. They have concentrated on high-rel products e.g. for the military market. They do not manufacture the "substrates" but they have specialised in assembly, C&W (Chip and Wire), COB and component placement. Combitech has produced several MCM-C, -D and -L circuits for both National and EU programmes.

A captive manufacturer for the medical market is Pacesetter. Today Pacesetter is producing the world's smallest pacemaker. The technologies used are HTCC (High Temperature Cofired Ceramic) with wirebonded ASIC's and adhesively joined passive components.
The military sector is served by Bofors Missiles and Celsius Tech Electronics from the same billion-dollar Celsius group. Thick film hybrids are manufactured for intelligent ammunition, missile systems and radars. MCM-C is produced for a memory application.
Ericsson Microwave has traditionally supplied airborne tracking radars, system computers and countermeasure systems. Thin film has traditionally had a strong point for the microwave circuits but today also thick film and LTCC (Low Temperature Cofired Ceramic) is used in the production of MCM-C circuits.
Ericsson is a multi-billion dollar company with 75,000 employees world-wide. Their Microelectronic Systems Group is providing the business groups with technology and application know-how. The same service is now offered for non-Ericsson customers. All microelectronics technologies are mastered. MCM prototypes have been produced.
The business group Access devices have concentrated on thick film technology for the mass production of CLIC's (Complete Line Interface Circuits), Cu-printed power modules for DC-DC converters and finally resistor networks for over-voltage protection on telephone lines. The business is ramping up dramatically which in return has called for new investments in automatic printing lines. Ericsson Mobile Phones produces the world's smallest GSM phone. SMT is used extensively with double sided mounting of large TQFP's (Thin Quad Flat Packs). Flip chips are used in an COG (Chip On Glass) application with display drivers.

The growth rate in the telecommunications sector is influencing the microelectronics market. This is reflected in massive investments in production equipment and it is of benefit to the subcontractors and to the infrastructure for microelectronics solutions. The military and high-rel sector has initiated many initiatives concerning the newest technologies such as MCM's.

The major technology is certainly thick film and the demand is growing rapidly, thanks to the needs in telecom subscriber lines. In addition to thick film all other technologies are used: thin film, silicon, micro-machining, MCM-C, -D and -L. Flip Chips seems to have found some niche applications. BGA's are investigated in national projects for automotive electronics.

6. FINLAND

6.1 General
Nokia is for Finland what Ericsson is for Sweden. This enormous industrial locomotive is in many ways controlling or influencing development, education and the nature and size of the many subcontractors.

6.2 National Programmes
The Finnish National Programmes has had a unique importance to the electronics industry in Finland. The main goal has been to expand and to make the electronics business competitive. A fundamental requirement has been that the results of the programmes should be implemented in real production during the project period or shortly after their closing.
With project managers from the involved companies defining and running the projects and with basic research at the very responsive research institute, VTT, and with partnership-like activities in research and education at the universities - the results have been very impressing.

6.3 Examples of microelectronics activities
VTT has a large wafer fab, micro machining, and wafer bumping for flip chips. They have thin films for electroluminiscent displays, PTF plus thick and thin film technologies. The universities in Oulu and Helsinki have supplementing application oriented research to fill the gaps and for future outlook.
In addition to VTT two other companies have wafer fabs. Micronas offers BeCMOS for mixed signal applications in the communications market with Nokia as the main customer. Micro machining of pressure sensors makes Micronas to the largest independent manufacturer in Europe of automotive pressure sensor systems. Besides the integrated circuits, Micronas has a facility for the production of hybrid circuits for hearing aids, proximity detectors, HF telecom circuits, optical circuits and for low and high power industrial systems.
The other company with a wafer fab line is VTI, now Breed Technologies. This line is used for the manufacture of silicon micromachined sensors.

Component manufacture takes place at VTI mentioned above. Here capacitive sensors for acceleration, pressure and angular rate are produced for the automotive industry. The main application is front and side airbags. Other areas are suspension systems, vibration monitoring, anti-skid braking systems, inclination and manifold absolute pressure sensing. Volumes are increasing due to the European demand for more and more safety devices and for easier and more smooth driving. The present generation uses a sensing element and an ASIC mounted on a thick film substrate functionally trimmed and calibrated. Next generation will be a two chip solution - just as described for Sensonor.
Another very large component manufacturer is LK-Products producing ceramic filters for mobile phone applications. Owned by Nokia the volumes are almost doubling each year. The production is very thick film like, but also thin film technology is used in a sister company making SAW filters.

Aspocomp Microelectronics is the largest hybrid sub-contractor in the Nordic countries. The main customers come from the Nordic telecom business. Picopak is another service provider but within quite a different sector. Picopak offers wafer bumping, TAB, and flip chip assembly. Volume production today is in contact-less smart cards with bumped and flip chip mounted memory chips on flex. Last but not least of the contract manufacturers is Elcoteq. 5 factories makes Elcoteq to the largest in the Nordic area for PCB assembly. Also TAB, heat seal, and flip chips are offered in microelectronics packaging solutions with dies bumped and inner and outer lead bonded.

The captive suppliers include Vaisala, producing products for environmental and industrial measurements in meteorology. Sensors for humidity and pressure are produced using thin film technology, micro machining and special proprietary polymers. Planar International produces high performance electroluminiscent flat panel displays based upon atomic layer epitaxy. They serve markets as the medical, instrumentation, IT (Information Technology) and military. With decreasing pitches in their TAB display drivers, hot seal and anisotropic conductive films are used for the interconnection to the glass substrates.

Nokia Research Center is the corporate R&D body of the Nokia group. They develop and assess new product concepts based upon emerging technologies, and find new and innovative solutions for Nokia's products. A strong multidisciplinary team carries out research projects within all relevant technologies and techniques.

Nokia Mobile Phones is one of the worlds largest manufacturers of mobile phones. They are carefully tracking the most economical ways to mass-produce products fulfilling the customer requirements. Packaging is a keyword, therefore many technologies are evaluated such as MCM's, Flip Chips, Chip Size Packages, PTF, vs. the use of finer and finer pitch TQFP's.

Again for Finland most sectors are served, but compared with Norway and Sweden, the military sector is negligible. Certainly the biggest volumes goes to the telecommunication area.

Thick film is far the most used technique. Direct chip attach and the various joining techniques have very high attention. Two companies are offering bumping service for TAB and for flip chips. Both flip chips and TAB are used. The national programmes and the national needs (Nokia) will to a great extent determine the winning technology. MCM production at present is mainly taking place at the research institutes. Some pilot series have been tested for telecom applications.

With the infrastructure created by the national programmes Finland has a very good starting point for fast reaction to coming new demands or new technologies.

7. CONCLUSION

The trends and a status for the microelectronics activities in the Nordic countries have been presented. The production sectors and markets in the Nordic countries have been described for a broad selection of technologies.

It is shown that the Nordic countries still have a huge interest in hybrid microelectronics for their telecommunications industry, for off-shore, for instrumentation, for automotive specialities, for power applications, for miniaturisation in hearing aids and pacemakers. The National programmes have played a very important role in the development of microelectronics activities and products now commercialised in Norway, Sweden and Finland.

The growing Nordic Microelectronics markets coupled together with new management philosophies resulting in reduced focus on non-core activities will require a base of competent sub-contractors. This could have interest for dynamic subcontractors.

The focus however will not as much as perhaps before only be on the price of the parts produced. More and more important is the total project economy including time-to-market. The future winners of sub-contracting orders will be the very responsive technologically competent suppliers who understands listening to the needs and to react accordingly.

8. REFERENCES

Soren Norlyng, *Nordic Microelectronics. Status & Trends*, ISHM-NORDIC, 1995.

IC PACKAGING FOR MINIATURISED CONSUMER ELECTRONICS

Co van Veen
Nederlandse Philips Bedrijven B.V.
CFT Centre for Manufacturing Technology
PO Box 218, 5600 MD Eindhoven, The Netherlands
Phone: +31-40 2733364; Fax: +31-40-2736815

ABSTRACT

The packaging of ICs for consumer electronics is dominated at present by single chip packaging. Developments of the past years in the single chip packaging field with to potential for miniaturisation in low cost consumer electronics are discussed. It turns out that the new developments as COB, TAB, BGA flip chip and chip size packages all have their limitations, partly from the package concept, partly from the printed circuit board technology. As a result it is expected that regular single chip packaging in the form of QFPs will remain mainstream for a long to come in this particular area.

Keywords: Flip Chip, Chip Size Package, BGA, COB, TAB, MCM

Introduction

The booming expansion of sales in personalised miniaturised electronics puts high demands on the capabilities of process engineers and the equipment they use. The proper choice of packages and the associated mounting technologies is decisive for reaching the so-called zero-defects goal in the assembly process. In the consumer electronics world, a distinction is made between technology limited and non technology limited products. Products with dimensions that are not limited by the size of the electronics are generally made in wave-soldering technology for cost reasons, whereas, (for example) personalised portable electronics, due to their small physical size in relation to the limitations in processability, are made in reflow-soldering technology

In this world of miniaturised portable electronics there is a continuous and fierce battle between the advocates for bare chip mounting technologies and the conservationists, who at any price want to continue on the well known technological road of squeezing the last bit out of the present packaging technology. It is well understood that the former are predominantly represented by the pre-developers, whereas the latter find more backing in the industry, where the actual products, comprising the new technology, are manufactured. In this contribution we will deal with a number bare chip technologies, mostly from the point of view of manufacturing products.

Different bare chip mounting technologies will be discussed: Chip on Board, Tape Automated Bonding, Flip Chip and the solution to the known good die problem which can be found in the "chip size", "chip scale" or near chip size package.

Chip on Board technology

Increasing the number of I/Os of an IC circuit implies smaller pitch of bonding areas on the crystal. With present, common, wire-bonding technology, bonding areas of 50 x 50 μm are feasible, but smaller areas give problems. For the wire bonding of consumer type ICs in leadframes, the pads are typically 80*80 μm at a pitch of 120 μm. The trend is to go down rapidly to 100 μm pitch, whereas in three years time a bondpad pitch of 80 micron is foreseen. Successful wire bonding at these pitches has been demonstrated among others by Kulicke and Soffa. The use of thinner bonding wires (about 25 μm) provides easier bonding, but these wires are more liable to bend against a neighbouring wire (wire sweep), or to cause parasitic interaction. Having more I/O connections means, moreover, more connecting wires, closely juxtaposed, with increased difficulty in the control of the whole bonding process with respect, for example, to lengths of individual wires and mutual distances.

The marriage with reflow soldering is an unfortunate one, since it leads to a mix of technologies. The start with the reflow part allows for printing of solder paste using stencilling. The

9

partly finished product is then subjected to a curing cycle for the die attach, a (plasma) cleaning cycle to eliminate the residues of the reflow process and a curing cycle for the glob. On the other hand if one starts with the COB step, the stencilling of solder paste is no longer possible and one is left with dispensing of solder paste. Furthermore the handling of bare ICs is always connected with the so-called 'Known Good Die' (KGD) problem.

Furthermore the use of COB is hampered by the fact that the obtained degree of miniaturisation, compared with a fine-pitch QFP, is rather limited, and the fact that additional precautions have to take place with respect to the printed circuit board surface (soft thick gold compared to flash gold for reflow soldering).

Tape Automated Bonding

The larger use of TAB is found in LCDs as the carrier for row and column drivers for those cases where only a very small edge of the glass is acceptable for the interconnect. TAB has the advantage that the bare chip is testable in the film. The use of TAB for circuit technological applications is limited to handheld devices, containing an LCD display, like remote controls, pager modules etc. The advantage is the extreme thinness of the package combined with the ability for burn in at the semiconductor plant. The board assembler user is left with the task to cut the component out of the tape, at the manufacturing site. The fact that TAB cannot be introduced in the manufacturing line with integral reflow processing i.e. placing into solder paste, is a great limiting factor. The additional equipment not only for the highly accurate placement but also the simultaneous soldering onto the board, which has to locally supplied with solder, of course, at a cost. The TAB foil occupies a relatively large board area, not much smaller than that of packaged ICs with a similar pincount. It may be expected that for the high pincount applications in stead of TAB the IBM's TBGA or similar packages will be used, as they are compatible with the present reflow technology.

Flip Chip technology

Contrary to COB and TAB, flip chip can be integrated into the fine-pitch reflow line. The interconnection is made of either high melting solder bumps (95/5 lead/tin) on the bondpads and eutectic solder on the footprint of the printed circuit board. Bump pitch can be as low as 175μm.

The bumps are made by evaporation or, as is presently more pursued, by electroplating or even stencilling, provided the bondpad pitch is around 250 μm and up. GE's Delco is exercising wafer bumping through stencilling at 200 μm. Delco is at present according to their own statements the worlds largest user of flip chip technology. The main driving factor is said to be cost. The statement that flip chip is an unpackaged die and that therefor it is cheaper is at least questionable. Already in a zero order approach it is found that the price of bumping the wafer is hardly cheaper than packaging into standard packages like SOs. Furthermore in automotive the flip chip is most probably mounted onto a ceramic substrate, rendering the use and tedious application and curing of underfill for mechanical reliability superfluous.

Low cost flip chip systems in consumer electronics contain either high temperature solder bumps and therefor need a depot of eutectic solder on the board. Specifically local solder is expensive in PCB technology. This system although expensive, guarantees a proper stand-off of the flip chip above the board for the application of the underfill. Specially the flow of the underfill can be hampered into too narrow capillary spaces due to the fact that the filler particles are left behind at the entrance. For real low cost application eutectic solder bumps, made by electroplating or stencilling are used. In this case there is no need to apply solder onto the board. Care has to be taken in the design of solder pad on the substrate in relation to the solder volume in the bump to assure a proper stand-off for the application of the underfill. Although in principle the copper pads can be designed in subtractive technology with the solder resist surrounding it. In practice it is easier to define the pad size through definition of the holes in the solder resist. A consequence of this is that the fiducials for placement have to be defined in an identical way. The surface tension of the molten is the driving force that keeps everything in place. Severe restrictions are put on the local flatness of the substrate at the flip chip location, furthermore this flatness has to be maintained during the reflow process, leading to requests for a substrate material with high Tg. The latter is also advantageous for the ability of fast curing of the underfill.

The thermal mismatch between IC die and substrate is proportional to the die dimensions. With increasing size, the reliability under thermal cycling will be a dominant factor, and the choice of suitable substrate will be the major issue. With ever decreasing die sizes, the dissipation will be the ultimate limit. Placement of the flip chip is done in a sticky flux, which at best, is supposed not to leave residues. In other cases, cleaning is necessary for the underfill, which is applied after soldering. Care has to be taken that little or no voids occur in the cured underfill. If the pitch on the IC is larger (300 μm and up), the solder on the printed circuit board can be applied by stencilling even in the regular solder paste application.

As with COB, flip chip generally is hampered with a deficient knowledge of the 'goodness' of the die. IBM does wafer testing of the bumped wafers, after which they are sawn and packed in either tape or waffle tray.

There is at present still a great reluctance by semiconductor manufacturers to release wafers for bare chip mounting applications. Requests to deliver bumped wafers to external customers are seldom honoured. Also manufacturers who want to deliver bumped dies will have to set up their own bumping facility, as at present no reliable high volume subcontractors for bumping services have been identified. This implies that flip chip at present is limited to a few larger vertically integrated companies, who have the complete processing in their own hands. This will also be lethal to the concept of flip chip as it will effectively prevent the spreading of this technology to small and medium size companies.

Chip Size Packages

In recent developments, the total size of the BGA package is reduced to about that of the IC die itself, by which means a 'chip-size package' is obtained. This package has characteristics in common with a flip chip, but the die-testing problem (KGD) has been solved, as the die can be adequately tested in the package. As can be seen from table 1, this type of IC package compares favourably with respect to fine pitch QFPs, with respect to occupied area and assembly weight (IC+package+substrate) in a ratio of about one to three or four. Furthermore the electrical performance of flip chip constructions is excellent due to short interconnects.

In a period of one to two years time, a lot of companies have published work on chip size packages. These packages make the miniaturisation achieved by flip chip assessable to everyone. Companies like Tessera, Sandia and Aptos, Mitsubishi, Matsushita, 3M, Toshiba, Hitachi, Shinko, NEC, Shellcase, Micro-SMT, Fujitsu, Motorola and Amkor Anam, all have come with their own or have participated or have licences in chip-size packaging activities. Vardaman and Crowley of TechSearch International have in their report on Chip Size Packages made a distiction into five different categories: Flex circuit interposer, rigid substrate interposer, transfer moulded, custom leadframe (for memories mainly), and wafer level assembly.

The packages solve the 'known good die' problem, and are therefore very well acceptable for semiconductor manufacturers. On the other hand they very well fit into the regular reflow line as a normal SMD component. However, only if they are equipped with sufficient strain relieve for the bump-

Package	pitch [mm]	substrate [mg]	die [mg]	package [mg]	weight % die
PLCC 68	1.27	1450	160	4500	3
VSO 40	0.76	450	35	600	3
QFP 80	0.80	1100	80	1300	3
SOL 24	1.27	400	32	500	4
BGA 169	1.27	700	100	1500	4
SSOP 24	0.50	100	17	200	5
QFP 80	0.50	400	80	500	8
COB	0.30	370	80	400	9
Flip Chip on FR4	0.30	100	80	10	40
μBGA	0.50	100	80	10	40
Flip Chip on flex	0.30	12	80	8	80

Table 1: Weight of packaged die assembly. All weights are indicative only and given in milligrams. All dies are taken to be 0.6 mm thick and the substrate thickness is variable, depending on the package type.

interconnect, rendering underfill superfluous. This appears only to be the case for the present version of the μBGA as developed by Tessera. This has been recognised by electronic industry and by subcontractors specifically. Amkor Anam , Shinko and Hitachi have taken licence on the Tessera technology.

Another point of attention is the fact that generally consumer ICs have small dimensions and the Tessera concept starts to become effective for ICs having dimensions of 4 mm and up. In practice this renders the technology out of reach for 80 to 90 percent of the consumer ICs.

Related to the array concept of the BGA, it is nearly always necessary to mount it on a multilayer board to be able to connect all its I/Os to other components. This makes the entire configuration at present too expensive for tabletop indoor consumer electronics applications.

Furthermore the Tessera concept will require filled and tented vias in the board at a pitch of 0.5 mm, which although technically feasible, drives the price of the boards at present too high for consumer applications. Therefore we foresee that the μBGA concept will be used for those low pincount applications, where the pitch can be released to 0.75 mm, in order to allow for surface routing.

Market potential

In a recent study (October 1995) BPA has projected the number of chipsize packages for the year 2005 at 3 billion compared to 6 billion flip chips, 2 billion BGAs and 20 billion PQFPs on a total of 100 billion packaged ICs. In their study they find a tendency in CSP towards higher pincounts, whereas the lower pincounts have better coverage from flip chip.

Subcontractors

Amkor Anam, worlds largest subcontractor for packaging, plans to ramp up the production for the μBGA to 250 K/wk or as needed by market in the third or fourth quarter of this year. Shinko, another major Japanese subcontractor who also has taken the Tessera licence, reports potential μBGA customers in Japan for various applications ranging from audio-visual, notebook computers, telecom till D-RAM with quantities from 100 K till 1000 K pieces/month. Pincounts range from 32 till 240.

Tessera will install equipment at the end of '95 and will package 20000 units/day by the end of 1996.

According to Tessera the pricing of the mBGA will drop from \$0.02 / IO in 1996 to \$0.0074 / IO in 1999. This compares well with the price characteristics of regular BGAs, which at present are offered at \$0.0167 / IO for a 225 pin package.

BGA

After many years of developments it appears that BGA for single chip packaging is finally taking off. For miniaturisation aspects, BGA has to compete with QFPs with pitch 0.5 and even 0.4 mm. The argument of the ability for inspection withstood the advent of this package for long time. QFP is a technology which has ripened over the years and is therefor hard to beat by any new technology. However, there appears to be a general understanding that BGA becomes effective at pincounts of 200 and above. This puts the BGA out of the scope of attention for consumer applications as for consumer applications the typical pincount equals twenty to thirty and only seldom surpasses hundred. For DRAM a BGA packages has appeared with 119 IOs and only recently during the MCM conference in Denver news broke that Amkor Anam works with a number of harddisk manufacturers to introduce a miniBGA with typically 80 pins or less for prices which are between a QFP and the regular BGA.

The future

Based on the aforementioned we assume that in the near future single chip packages will predominate the packaging world for consumer electronics. Flip chip and Chip Size packages will remain in the focus of attention. Work will be invested in the development of printed circuit boards which are better than today suitable for accepting flip chips and chip size packages. Whether this will be at the level of an MCM-L or that the technology can be made so low cost that it will be affordable for motherboard applications can not be foreseen at present.

Literature

Meehan; Delco electronics; Area Array Packaging Workshop; November 1995, Berlin, Germany

Vardaman and Crowley; Techsearch International; Austin TX

BPA Ltd.; Known Good die and Chip Scale Packaging; Dorking;Surrey; U.K.

THE MOVE TOWARDS FURTHER MINIATURISATION

By

M G SAGE

BPA (Technology & Management) Ltd

BPA House, 250-256 High Street, DORKING, Surrey, RH4 1QT, UK
Tel: +44 1306 875500   Fax: +44 1306 888179

Currently 40% of all electronic products are portable and this figure will reach nearly 60% by the year 2000. While many of these products contain only a limited amount of electronics eg smart cards, the growth in portables eg phones, laptop computers, PDAs is influencing a major shift to new component technologies and their interconnection and packaging.

This move to portable electronics is the third stage in the move from firstly standalone to secondly desktop products and is further complicated by the amalgamation of communications and computing capability. The implications for interconnection and packaging are covered in this paper.

MAJOR CHANGES

The electronics industry is being complicated due to the integration of the traditional market sectors of Telecommunications, Computer, Communications and Consumer opening up a whole new range of products and services. Already we are seeing the integration of handheld computers into portable phones creating a powerful combination of data processing and person to person communication. At the same time there is a strong move to portability which demands further miniaturisation in electronic hardware; how significant is this trend? In order to evaluate the hardware technologies used in various products BPA has created a series of roadmaps that cut across the traditional market sectors, see Figure 1.

13

G. Harman and P. Mach (eds.), Microelectronic Interconnections and Assembly, 13-19.
© 1998 *Kluwer Academic Publishers.*

Figure 1

GLOBAL TECHNOLOGY ROADMAPS

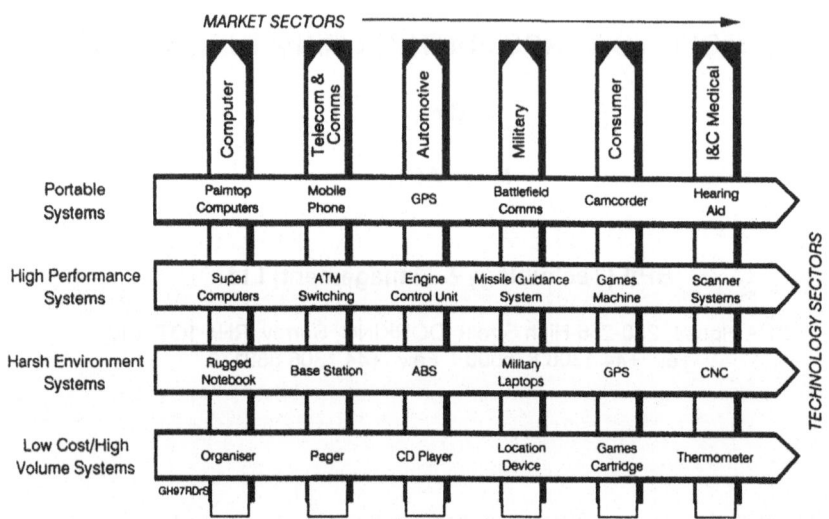

From the technology groups selected the number of portable products emerged as the largest. This included smart cards, hearing aids etc but even so the number of products such as laptops, phones, camcorders, were not insignificant. From the forecasts of portable growth the year 2000 indicated that around 60% of all end products would be in this category, see Table 1 below.

Table 1

TOTAL EQUIPMENT QUANTITIES BY
GLOBAL TECHNOLOGY ROADMAP SECTORS 1995-2005

Roadmap Sector	1995		2000		2005	
	M Units	%	M Units	%	M Units	%
Portables	1288	41.4%	3539	55.6%	6007	61.8%
High Performance	426	13.7%	732	11.5%	875	9.0%
Harsh Environment	246	7.9%	579	9.1%	1118	11.5%
Low Cost High Volume	1150	37.0%	1515	23.8%	1720	17.7%
Total	3110	100.0%	6365	100.0%	9720	100.0%

Source: BPA   [GJ81RR-I]

Note: Based on BPA's analysis of shipments by the various
                         historic methods of electronic industry shipments analysis
                         eg EPD, Comms etc and other data

IC PACKAGING

An analysis of integrated circuit packages gives an indication of the changes that are taking place, see Figure 2 below. While surface mount dominates the current period direct attach types are starting to emerge. Direct attach devices include chip-on-board, tape automated bonding and flip chip.

Figure 2

WORLD IC PACKAGE TYPES 1995-2005

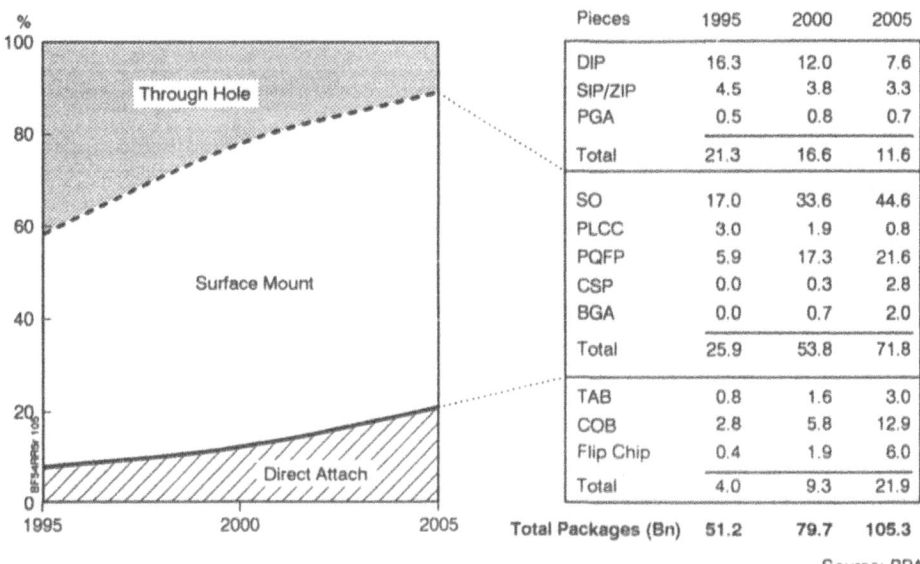

Pieces	1995	2000	2005
DIP	16.3	12.0	7.6
SIP/ZIP	4.5	3.8	3.3
PGA	0.5	0.8	0.7
Total	21.3	16.6	11.6
SO	17.0	33.6	44.6
PLCC	3.0	1.9	0.8
PQFP	5.9	17.3	21.6
CSP	0.0	0.3	2.8
BGA	0.0	0.7	2.0
Total	25.9	53.8	71.8
TAB	0.8	1.6	3.0
COB	2.8	5.8	12.9
Flip Chip	0.4	1.9	6.0
Total	4.0	9.3	21.9
Total Packages (Bn)	51.2	79.7	105.3

Source: BPA

The three groups of integrated circuits can be largely grouped into insertion, surface mount and area array, as shown in Figure 3. Area array types are already in use as ball grid arrays (BGAs) with micro BGAs sometimes called chip scale packages (CSPs) - being only 10% larger than the die. CSPs are starting to be produced in significant numbers with a 288 bump 0.5mm type already in use in a portable phone. The use of these very small packaged ICs will place new demands on substrate interconnection density.

<u>Figure 3</u>

THE MOVE TO PORTABILITY

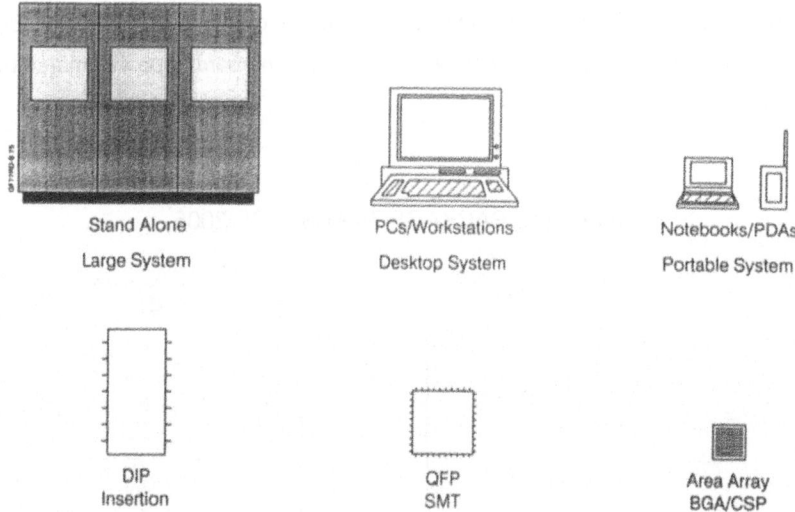

| Stand Alone | PCs/Workstations | Notebooks/PDAs |
| Large System | Desktop System | Portable System |

| DIP | QFP | Area Array |
| Insertion | SMT | BGA/CSP |

MULTICHIP MODULES (MCMs)

A great deal of publicity has been given to MCM technology over the past ten years. They have been generally defined as more than one bare die (or near bare die) mounted on a high density substrate. Typically they have been `islands' of high performance in mainframes, and workstations, but now the technology, (as overall interconnection and packaging), is finding increasing use in portable products such as phones and PCMCIA cards. The forecast of MCM technology (including silicon) is conservatively forecast to be $23 B by the year 2000 and will continue to grow rapidly and find use in all forms of electronic products.

SUBSTRATES

The drive to further miniaturisation, particularly for portable products, is demanding more components per unit area which combined with direct attach ICs and a smaller board, gives a significant increase in the required interconnection density. Current substrate technology such as the PCB will find it increasingly difficult to reach the high density interconnection demands both from a technical and cost point of view. Hole diameters are the major blockage to higher density and substitutes for mechanical drilling, using photo imaging, lasers and plasma etching are coming into use with a growing number of different buildups.

It is of note that nearly all the major developments and use of the new substrates is taking place in the system houses, due to the recognised need, capital investment and development time; the latter averaging about five years. One of the many new substrates that illustrates the degree of change taking place is the ALIVH technology from Matsushita. This board is designed to maximise component count and interconnection density by moving more surface interconnect inside the substrate. Buried vias are made by lasers with copper paste infill, with post lamination, circuit formation and blind vias; an innovative development, see Figure 4.

Figure 4

"ALIVH" BY MATSUSHITA
["ALIVH": Any Layer IVH Structure Multilayered PCB]

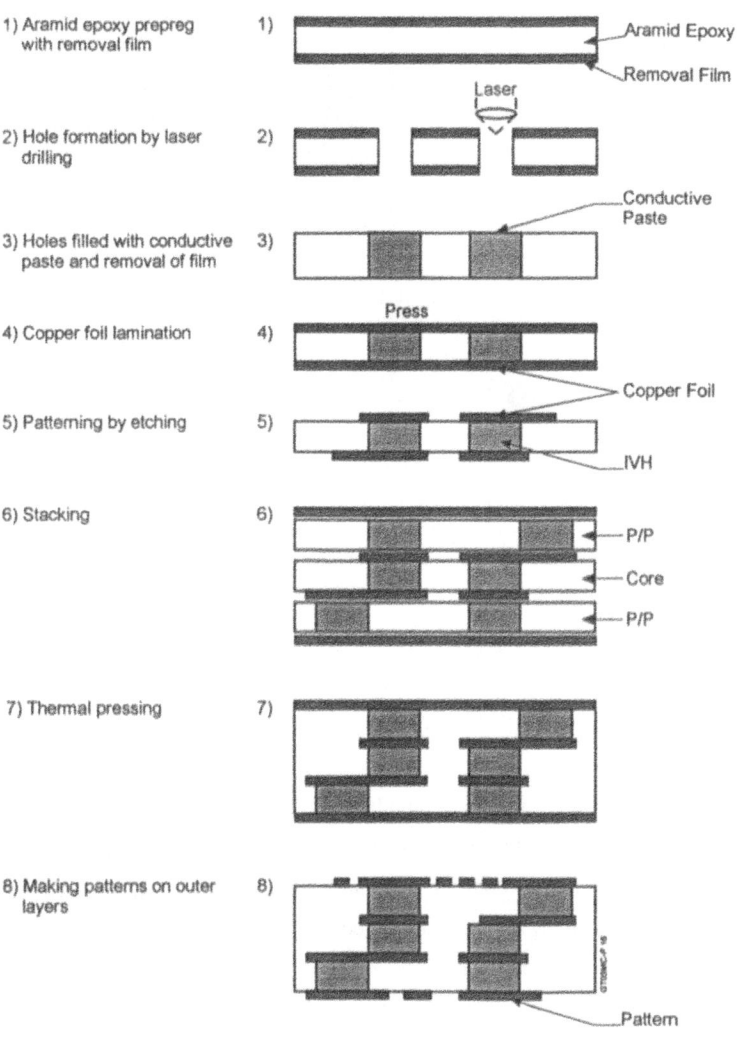

1) Aramid epoxy prepreg with removal film

2) Hole formation by laser drilling

3) Holes filled with conductive paste and removal of film

4) Copper foil lamination

5) Patterning by etching

6) Stacking

7) Thermal pressing

8) Making patterns on outer layers

Approximately thirty new substrate technologies are in development and only a small number of these will survive. Cost is a major driver and forecasts for the new substrate technologies are given in Figure 5. The move to a new substrate industry in terms of technology and structure has started.

Figure 5

COST COMPARISON

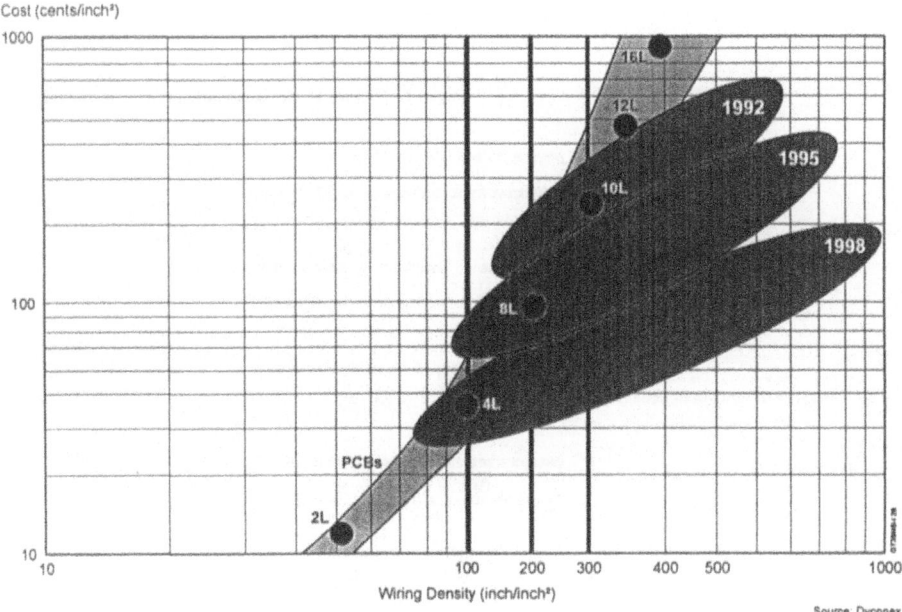

CONNECTORS

The new interconnection and packaging demands from ICs and substrates will be reflected in new connector technology and already we are seeing flexible and elastomer solutions providing both edge and area connections. The use of palladium dendrites is providing an interesting interconnect button. Developed by IBM this form of connector could well find major use to meet the new interconnection and packaging needs.

CONCLUSION

The integration of the traditional market sectors eg telecom and computer, combined with the trend to portable electronic products is creating a demand for further miniaturisation. The effect on all components and their interconnection and packaging is significant. New design tools and assembly methods will also be required to ensure success with what can be identified as the third era of the electronics industry.

PLASTIC PACKAGING IS HIGHLY RELIABLE

Prof. Nihal Sinnadurai, Senior Member IEEE, Fellow Institute of Physics
Principal Consultant, Electronics Technology and Reliability,
Hill House, 83-87 Anglesea Road, Ipswich IP1 3PJ, England .
[Fax:(+44) 1473 211576]

ABSTRACT

Plastic encapsulation is now a highly reliable method of packaging active semiconductor devices and microelectronics in general. In this new work, tropical climates have been analysed and shown to be significantly more severe than those safeguarded by current hermeticity and accelerated test standards in Mil-Std-883, Test Methods & Procedures for Microcircuits. Field failure returns have revealed that Mil-Std-883 approved hermetic CerDIP (ceramic dual-in-line) packaged ICs, installed in digital switching systems, failed catastrophically with a failure rate of 1755 FITs. The cause of failure was severe ingress of moisture, resulting in dewpoints up to 30°C. Alternative indigenous developments of modular digital switches for widespread rural use in India have incorporated plastic encapsulated components selected according to the criteria developed from earlier extensive and successful reliability work by British Telecommunications (BT). Such criteria include the use of HAST (highly accelerated stress test, invented at BT Labs). The pioneering work at BT Labs demonstrated that commercial plastic packaged devices from certain sources were more reliable than their hermetic counterparts, and that both plastic and hermetic packaged devices should be subjected to similar environmental tests, including HAST, in order to separate the robust from the vulnerable packages.
Other researchers have since confirmed that polymers endure as high reliability materials for packaging and interconnecting microelectronics, including pioneering work in the USA by: the IEEE Gel Task Force, AT&T, CALCE University of Maryland, as well as continuing work by BT Labs on optoelectronics, and analyses of MCM applications for satellites. The evidence is that plastic encapsulation is now a very reliable and cost-effective option.

1. THE NEED FOR A PACKAGE

The history of integrated circuits contains many chapters of unreliability arising from the exposure of the integrated circuit chip to harmful effects of the external environment. Early problems with gel filled packages led to the defined need for hermetic packages, and detailed reliability physics studies identified a hermeticity requirement of 5×10^{-11} millibar.litres/sec. However the inability of commercial leak testers to measure such low leak rates caused the hermeticity specifications to be set at much higher levels, which have progressively been tightened to the present level of 10^{-8} millibar.litres/sec. The inadequacy of such a requirement is demonstrated later in the paper. Meanwhile the inevitable pressure for cost reduction and increasingly automated manufacture, led to the development of plastic encapsulation by transfer moulding and other methods. Early suspicions that contaminants in the plastic and the permeation of moisture would cause failures were amply confirmed and led to an enthusiastic condemnation of the use of plastics by defence and telecommunications reliability authorities. The condemnation was entrenched in military specifications for two decades without contemporary evidence. Meanwhile, non-military organisations proved and adopted plastic encapsulated microelectronics (PEMs) and have benefited commercially. Today more than 85% of IC packages in the world are plastic and many are used in high reliability applications, with considerable evidence not only to support the high reliability obtainable with plastic encapsulations, but also the increasing evidence of many field failures of hermetic packages constructed and evaluated to Mil-Std-883 hermeticity standards.

G. Harman and P. Mach (eds.), Microelectronic Interconnections and Assembly, 21-32.
© 1998 *Kluwer Academic Publishers.*

2. LABORATORY EVIDENCE OF RELATIVE RELIABILITY OF HERMETIC AND PLASTIC PACKAGED DEVICES

2.1 Pioneering Work

Reliability engineering of plastic encapsulation was pioneered during 1966-1985 at British Telecommunications Laboratories (BT Labs), England, with the active participation of major plastics materials suppliers. The evaluations were facilitated by the invention of the non saturating autoclave (later named "HAST") reliability assessment technique[1] at BT Labs in 1968. The high reliability capabilities were published in the early eighties, and plastic encapsulation was adopted for terrestrial telecommunications use at that time. The outcome of the work over many years, led to two developments:
(i) proven relationships for accelerated ageing by damp (with non-saturated water vapour) and heat within distinct validation limits for systematic acceleration[2].
(ii) strong evidence that discrete devices junction coated with particular plastic encapsulants were highly reliable[3].

2.2 Review of Pioneering Work

The first experiments which produced convincing evidence of high reliability were conducted on test vehicles comprising an assembly of npn and pnp transistors and specially designed moisture sensors, assembled onto ceramic substrates[3] (Figure 1). These were then coated with one of 15 different plastic coatings. The test vehicles were then subjected to high temperature and separate high humidity tests with electrical bias applied to the devices. Evidence from over 4000 hours of overstress testing on 500 test vehicles (i.e. 2 million device hours) showed that reliability equivalent to 25 years in tropical climates was achieved by four silicone plastic encapsulations, while most of the other materials continued to be hazardous to device reliability. A second series of experiments used the better junction coating materials from the first experiment, selected for their compatibility with a range of top coatings which were added to provide "screwdriver" protection, i.e. protection from mechanical and handling damage. The results from the dual coated test vehicles showed that, of the materials tested, two combinations - a silicone resin mechanically protected by a filled silicone, and the silicone resin protected by a filled phenolic - demonstrated high reliability also equivalent to over 25 years in tropical climates. On the basis of such work, British Telecommunications adopted the use of plastic encapsulated integrated circuits (ICs) in high reliability applications and opened the door to the acceptance of plastic encapsulations for high reliability applications. Subsequent publications have confirmed the promise of the reliability of PEMs[4,5,6,7,8].

The identification of high reliability plastic coatings led to the invention of a low cost, high performance, plastic chip carrier given the trade name EPIC (Figure 2) which was fabricated by standard printed wiring board techniques[9]. Reliability assessments of the EPIC were carried out alongside other commercially available packages containing the same species of IC[10]. An example of the test vehicles subjected to damp heat tests is given in Table 1. The results of both dry heat with bias and damp heat tests with bias were that commercially manufactured surface mount micropackages from three sources were capable of achieving high reliabilities in both temperate and tropical climatic conditions, while there still remained a number of suppliers unable to achieve adequate reliability even in temperate climates (Figure 3). It is particularly important to note that the silicone junction coated ICs and one manufacturer's plastic moulded small outline IC package (SOIC) outperformed the CerDIP "hermetic" packaged ICs. The mechanism of chemisorption of silicones onto the silicon passivation surfaces was identified as the adhesion and protection mechanism with silicone encapsulations, while Van der Waal adhesion forces provided the adhesion mechanism for epoxy encapsulations. The fact that hermetically packaged ICs failed in the high humidity tests clearly showed up the false assumption that Mil-Std- 883 tested hermetic packages do not need to be subjected to humidity tests and forewarned of problems to come (the vulnerability of Mil-Std-883 "hermetic" packages to external moisture is described later in the paper). Therefore, all packages must be subjected to all the tests if they are to prove their ability to survive in the climates represented by the accelerated test conditions (the accelerated test conditions are described later in this paper). The reliability tests were accompanied by full failure analyses and showed weaknesses in some of the plastics materials, such as ionic and halide contamination, and also the excellent and intact survival of the silicone junction coated ICs. The reliability tests also established the credentials of the BT specification M219F[11] for Type Approval of Encapsulating Resins, which was devised and improved into a simple set of materials properties relevant to the reliability interaction of plastics and microcircuits.

TABLE 1. EXAMPLE OF RELIABILITY TEST PROGRAMME

HIGH HUMIDITY TESTING
IN NON-SATURATING AUTOCLAVE (108°C, 90%RH)

Source Code	Batch Size	Package Type	IC Type	IC Source	Metalisation + Passivation
A	40	SOIC	741	B	Al + Oxide
	40	SOIC	748	B	Al + Oxide
D	40	SOIC	741	B	$Ti/Pt/Au + Si_3N_4$
	40	SOIC	748	B	$Ti/Pt/Au + Si_3N_4$
P	40	SOIC	741	E	Al + Oxide
F	40	SLCC	741	B	Al + Oxide
J	40	SLCC	741	B	$Ti/Pt/Au + Si_3N_4$
N	20	SLCC	741	L	Al + Oxide
	20	SLCC	748	L	Al + Oxide
T	20	3LCC	348	V	Al + Oxide
U	20	EPIC	348	V	Al + Oxide
W	20	3LCC	3096	Y	Al + none
X	20	EPIC	3096	Y	Al + none
M	20	CerDIP	741	L	Al + Oxide
	20	CerDIP	748	L	Al + Oxide
Z	15	EPIC	348	V	Al + Oxide

In the USA, evidence of the inadequacy of military specifications and the reliability of plastic compared with hermetic packaging has been produced by Condra et al of CALCE, University of Maryland[12]. They conducted studies of simple logic circuits and showed that the reliability achieved in temperate climatic conditions is three times greater than that predicted for military parts. This study was reinforced by the findings from an experimental programme to compare the reliability performance of similar ICs packaged in hermetic and plastic packages under steady and cycled damp heat representative long temperate lifetimes, which showed that hermetic packages had no lifetime advantages over plastic packages. In exploring opportunities for soft encapsulations, Wong of Bell Labs[13] undertook comprehensive materials analyses and showed that silicone gels possess remarkably useful properties to safeguard high reliability integrated circuits. The IEEE Computer Society "Gel Task Force" with representatives from 24 companies, pursued this opportunity and followed the earlier BT initiatives by undertaking evaluations of polymer gel coatings for integrated circuits[14]. Some 1440 IC chips containing specific test patterns and protected by different glassivations, were tested with five gel types and one silicone RTV. The test patterns were intended to reveal simple corrosion of interconnections and breaks in the wires. The evaluation tests comprised thermal shock, salt spray and HAST. The outcome was that the more rigid and thicker coatings caused wires to be broken. It was clear that thick silicone RTV coatings, applied without dilution, caused strain and damage to the wires (Figure 4). This was consistent with the recommendation from the original work[3] that junction coatings should be applied thin, barely to cover the wirebonds on the chip, i.e. about 25 microns thick. Three of the thinly applied gels did achieve good protection of the test vehicles (Figures 5), the test vehicles without any glassivation over the metalisations performing the poorest of these. The Task Force undertook no determination of the physical or chemical adhesion or protection mechanisms of the encapsulations, but did establish that gel coated devices would withstand acceleration forces up to 15000 G.

More recent work by BT on low cost packaging for optoelectronic components[15] adopted the observations from the IEEE Task Force and the previous work at BT[3] and proceeded with an evaluation of the use of the better gels as reported by the Task Force, applied to PIN and laser diodes and GaAs ICs. The accelerated ageing conditions and the acceleration factors to predict longevity were based on the earlier work[3]. The PIN diodes assembled onto "light-to-logic" circuit substrates and subsequently junction-coated and tested at 85ºC/85% RH temperature, humidity and bias (THB) conditions, showed that three of the four gels provided excellent reliability protection (Figure 6) equivalent to many decades of longevity, provided that the acceleration factors for junction-coated silicon devices from the earlier tests were applicable. Under the same overstress conditions, the lasers showed inconsistent behaviour (Figure 7), with the degradation occurring cumulatively in the small population of devices. The investigator was encouraged by the survival of 3 of the 4 gel coated lasers beyond 3500 hrs. The damp heat tests on gel coated GaAs ICs showed a remarkable stability of the ICs up to 6000 hours (Figure 8). These encouraging observations led to the development of low cost assembly of an optoelectronic circuit board with the individual active devices (PIN, Laser diode and GaAs logic ICs) joined using Light Cured Resin (LCR) adhesives based on oligomeric aromatic methacrylate (OAM) encapsulated in Wacker 905 silicone gel.

2.3 Polymer materials in Space

As already stated, silicone and some epoxy encapsulations of semiconductor dice provide high reliability protection not only for longevity in temperate climates but also in tropical climates. This reliability benefit can be extended also to the high reliability demands of satellite communications[16]. Space adds additional hazards, arising from launch and manoeuvre stresses and from radiation. Nevertheless, plastic encapsulation by the use of junction coatings (compliant thin coatings directly applied to the surfaces of the dice) offer very high protection of semiconductor devices. Glob Top is an unfortunate terminology, because it has been associated with low technology low reliability applications. While the major providers of space vehicles adopted the ponderous and costly route to component approval, such a luxury was not available to amateur communicators. Nevertheless amateur radio does require communications satellites that are dependable and expedient but not expensive. This has led to an innovative path to component and system approval[17] and a determined appraisal and adoption of low cost components including plastic packaging. The Amsat approach is to "Hit the middle of the line" i.e. concentrate on robust design, reliability commensurate with cost, gain experience over the whole line and build up a database of experience. As a result of their rapid learning curve and experience, currently there is in excess of 60 satellite years of orbit experience with amateur satellites and UoSat-OSCAR-11 remains in operational use after 11 years in orbit.

The potential for applications in space arises because reliable dice can be protected by an intimate application of a junction coating. There are now many excellent silicones which form intimate bonds with the surfaces of semiconductor dice by chemisorption[22]. This requires the surfaces of the dice to be chemically clean, which can be achieved by either wet processes or ideally by a final plasma clean[22]. Application in vacuum facilitates intimate contact. Coatings must be dispensed by controlled release from a syringe, allowed to level out under gravity and then thoroughly cured, which means beyond the manufacturers' recommended curing times (which are defined for a high percentage of polymerisation). Effective curing times are, e.g. 150°C for 30 mins + 3 hours at 180°C. Such a coating achieves an intimate and robust seal which will safeguard the existing reliability of the dice against external climatic attack. For space applications, the encapsulation adds mechanical support of wire or TAB bonds, and adds the major advantage of attenuating particles which can give rise to soft errors. α particle activity of silicones is very low, of the order of 0.002 per cm^2 per hr for 1-8 MeV energies. On the other hand, silicone effectively attenuates α particles, for instance, resulting in zero soft errors in 10^6 device hours of operation, compared with 200 soft errors in 10^6 device hours if the silicone protection was not present[22]. As α particles can be sourced also from alumina substrates and lead based solder terminations, the silicone adds versatility to the range of materials that may be used in the assembly of the MCMs.

The options for MCM materials for use in space are still wide open. The payload advantage currently remains with silicon substrates using thin film multilayers with either SiO_2 or polyimide dielectrics. Silicon substrates also provide advantages of heat removal. Nevertheless, a number of low dielectric constant polymers are providing exciting opportunities for high performance. The Radiation Hardness of a number of MCM polymer materials have been tested to differing degrees of severity. For instance, polyimide has been found to retain essential properties with dosages in excess of 100 MRads of γ radiation, while benzo-cyclo-butane (BCB) materials have been tested successfully with dosages in excess of 100 MRads of γ radiation[16]. Therefore there are few fears for the inherent useability of polymers in satellite communications applications [16].

3. FIELD EVIDENCE OF THE RELIABILITY OF HERMETIC AND PLASTIC PACKAGED COMPONENTS

3.1 Failures of installed telecommunications equipment in India due to hermetic packaged IC failures.

The major investment by the Indian Department of Telecommunications (DoT) in acquiring digital switching systems for the planned expansion and modernisation of their telecommunications network was achieved by importing systems and by large scale technology transfer. The digital system was supplied as a mature technology using components, including mandatory hermetically packaged ICs, to MIL 883. However, no attempt was to adapt the technology to the conditions of supply and use in India. Fortunately, an United Nations Development Programme (UNDP) expert advisor scheme was set in place at the same time, which led to an effective scheme being set up in the city of Bangalore for reliability assessment, field failure data retrieval, failure analysis, and analysis of the climatic conditions relevant to the use of electronics in India.

DoT set up SCRAM (System for Component Reliability Analysis and Measurement)[18]. Controlled retrieval of information from the installed base of digital telephone exchanges analysed by SCRAM, has produced the distribution of failures by component type shown in Figure 9, in which it can be seen that the predominant contributor to failures were ICs, followed by hybrid modules, transistors and capacitors. The data was further analysed into classifications of defects for all components, and an example is shown for ICs (Figure 10). For the period in question, the IC failure rate alone amounted to 1755 FITs (1 Fit = 1 failure in 10^9 device hours) in a population of 2.35 million ICs.

Some 10,000 components including 2000 ICs had failed over 5 years and were retrieved by the DoT failure analysis laboratory. Failure analyses of almost 4000 devices confirmed that the IC hermeticity met the Mil-Std-883 criterion of 10^{-8} millibar.litres/sec with only 3 exceptions. The analyses also showed that the dewpoints within the packages ranged from -20°C to +30°C, corresponding to relative humidities that will have reached as high as 95% during operation. The failure mechanisms of the active devices (ICs, hybrid microcircuits, transistors, diodes) were distributed as shown in Figure 11. Apart from those failing due to electrical overstress, attributable to a poorly suppressed electrical environment, the most consistent failure mechanism observed was corrosion of the aluminium metalisations.

The analyses of the devices and the environment showed that, after manufacture, the packages gained moisture sourced from the high levels encountered during transport and storage in India, entering through the "acceptable" leaks in the package body. The moisture thus retained within the package will have started a train of chemical corrosion even during storage, as observed during the earlier BT work. The increased humidity as the equipment was moved into an air conditioned environment and the presence of bias when the equipment was electrically operated will have accelerated the corrosion as the equipment was brought into service.

The manner in which hermetic packages can gain moisture is illustrated by calculations of the water penetration into packages having a leak rate equal to the Mil-Std-883 specified 10^{-8} millibar.litres/sec. Even a modest external ambient moisture of 25°C, 70%RH can cause a gain of 5000 parts per million of water into such a hermetic package within 2 months[19] (Figure 12). The analysis shows that the hermeticity required to properly safeguard active devices for high reliability applications must be less than 10^{-11} millibar.litres/sec.

3.2 Analysis of the climates and conditions for assuring reliability in severe climates

The problems of the more severe climatic conditions were anticipated well before the failures actually occurred. A programme to analyse the climatic distribution, average climatic conditions, and conditions representing reliability safeguards covering 95% and 99% of the occurrences likely in India, was therefore already well in hand. Meteorological data obtained over 30 years from 162 locations, weighted according to the geographic area they represent, has been analysed. The outcome of this work has been reported[20]. This data converts into Table 2.

Table 2. Conditions for reliable operation, within specification, averaged over the required life

Climatic Condition	Covering 95% of Times & Places	Covering 99% of Times & Places
Hot Humid	29°C, 83%RH (33 mbar)	29°C, 86%RH (34 mbar)
Hot Dry	40°C	42°C

The information from the climatic analysis has been applied to determine test conditions for both equipment and components. The particular specification for environmental testing conditions for telecommunications equipment is the DoT specification QM-333[21].

The climatic conditions given in Table 2 have been applied to the equation for reliability assessment by damp heat testing of non-hermetically packaged ICs[22] which has been solidly established as definitive by comprehensive reliability testing and the physics of failure model of isothermal absorption and permeation.

The generic Sinnadurai-HAST (S-H) model is:

$$t_s = 175000/\{\exp\{X[(RH_s)^n-(RH_{amb})^n] + 7000[1/T_{amb}-1/T_s]\}$$

where:
- t_s is the required duration of the stress test
- n is dependent on the physics of failure related to the type of component
- X is the humidity activation coefficient
- Y is the activation energy divided by Boltzmann's constant
- RH_s is the overstress humidity
- RH_{amb} is the ambient humidity (e.g. from Table 2)
- T_s is the overstress temperature in absolute units
- T_{amb} is the ambient temperature (e.g. from Table 2) in absolute units

Whence detailed stress test conditions to simulate 20 years operation have been calculated. Such accelerated tests must be conducted with the electrical bias of the devices representing, as nearly as possible, the worst case operational condition. Other models for damp heat acceleration of ageing have been derived from this original work which is soundly based on a physics of failure model. Other expressions have been derived simply by regression analyses and are prone to serious error. Such an expression for accelerated ageing by moisture was produced by Peck [23] who adapted the S-H RH^n expression by undertaking a statistical regression exercise and produced:

$$t= A.RH^{n.} \exp(E_A/kT)$$

in which n was found to be -2.66 and E_A was 0.79eV

The disagreement between the interpretations of Peck and Sinnadurai arise because Peck used the S-H expression and then applied it to both unsaturated and saturated vapour pressures without consideration of the physics of failure and therefore the resultant expression was not constrained by the realities of the effects of the change of state of moisture from unsaturated to saturated. On the other hand, Sinnadurai established a physics of failure model for damp heat ageing, which was based on isothermal adsorption for plastic packaged ICs and diffusion for thick-film resistors. The S-H expression is restricted by the physical processes on which it is based and has been proven by many hundreds of millions of device test hours over 20 years in the non-saturating HAST. An upper limit of 130°C and unsaturated vapour up to 99% RH is advocated as the range of Sinnadurai-HAST testing of plastic packaged ICs. The absurdity of unconstrained statistical manipulation of data is revealed in the proposition derived from Peck's work that a test at 140°C & 100% RH for a duration of 20 hours is equivalent to 1000 hours at 85°C & 85% RH. At 140°C & 100% RH, the polymer of a plastic package would progressively de-bond and the exterior terminations of the package would suffer electrolytic damage. Unfettered acceleration can be extrapolated to produce an absurd few seconds of testing at high temperatures and 100% RH being equivalent to 100 hours at 85°C & 85% RH. These are good reasons to discount and therefore discard the Peck expression.

The S-H model recognises these constraints to validity. Inserting the HAST stress test conditions used at BT Labs into the S-H equation, gives:

$$t_s = 175000/\exp\{0.00044[(90)^2-(RH_{amb})^2] + 7000(1/T_{amb}-1/381)\}$$

The applied stress condition of 108°C, 90%RH, in a non-saturating autoclave HAST test[2] gives more dependable but more severe damp heat acceleration of ageing than is obtainable with humidity chambers. The calculation has also been extended to higher temperatures of 125°C and 95%RH.

Hence, the test durations for THB (Temperature and Humidity and Bias) overstress tests to simulate 20 years (175000 hrs) operation in various climates have been calculated (Table 3).

Table 3. THB Reliability Tests for 20 Year Survival in Different Climatic Conditions

	TEST	COND	ITIONS	TO BE	APPLIED
	85°C & 85%RH	95°C & 95%RH	108°C & 90%RH	125°C & 90%RH	125°C & 95%RH
CLIMATE	TIME (hours)	TIME (hours)	TIME (hours)	TIME (hours)	TIME (hours)
Temperate General (12°C & 72%RH)	500	130	100	50	TOO SHORT
Temperate Equipment Room (30°C & 25%RH)	300	80	60	TOO SHORT	TOO SHORT
Tropical Coverage of 95% of India (29°C & 83%RH)	4100	1050	850	400	260
Tropical Coverage of 99% of India (29°C & 86%RH)	5100	1300	1000	500	320
Tropical Severe (35°C & 90%RH)	10000	2600	2000	950	630

Clearly the non-saturating autoclave HAST test is considerably more efficient than the 85°C, 85%RH humidity test. The extensive base of experimental work that has been conducted with the apparatus has also established its credentials for robustness and precision of control.

It is an unfortunate fact that the nations which can least afford the problems of severe climatic conditions are those which require greater and therefore more costly safeguards of component and equipment reliability. The evidence presented here, accumulated over many years, has clearly revealed the vulnerability of supposedly reliable components used in modern equipment. The Indian (and other tropical) climate requires more severe testing of electronics components than is applied abroad.

3.3. Field results from indigenous equipment using plastic packaged microcircuits.

In order to establish systems more relevant to the dispersed rural needs of India, its diverse and severe climatic conditions, and the need to achieve a high storage and operational reliability of its equipment, the Indian Government established the Centre for Development of Telematics (C-DOT)

With significant determination and ability, C-DOT developed a modular digital telecommunications switching system based on a small switch building block suitable for rural locations not having air conditioning. As a result of the reliability problems the DoT encountered with the Mil-Std-883 hermetic packages in imported systems and because of the ponderous nature of the Mil-Std-883 approval route and the greater availability of high quality plastic packages, the C-DOT developers made the bold decision to use plastic packaged ICs.

Components were evaluated using accelerated ageing tests appropriate to assuring 20 years reliability in uncontrolled climatic conditions, determined during the course of the work described herein. The equipment was tested in walk-in chambers to the conditions specified in QM-333[21]. The choice made by C-DOT thereafter was to approve the manufacturers of the components that met the evaluation criteria.

The switching systems were successfully developed and have been in the field for two years. Modularity has enabled switches with 10000 line capacity to be built. The reliability results in the field have been promising with one serious exception.
- IC failure rates were less than 1 Fit out of 154000 line circuits.
- hybrid microcircuit failure rates amounted to 600 FITs during 3×10^9 device hours
- the alarming exception was the rogue failures of a particular type of memory IC which passed the functional tests upon assembly into the memory cards, but then failed on installation into the equipment. Allowing a generous amount of 10 hours for survival upon installation, yields a component failure rate of 8×10^6 FITs! Because a single IC failure caused the memory card to fail, 40% of the memory cards failed. The manufacturer of these ICs was ostensibly a major international manufacturer, as the package logo and code showed. However, tracking down the source proved otherwise! The logo was an imitation, and the components, while having the same function, were fake. The finding was a severe lesson that manufacturer approval alone, and the lack of routine vigilance and assessment, is a dangerous gamble. Nevertheless, the excellent behaviour of the authentic ICs proved the bold decisions to use known good plastic packaging, was wise. Thus there is now an awareness, alertness and responsiveness to the needs of more demanding climates, and for cost-effective ways of achieving them.

3.4. Field data from non-telecommunications installations using plastic packaged microcircuits.

Benefits are already being realised, for instance, in an 8 billion Franc project in France in which 1 billion Francs is being saved by using reliable plastic packaged ICs in portable military communications[24]. Defence applications already have a 5 year history of the use of plastic packaged devices in missiles with no field failures. The reported[25] increasing military use shows that the final bastion of resistance to the use of low cost high reliability plastic packaging, has been penetrated.

The relative reliability of plastic packages at 55°C can be scaled as follows[26]:

simple devices	1 Fit
memory devices	10 FITs
ULSI logic & µP	100 FITs

Packages using the purer[11] moulding compounds have an average reliability of 40 FITs. This can be improved to 30 FITs by the use of good quality Si_3N_4 passivation.

Automotives are also increasingly using plastic encapsulation even in the hostile environment under the bonnet. A good example is the Ignition Control hybrid microcircuit of the Fiat Cinquecento[27]. The active integrated circuit is a bare-chip-and-wire assembly which is junction coated with a silicone encapsulant. Some 500000 vehicles have had no system failures in the field in two years.

4. DISCUSSION

The traditional requirement for hermetic packaging may now be overturned in favour of plastic packaging on many grounds:
(i) the full range of laboratory evidence of high reliability of silicone junction coated ICs undertaken by many researchers, showing that plastic encapsulations may now be used for microelectronics and optoelectronics in telecommunications, automotive, military and space applications as the better option in many instances.
(ii) the inadequacy of the hermeticity specification.
(iii) the demonstrated field failures of Mil-Std-883 "hermetic" packaged devices, and the massive ingress of moisture, which caused many failures of telecommunications switching systems.
(iv) the demonstrated better field reliability in rural tropical climates, of assessed, standard plastic packaged ICs from major manufacturers.

The invention of the non-saturating autoclave test (HAST) has proved to be an invaluable tool in the drive to develop high reliability plastic packaging. Combined with the climatic analyses, there are now clear methods for evaluating plastic packaged device reliability for various climates. Of course, tests only provide the basis for selecting the right technologies. The essence of reliability engineering is to develop the technologies and build in the reliability improvements.

Fortunately the good prospects of achieving the required reliability at low cost followed from the initiatives described earlier in this paper. These initiatives have been recognised in the later influential work[14,15].

It is a profound fact that each day's production of commercial plastic packaged ICs provides sufficient data to obtain information on parts per million defects. By contrast, the ponderous "qualified components" route for hermetic package evaluation can take three years to obtain the same magnitude of data.

The inexorable momentum resulting from production throughput, product quality, and good reliability for global applications will make plastic packaging the undisputed dominant encapsulation of future high reliability microelectronics.

REFERENCES

1. F. N. Sinnadurai, "The Accelerated Ageing of Semiconductor Devices in Environments Containing a High Vapour Pressure of Water", Microelectronics and Reliability, Vol. 13, pp. 23-27, (1974).

2. N. Sinnadurai, "More Than a Decade of the Non-Saturating Autoclave as a Highly Accelerated Stress Technique for Evaluating the Reliability of Non-Hermetic Microelectronics Components", Microelectronics and Reliability, pp. 833-836, Vol. 24, (1983).

3. N. Sinnadurai, "An Evaluation of Plastic Coatings for High Reliability Microcircuits", Microelectronics. Journal, Vol 12, pp. 30-38, (1981).

4. H. Baudry and G. Kersuzan, "Individual Encapsulation for Integrated Circuits: How to Use Thermal Analysis to Optimise Curing Conditions Towards Reliability", Proc. 4th European Hybrid Microelectronics Conference, pp. 111-121, (1983).

5. S. M. Boyer, "Plastic Encapsulation of Active Devices on Hybrid Microcircuits", Proc. 4th European Hybrid Microelectronics Conference, pp. 122-132, (1983).

6. J. J. Granger, J. C. Basset, N. Vimont and P. Viret, "Wire Bonded Chips Encapsulation Using Resin Droplets", Proc. 4th European Hybrid Microelectronics Conference, pp. 146-154, (1983).

7. O. Mallem, J. Lantaires, "The Plastic Composite Package Provides an Economic Alternative to LSI-VLSI Packaging for High Density Microcircuits", Hybrid Circuits, Vol. 8, pp. 5-12, (1985)

8. P. Collander, M. K. H. Huhtanen, R. A. Mäkelä, S. M. Palo, "Humidity Testing of Plastic Coated Integrated Test Circuits", Proc. International Microelectronics Symposium, pp. 249-254, (1987).

9. N. Sinnadurai, "EPIC: A Cost-Effective Plastic Chip Carrier for VLSI Packaging", IEEE Trans. Components, Hybrids and Manufacturing Technology, Vol CHMT-8, pp. 386-390, (1985).

10. N. Sinnadurai, D. Roberts, "Assessment of Micropackaged Integrated Circuits in High Reliability Applications", Microelectronics Journal, Vol 14, pp. 5-25, (1983).

11. British Telecommunications Specification M219F for "Type Approval of Encapsulating Resins, Coatings and Adhesive Materials for Semiconductor Devices and Microcircuits", BT Quality and Reliability Centre, Bordesley Green, Birmingham B9, England.

12. L. Condra, S. O'Rear, T. Freedman, L. Flancia, M. Pecht, D. Barker, "Comparison of Plastic and Hermetic Microcircuits Under Temperature Cycling and Temperature Humidity Bias", IEEE Transactions on Components, Hybrids and Manufacturing Technology, Vol 15, pp. 1-11, (1992)

13. C. P. Wong, "High Performance Silicone Gel as IC Device Chip Protection", Mat. Res. Soc. Symp. Vol. 108, pp. 175-187, (1988)

14. J. W. Balde, "The Effectiveness of Silicone Gels", IEEE Transactions on Components, Hybrids and Manufacturing Technology, Vol 14, 1991.

15. I. P. Hall "Novel Low Cost Techniques for Low Cost, High Performance, Optoelectronic Component Assembly", Hybrid Circuits, Vol. 34, pp. 44-47, May (1994).

16. N. Sinnadurai, "MultiChip Module Applications in Communications Satellites", NATO Advanced Research Workshop on MCM/Mixed Technologies, Islamorada, Florida, ISBN 0-7923-3460-4, pp. 169-176, (1994), and Microelectronics International, Vol. 37, pp. 31-32, (1995).

17. J. A. King "Living in Two Worlds", JC 13.5 meeting, USA, 6 May (1992).

18. T. K. Ramaswamy, T. S. Kuppuswamy, "Quality Aspects of Telecom Equipment", Telecommunications (the Indian Journal of Telecommunications), pp. 14-18, (1991).

19. D. Stroehle, "On the Penetration of Gases and Water Vapour into Packages with Cavities, and on Maximum Allowable Leak Rates", Proc. IEEE 15th Annual Reliability Physics Symposium, pp. 101-104, (1977).

20. "Report on Climatological Conditions Encountered by Telecom Equipments in India", General Manager TQA Circle, Department of Telecommunications, Bangalore, India, Issue 1, (1988)

21. "Specification for Environmental Testing of Electronic Equipment for Transmission and Switching Use", Telecom Quality Assurance Circle, Department of Telecommunications, Bangalore, India, QM-333, Issue 1, (1990).

22. F. N. Sinnadurai, "Handbook of Microelectronics Packaging and Interconnection Technologies", Electrochemical Publications Limited, 1st ed., pp. 6-7, 150, 153-155, (1985)

23. D S Peck, "Comprehensive Model for Humidity Testing Correlation", IEEE International Reliability Physics Symposium, pp. 44-50, (1986)

24. Private communication to N. Sinnadurai, 1993.

25. "Plastic-packaged ICs in military equipment", IEEE Spectrum, pp. 62-63, Feb., (1991).

26. M. Barre, "Prsentation at "Workshop on Plastic Packaging - Present Reliability and Procurement Evaluation Programmes and Existing Experience in Plastic DEvice Utilisation for Military and Space Applications", Loughborough UNiversity, England 19-20 July 1994.

27. Private communication to N. Sinnadurai during visit to automotive electronics plant, May 1994.

Thermal Simulation and Characterization of Single Chip Packages

Orla Slattery, Ciaran Cahill, John Barrett, Martin O'Flaherty and Kenneth Rodgers
National Microelectronics Research Centre,
University College,
Cork,
Ireland.

Telephone: +353 21 904174, Fax: +353 21 270271
e-mail:slattery@nmrc.ucc.ie

Abstract

Electronics systems are continuing to strive for higher speeds and reduced size and weight. This results in increased packaging and power density, with a requirement for enhanced thermal performance and more sophisticated techniques for heat removal at the micropackage level. This, in turn, has a significant influence on the interconnection and first level assembly techniques that are used and leads to a requirement for accurate thermal design and validation tools in order to make technology and material choices at the micropackaging and assembly level. Validated thermal modelling tools are proving to be a fast and cost effective method of predicting thermal behaviour and optimising package design to enhance thermal performance. Thus, they can be quickly used to assess different package designs, heat flow strategies, and IC die attach/encapsulation materials without the need to build prototypes.

This paper details the use of thermal simulation techniques to carry out thermal performance analysis of various single chip package types including CPGA, PQFP and BGA packages. Corresponding thermal characterization of these package types is also discussed. The work will be presented in the context of two ESPRIT projects: DELPHI and CHIPPAC.

I. INTRODUCTION

Thermal management is becoming an increasingly more important issue in package design and qualification. Higher speeds and reduced size and weight results in increased packaging and power density. Increased operating temperatures enhance most failure mechanisms and thus can have a serious effect on package reliability. Development of reliable packages requires an ability to accurately predict the device thermal performance in its operating environment. The use of thermal simulation techniques as a means of package characterization is becoming more widespread as a means of predicting package temperature range and subsequently as a design tool to optimize materials and geometry for optimum heat removal at the package level and also at the interconnection and assembly level.

Effective use of thermal simulation techniques requires validated package models and is the main subject of this paper. The paper presents an update of an ongoing Esprit Project, DELPHI whose aim is to develop a library of validated physical models of an integrated design environment. Results from Esprit project CHIPPAC are then presented as a case study into the use and application of simulation techniques.

G. Harman and P. Much (eds.), Microelectronic Interconnections and Assembly, 33-43.
© 1998 *Kluwer Academic Publishers.*

II.   INTRODUCTION - THE DELPHI PROJECT

Traditionally, package manufacturers data give values of thermal resistance measured under idealised conditions. These values are not always accurate under different operating conditions. Package thermal management is becoming increasingly more important and thus it is necessary for package suppliers and users to have accurate thermal resistance values. One method of insuring this is for suppliers to provide a model of their components to users which may be used to predict thermal behaviour under any operating conditions. This is the basis for the DELPHI project.

The project name DELPHI stands for Development of Libraries of Physical models for an Integrated design environment. The DELPHI Esprit III funded project is concerned with the design of validated compact models for electronic components. The project aim is to develop a methodology of both validating thermal models and generating compact models for electronic components.

The general procedure for the validation of a model and generation of a compact model is shown below.

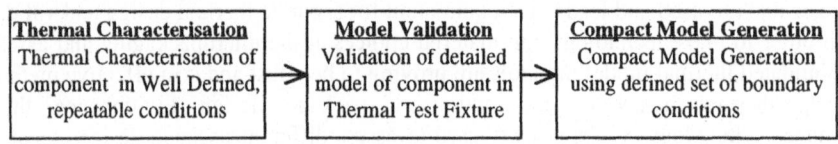

The project is researching improved techniques in both the thermal characterisation [1] and the compact model generation [2, 3]. Figure 1 shows an example of a compact component model, indicating thermal heat flow paths from junction to board. The DELPHI project is concerned only with validated package models and so the compact component model is a conduction only model of the component. Heat transfer between from package to board and convective flow paths are not considered.

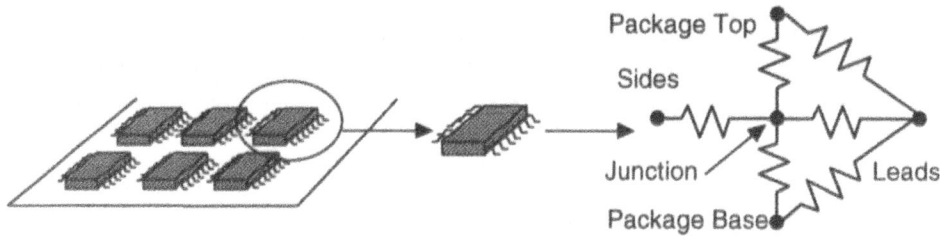

Figure 1: Compact Component Model

Existing thermal measurement standards were investigated from the point of repeatability, reproducibility and model validation. The current SEMI, JEDEC and MIL Standards do not meet all the above criteria.

In [4] a round robin junction to ambient thermal test experiment, designed to evaluate a preliminary test method, tests were carried out in eight thermal test labs. The component, mounting and entire kit was supplied to the participants. The component was a 208 PQFP assembled onto test boards. Each lab was given written test methods and steady state and transient test were carried out. The results of the experiment, which showed variation between measurement results from different labs highlight the need for standardisation of measurement techniques and detailed data recording and reporting. In an effort to address all the above issues, special measurement test fixtures were developed as part of the DELPHI project.

III.   DELPHI MEASUREMENT TEST FIXTURE

In order to validate the thermal models, the test system needed to comply with the following criteria;

- Boundary conditions must be well defined,
- All important heat flow paths must be addressed,
- Tests must be repeatable,
- Accurate device calibration must be achievable.

A dual cold plate (DCP) test fixture was developed for the measurements. This fixture was designed to allow for junction to case thermal resistance measurements of the package according to the above criteria. The system is a structure in which both plates are cooled by the same re-circulating fluid. Equal temperature on top and bottom cold plate (DCP1) allows for measurement of a single junction to case thermal resistance value. To fully characterize the package, measurements were performed whereby heat is extracted through;

- both the top and base of the package (DCP 1)
- the package base with the top insulated (DCP 2)
- the top of the package with the base insulated (DCP 3)
- through the package leads to the PCB (DCP 4)

The test fixture set ups are shown in figures 2 and 3. Figure 2 shows the test fixture configuration for the DCP1, DCP2 and DCP3 cases, heat is extracted by direct contact of the package top and bottom with either a heat sink or insulating material. In the DCP4 configuration, the package itself is not directly contacted but heat is extracted through the package leads and the PCB is contacting an aluminium board.

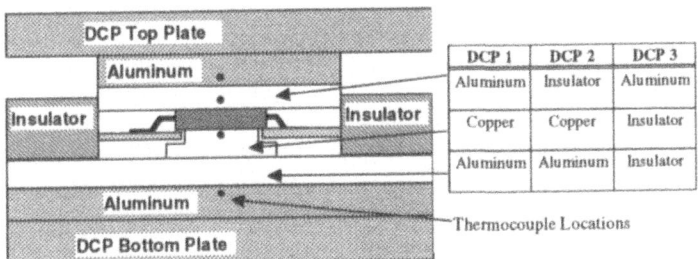

	DCP 1	DCP 2	DCP 3
	Aluminum	Insulator	Aluminum
	Copper	Copper	Insulator
	Aluminum	Aluminum	Insulator

Thermocouple Locations

Figure 2: DCP1, 2 and 3 test fixture configuration

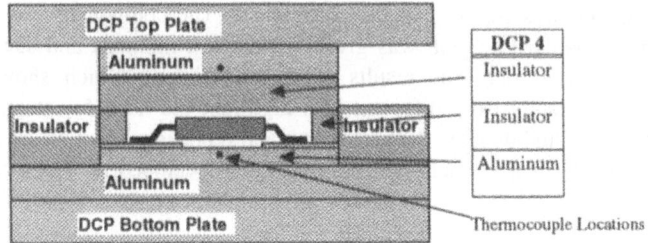

Figure 3: DCP4 Test Fixture Configuration

These four test set ups are designed for conduction only heat flow. It was decided to use only conductive heat flow in the measurement/modelling validation as convective flow would involve parameters such as heat transfer coefficients which are difficult to exactly quantify and do not comply with the test system criteria. Natural or forced convective flow would introduce boundary conditions that are not well defined and repeatability and reproducibility of measurement results would be undermined.

The topplate and baseplate reference temperatures were continuously monitored during all tests using thermocouples embedded 1mm from the relevant surface. Pressure is applied via a screw/clamp mechanism to the assembly. Layers of thermal grease compound were applied to all interfaces to minimise their resistance.

The test die used in the characterization included heater resistors and temperature sensing diodes. The test procedure allows simultaneous calibration of diode forward voltage (Vf) along with thermal measurement. This provides accurate calibration and reduces measurement uncertainty as a single setup is used for both calibration and measurement. The criteria used for correlation of measured to modelled results is junction to case thermal resistance defined as;

$$Rth(j-r) = \frac{Tjunction - Tref}{Power}$$

Where $T\,ref$ is the temperature at one of the known boundary conditions. This temperature is measured by thermocouple as indicated in figures 2 and 3. Junction temperature is measured by the temperature sensing diodes.

IV.   DELPHI SIMULATION

The simulation tool used is FLOTHERM a CFD based thermal modelling package from Flomerics Limited. A generic model was constructed of the test fixture previously described. The same basic model was used to simulate each of the measurement set ups - DCP1 to DCP4, and material conductivities were amended depending on whether the top and base plates were heat sinks or insulators. A detailed model was built for each of the package types characterized and the package model was incorporated into the test fixture model. Here, the results of the measurement/modelling correlation will be presented for three package types. Firstly, thermal analysis of a PQFP package will be presented and then validation results for a TQFP and CPGA packages will be discussed.

V.   THERMAL CHARACTERIZATION OF PQFP PACKAGE

A 208 lead PQPP package used in this evaluation. The PQFP208 was a 28mm square package with a thickness of 3.6mm, die size was 9.5mm square and 635μm thick.

As the aim of the DELPHI project is to produce accurate validated package models, measurements were carried out at two independent sites; the NMRC and Alcatel Bell in Antwerp. The thermal resistance was evaluated in a DCP 1 configuration over three cold plate temperatures 20°C, 30°C and 40°C. Simulation of both the NMRC and Bell test fixtures were carried out for each of the three cold plate temperatures and the results are shown in figure 4. The results show good correlation between all cases with overall variation less than 3%.

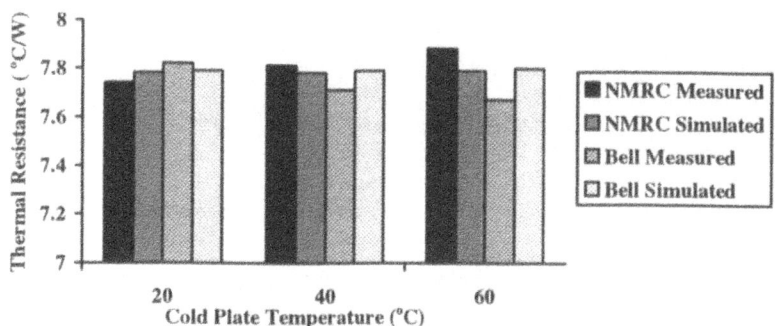

Figure 4 Comparison of Measured and Simulated Results for PQFP 208 Package

VI.   THERMAL CHARACTERIZATION OF TQFP PACKAGE

Characterization of TQFP packages was carried out for an 80 lead TQFP. The package is a 14mm square package, 1.4mm thick with a 5.2mm square thermal test die of thickness 300μm. The thermal test die used for this work incorporates a heater resistor in one quadrant of the chip and the diode is located in the centre of the adjacent quadrant. Thus, accurate prediction of junction temperature requires detailed modelling of the chip surface. During measurements exact alignment of the heat sink with the package is required to achieve accurate and repeatable results. Agreement between measured and modelled results for DCP1,DCP2 and DCP3 cases was found to be very good as shown in Table 1. The board and wiring conductivity greatly impact the DCP4 thermal resistance and work is ongoing to improve the accuracy of the model in this case.

	DCP -1	DCP -2	DCP -3	DCP -4
	No Insulator	Insulator on Top	Insulator on Base	PCB on Heatsink
Power	30C , 5W	30C , 5W	30C , 5W	30C , 1W
Measured	4.65	7.35	8.14	46.3
Simulation	4.66	7.2	8.0	30.8

Table 1 Measured and simulated thermal resistance results (°C/W) for 80 lead TQFP.

A contour plot showing the modelled temperature variation across the die is shown in figure 5. The case simulated is a DCP1 set up with an applied power of 5W. Heatsink temperature is 31.5°C. The measured diode temperature is 54.7°C and simulated diode temperature is 54.9°C. The plot shows maximum junction temperature of 81°C at the heater location. Temperature variation across the die is considerable in this case ranging from 81°C at the heater location to 53°C at the die edge. The temperature distribution across the die surface is due to the location of heater resistor in one quadrant of the chip only. Die surface temperature would be expected to be more uniform if the power dissipation were spread more evenly across the die. Standard thermal modelling techniques generally assume die power dissipation to be evenly distributed along the die surface. The results of this simulation particularly highlight the requirement for accurate detailed modelling of die surface parameters.

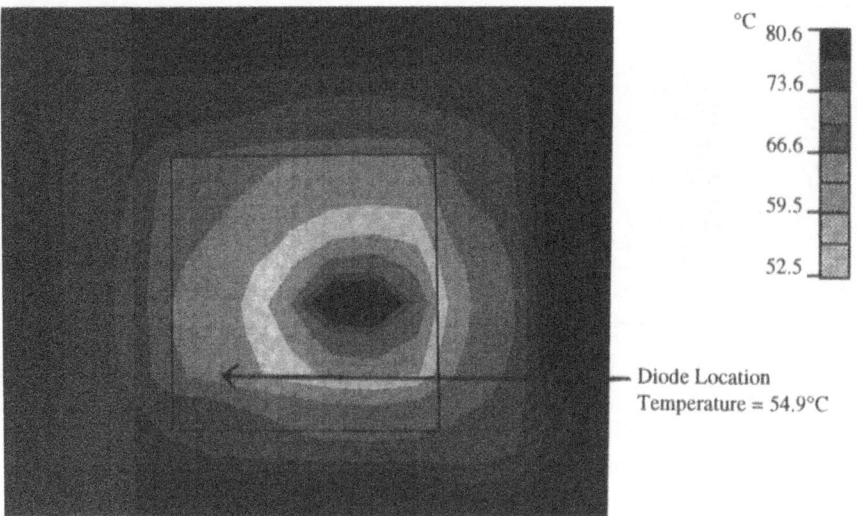

°C
80.6
73.6
66.6
59.5
52.5

Diode Location
Temperature = 54.9°C

Figure 5. Contour plot of predicted temperature on the die surface for the 80 lead TQFP.

VII. THERMAL CHARACTERIZATION OF CPGA PACKAGE

The package types used in the CPGA evaluation were a 224pin CPGA and a 180pin CPGA; each package had either a 6mm square die or a 10mm square die. The 180 CPGA was a 40mm package and 2.5mm high. The 224 CPGA was a 45mm package and 3.2mm high. The lid of the CGPA was gold plated nickel. The die used were NMRC designed PMOS3 and PMOS4 thermal test die incorporating heater resistors across the die surface and temperature sensing diodes at the centre of the chips. Three of the dual cold plate test set ups were used in this evaluation:

- heat extracted through base and top (DCP 1);
- heat extracted through base only (DCP 2) and
- heat extracted through top only (DCP 3).

Characterization of package thermal behaviour with heat extracted through the pins (DCP4) is ongoing

The FLOTHERM model assumed that the ceramic and silicon materials had a temperature dependent thermal conductivity. The correlation between the measured and modelled results for 180CPGA in figures 6 and 7. A comparison of the agreement achieved for the 224 pin CPGA is shown in figure 8 for the DCP 1,2 and 3 cases. A high level of agreement between the measured and modelled results was found. Agreement within 3% was achieved between measured and modelled results for both the 180CPGA and 224CPGA in the DCP1 and DCP2 configurations. There was some discrepancy with the 224CPGA for the DCP3 case measured and modelled thermal resistances varied by up to 14% in the package with the 6mm die. Physical analysis found this package to be warped and thus the resulting increase in interfacial contact resistance at the warped surface accounts for the larger than expected variation.

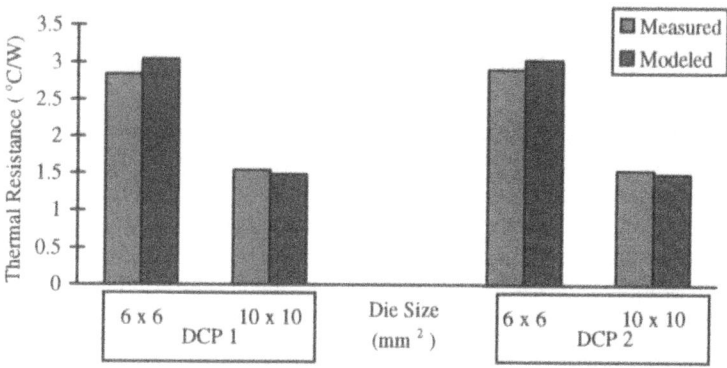

Figure 6 Comparison of measured and simulated results for 180 pin CPGA package analysed in the DCP1 and DCP2 configurations for two die sizes.

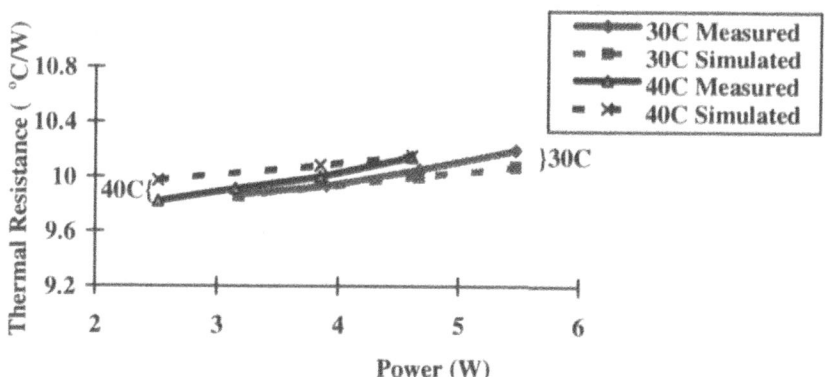

Figure 7 Comparison of measured and simulated results for 180 pin CPGA package analysed in the DCP1 and DCP2 configurations.

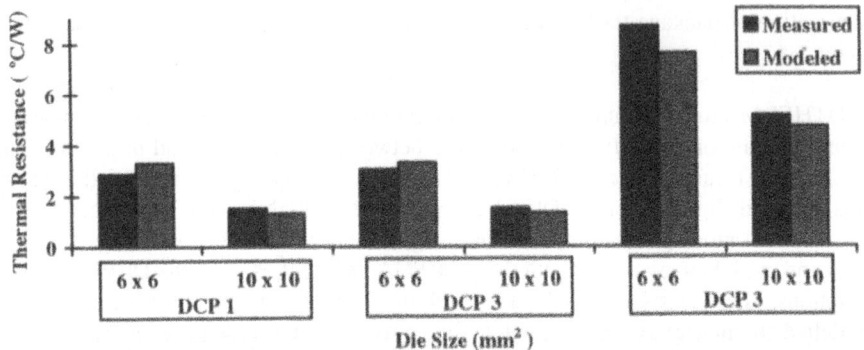

Figure 8 Comparison of measured and simulated results for 224 pin CPGA package analysed in the DCP1, DCP2 and DCP3 configurations using two die sizes.

VIII. INTRODUCTION - THE CHIPPAC PROJECT

In the ESPRIT Project CHIPPAC, finite element simulation techniques were used to evaluate the limitations of package designs and optimize package materials and geometry to comply with specified thermal requirements. The objective of CHIPPAC - Cost effective, High Performance PACkaging, was to develop technologies for efficient packaging of VLSI devices. As part of the thermal simulations work, a study was carried out of a BGA package with a ceramic base with 3 different lid materials, aluminium, ceramic and globtop. This section of the paper discusses the simulation of the CBGA package.

IX. ANALYSIS OF BGA PACKAGES

Operating conditions required the package to dissipate 3W in an ambient of 50°C. The maximum allowable junction temperature was 85°C. The lid types considered were aluminium, ceramic and globtop. These materials have considerably different thermal conductivities varying from 200W/mK for aluminium to 3W/mK for ceramic and 0.5W/mK for glob top. Finite element (FE) techniques were used to simulate BGA behaviour under forced convection. The simulations had two objectives: firstly to compare the package thermal behaviour with each of the three lid types and secondly to compare two different designs for the aluminium lid. The BGA package was 34mm square with a 15mm die (550μm thick).

Figure 9 shows the FE model of the simulation case for a CBGA with an aluminium lid. Owing to symmetry of the structure it was sufficient to create a 3D finite element model of one quarter of the structure only. The solder balls were modelled as a solid block and this layer was given an effective conductivity to take account of the actual area occupied by the solder. A layer of thermal grease was incorporated into the cavity of the package to enhance heat flow.

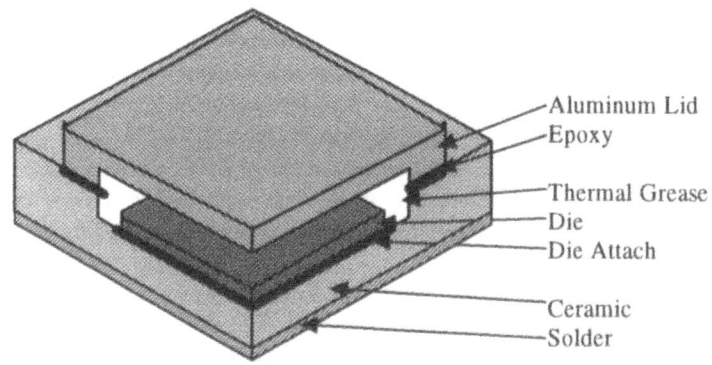

Figure 9 Schematic of FE model of CBGA package with an aluminum lid

Forced convection air flow was modelled by applying a heat transfer coefficient to the model boundaries. A value of 65 W/m²K was used to on 5 of the 6 boundaries and a value of 10 W/m²K was used along the bottom surface. The simulated maximum junction temperature achieved for the case with the aluminum was 76.8 °C.

Figure 10 shows a plot of temperature distribution for a CBGA with a ceramic lid simulated under the above conditions. The plot shows a junction temperature of 80.8°C and the temperature range across the package varies from to 67.5 C at the lid edge. Junction temperatures for each of the three lid materials are shown in table 2.

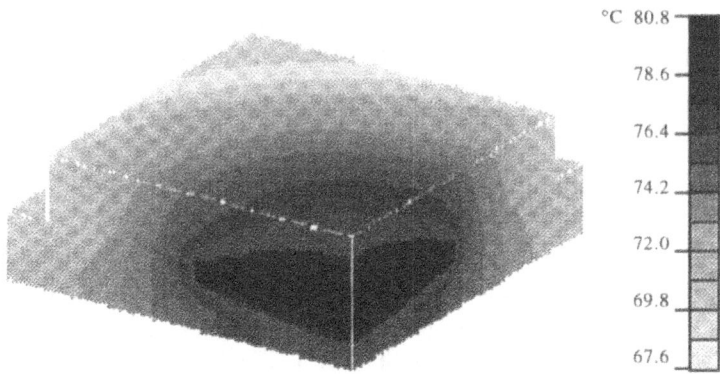

Figure 10 Simulated temperature distribution for a CBGA with a ceramic lid

Lid type	Simulated junction temperature °C
Aluminium	76.8
Ceramic	80.8
Globtop	88.9

Table 2 Simulated junction temperature for Ceramic BGA for different lid materials

The thermal conductivity of the aluminium is 200 W/mK compared to 3 W/mK for the ceramic and 0.5W/mK for the globtop material. As expected, the junction temperatures are lowest for the package with an aluminium lid. These results show that for the aluminium and ceramic lids, junction temperatures are lower than the maximum allowable temperature of 85°C. However, the predicted junction temperature for the CBGA with globtop lid is approximately 3°C higher than the specified maximum of 85°C. Current trends in globtop materials are concerned with development of globtops with higher thermal conductivity. Globtops with conductivities up to 2 W/mK are now commercially available.

The second objective of the simulation work was to compare two different designs for the aluminium lid. The first design consisted of a flat lid as shown in figure 9. The second design consisted of a T shaped lid , similar to the flat lid but incorporating an additional aluminium block extending into the package cavity. Owing, to the very high thermal resistance of the cavity grease material simulation results found that there was little variation in junction temperature between the two lids. In order to improve heat flow with the T shaped lid it would be necessary to bring the lid into direct contact with the chip and this is not practical

X. SUMMARY

This paper outlines the importance of thermal management and the need for accurate thermal modelling tools to accurately predict thermal behaviour in an increasingly more demanding environment. An overview of an ongoing Esprit , DELPHI is presented. The project aim is to develop libraries of physical models for an integrated design environment. The thermal test system designed for measurement of package junction to case thermal resistance is described. The four test set up configurations designed to extract heat from package top, bottom and leads are described. Correlation between measured and modelled results is presented for three package types: PQFP208,; TQFP80 and 180 and 224 pin CPGA.

Thermal simulations from Esprit project CHIPPAC are presented as a case study into the use of thermal simulation techniques in IC package thermal characterization.

ACKNOWLEDGMENT

The contributions of the project partners from the DELPHI and CHIPPAC project are appreciated. The DELPHI project partners are : Flomerics Limited; Alcatel Bell; Alcatel Espace; Philips CFT and Thomson CSF. The CHIPPAC consortium include: Bull; GEC Marconi; IBM; Mietec; IMC; SGS Thomson; ES2; Framatome; Montpellier Technologies; and Telefonica I & D. The authors wish to thank John Rea of the mechanical workshop at NMRC for manufacture the dual cold plate and thermal test fixtures and acknowledge Analog Devices BV who supplied the TQFP packages with thermal test die.

REFERENCES

[1] W.Temmerman, W.Nelemans, T.Goossens, E.Lauwers, C.Lacaze, "Validation of Thermal Models for electronics Components", IEPS 1995, San Diego, September 1995

[2] Clemens Lasance, Heinz Vinke, Harvey Rosten, Karl-Ludwig Weiner, "A Novel Approach for the Thermal Characterisation of Electronic Parts" Eleventh Annual IEEE Semiconductor Thermal Measurement and Management Symposium, February 7-9 1995.

3 H.Vinke, C.Lasance, "Thermal Characterization of Electronic Devices by means of Improved Boundary Condition Independent Compact Models", 45th. EUROTHERM Conference, Leuven, September 1995

[4] B. Joiner, B Siegal, T Tarter, B Bright, "Use of Experimental Data in Guiding Thermal Specification Development", Twelfth Annual IEEE Semiconductor Thermal Measurement and Management Symposium, March 5-7 1996, Austin Texas.

Current Trends and Future Issues in Solderability

Gary J. Ewell
The Aerospace Corporation
M4-987, P. O. Box 92957
Los Angeles, CA USA 90009
(310) 336-6003; Fax (310) 336-6914
ewell@courier4.aero.org

Abstract

This paper discusses solderability of solder joints. It introduces the issue in terms of the metallurgical and chemical requirements to form a good joint and focuses upon recent changes in Europe and the USA. Changes discussed include application of a variety of no-clean solder fluxes, environmental protection limitations on cleaning solvents, use of very-fine-pitch packages, extended shelf times of purchased components before use, and the recent advent of steam-aging and automated testing of solderability.

Finally, new approaches to maintaining solderability are discussed with respect to process limitations and possible further restrictions on design of SMT assemblies, including: use of inert soldering environments during joint formation, the recent development of organic solderability protective coatings (OSPs), and the development of Pb-free soldering alloys.

Introduction

Solderability is traditionally defined as the degree to which a surface can be wet by molten solder. Until a few years ago, solderability was usually determined by degree of solder coverage, via visual inspection after a dip and look type test, float test, or by measuring wetting angles on a globule or similar test [1]. As a result of years of studies, various authors developed lists of the causes of inadequate solderability. In 1984, R. Wild of IBM [2] estimated that 28% of all as-received components were unacceptable for storage or use without rework or at least retinning.

Recently the focus on solderability has become more intense. In part, the intensity resulted from the near elimination of many other causes of solder joint defects. The major cause of the intensity, however, arises from the rapid spread of automated surface mount technology (SMT). The use of SMT has had several impacts. First, there has been an increasing emphasis on tighter control of surface planarity. Second, the emphasis on adequate strength and fatigue resistance of solder joints has been heightened as joints moved from the stronger ring-and-plug geometry of leads in plated-through holes to a lap shear type of joint configuration. Third, new packaging schemes utilizing ball grid arrays (BGA) of solder balls or chip-on-board (COB) technology require more insurance of solderability. Finally, automated, surface-reflow techniques result in less mechanical and thermal solder agitation of the joint compared with hand soldering and thus are more prone to oxide film formation, contamination, and intermetallic compound (IMC) layer buildup in the joint.

Solderability

A more complete look at solderability shows that it results from an interplay of a variety of factors, including surface tension, wetting force and angle, surface cleanliness, the porosity, purity and adhesion of external coatings, the nature of surface oxides, sulfides, etc. that form, and the time and temperature of solder exposure, as well as the compositions of the flux and solder utilized. Thus for a manufacturer, solderability might be defined as the ability to form a quality solder joint using well-controlled soldering and fluxing practices and a good joint design.

G. Harman and P. Mach (eds.), Microelectronic Interconnections and Assembly, 45-51.

The metallurgical structure of a solder joint on a printed wiring assembly (PWA) is quite complex. It joins the copper on the printed circuit board (PCB) to the component soldering surfaces (termination, pad, lead, etc.) with a solder alloy and a flux. The structure is made more complex by coatings added to the component and PCB to preserve or ensure solderability, by elements added to the solder to minimize leaching, and by recent changes to improve the environmental compatibility of all the materials utilized. Coatings and platings are added to either establish a more solderable surface on components or to preserve a solderable surface by applying a protective coating.

Intermetallic compounds (IMC) that form by the metallurgical interaction of metallic elements, according to arrangements visible in phase diagrams, are usually spoken about as troublesome and problem-causing. Their formation, however, indicates that the metals involved have wet and successfully interacted; thus, the presence of IMC layer(s) in a solder joint is normal and to be expected. The problems arise when IMC layers become too thick, consuming much of the tin required for interaction in tin-lead solders, when the IMCs that form are unusually brittle, such as gold-tin intermetallics, or when the IMCs oxidize and are unsolderable. R. Bowlby [3] has stated that when copper-tin intermetallic compounds comprise more that 50% of the total solder coating thickness that solderability is significantly impaired.

Besides containing IMC layers (usually quite thin), a solder joint usually is a eutectic structure, sometimes containing dendritic structures of solid solutions, with surface oxides or other films. For tin-lead solder joints, the tin is partially consumed in IMCs, thus causing the eutectic structure and oxide to be lead rich. There have been several reports of cracks following the lead-rich portion of the eutectic structure.[4]

The complex nature of solder joints, as well as the number of factors influencing solderability, are the main reason that until recently measurements of solderability have been mostly qualitative.

Quantitative Measurements of Solderability

Two techniques are now available to provide quantitative, repeatable measurements of solderability. The first technique is the wetting balance or meniscograph. Developed over the last 15 years [5-7], this method is documented in IEC 68-2-44 and IPC-S-805, among other locations. The wetting balance measures weight change with time, allowing one to determine the time until wetting begins and then both the degree of wetting and time over which wetting (solder wicking) occurs. It has been used to compare coating, fluxes, solder compositions, tinning procedures and a variety of other solderability-related material and process variables. There have been questions raised, however, about repeatability of measurements with very small components and with flat pack packages.

Much newer is the sequential electrochemical reduction analysis (SERA) that determines both the nature of surface oxides, as well as their thickness[8-10]. SERA monitors the electrode potential during oxide reduction. Testing starts by inserting the PCB or component into a sealed cell with a buffered borate electrolyte. Passage of a current through a counterelectrode reduces surface oxides. Monitoring the electrode potential during the time that surface films are reduced gives information relating to the nature of the surface coating and its thickness.

Both of these measurement techniques, used in conjunction with steam aging, offer a means to quantitatively measure solderability as a function of exposure. Steam aging has been developed to simulate storage or shelf life exposure, with 20 hours of steam aging believed to represent 1 year of natural exposure.[11]

Developing and Maintaining Solderability: Current Status

Maintaining solderability on both PCBs and components continues to receive a high level of concern. The focus has been divided into five major areas: (a) retinning components, (b) hot air solder leveling (HASL) on PCBs, (c) storage in controlled environments (d) assembly in contamination-controlled working areas and (e) soldering in low-oxygen environments. Component issues relate to (1) retinning after burn-in and testing, i.e., whether third parties can perform this or if the original manufacturer must retin, and (2) concerns about the possibility of tin whisker formation for all-tin coatings. On PCBs, the issues relate to growing dissatisfaction with HASL, discussed later.

Hot air solder leveling applies a thin layer of molten tin-lead solder to a PCB following solder mask formation. Application is by passing the board through a molten wave or dipping it into a solder pot. Excess molten solder is then quickly blasted from the surface using high-velocity hot air. HASL has appeared to be the only method of preserving solderability that was robust enough to handle the multiple thermal exposures given a PWA during its assembly. Wave soldering, hand soldering, adhesive bonding and curing, etc. require coatings that can endure through extended periods of high temperature exposure. HASL produces solder coatings that meet those requirements. In addition, HASL is compatible with the multitude of processes and quality issues experienced during PWA assembly (Table I).

Table I. Solderability Coating Compatibility Requirements*

Pad and through-hole wettability
Solder joints free of defects
Wave soldering free of solder balls
Solder masks free of discoloration or lifting
Ionic cleanliness levels acceptable
Conformal coatings with good adhesion, curing
Chip adhesives with good adhesion, curing
Wire bondability

* Table modified after list presented by DeBiase and coworkers[12]

More recently, however, deficiencies and shortcomings of HASL have become quite apparent. Those deficiencies include:

- Relatively high cost, especially in equipment maintenance.
- Inability to hold tight solder thickness tolerances needed by very high pitch (<.05 mm spacing) packages.
- Environmental issues associated with waste disposal.
- Worker safety issues associated with lead exposure.
- Warping of very thin substrates, such as the 0.42 to 0.48 mm boards used to produce PCMCIA (portable computer) cards.

Conditions for component storage now receive greater scrutiny. Some facilities, worried about maintaining parts for very extended time periods, are instituting storage in dry nitrogen environments. The majority of firms are carefully controlling temperature and humidity, as well as instituting a first-in, first-out procedure or requiring just-in-time deliveries. Controlling contamination in assembly areas continues to receive attention, with everything from food and drink to hair sprays, hand creams, and adjacent manufacturing operations being restricted.

Performing soldering and component burn-in in low-oxygen environments has been a major thrust for a large number of manufacturers. Table II shows the effect of oxygen levels on solder joint quality. Correspondingly, the authors reported that solder defects due to nonwetting were reduced from 22% to 2% after institution of a nitrogen insertion soldering operation, while overall defects were reduced by more than 75 % [13].

Table II. Oxygen Level During Soldering and Solder Joint Quality*

Oxygen Level	Wetting Angle	Joint Rating
4 ppm	20 degrees	excellent
400 ppm	20 degrees	excellent
1000 ppm	20 degrees	excellent
2000 ppm	27 degrees	very good
4000 ppm	35 degrees	good
10,000 ppm	>35 degrees	good
41,000 ppm	36 degrees	good

* Ivankovits and Adams[1]

New Approaches to Maintaining PCB Solderability

At least six (6) alternatives to HASL for maintaining printed circuit board solderability have been proposed, ranging from organic coatings to a variety of immersion, electrolytic, and electroless platings. The alternative receiving the most publicity has been the use of what in the past has been called antitarnish coatings, but are now called organic solderability preservatives (OSPs) or organic protective coatings (OPCs).

OSPs come in two families. The first, older group are rosin-based; the newer group are azole-based, using water rather than organic-based solvents. Originally based on benzotriazole, a second generation of the new group uses substituted benzimidazoles. One example is the "Ronacoat OSP" marketed by Lea-Ronal. Table III briefly highlights the advantages and disadvantages of both families of products[14-16].

Table III. Advantages and Disadvantages of Several OSPs

	Rosin-Based	Azole-Based
Advantages	Coats entire board prior to fluxing	Forms very thin coating, nontacky
	Very resistant to heat	Water-based system
		Coating vaporizes or dissolves in flux
Disadvantages	Solvent-based, requires solvent cleaning	Difficult to apply
	Tacky surface	Shelf-life limited
		Handling problems: poor scratch resistance

The newer OSPs selectively bond with copper, resulting in coatings that are between 0.1 to 0.5 micrometer thick. Use of these materials, however, currently requires the manufacturer to modify his operation to limit the number and duration of high temperature operations before coating removal.

A number of alternatives to OSPs have been proposed, but there appears to have been only minimal process definition and development to date. Some of the other alternatives to HASL are listed below in Table IV.

Table IV. Other Alternatives to HASL

Coating Proposed	Reference
Immersion bismuth	17
Immersion silver with organic oxidation inhibitor	17
Electrolytic tin (matte)	15
Electroless nickel/immersion gold	15
Electroless palladium or immersion gold	18

Near-Term Future

The near-term future will probably bring changes that reflect increasing concerns about worker safety and environmental degradation. These changes are going to encourage continued movement away from cleaning compounds harmful to the environment and from lead and other metallic elements harmful to workers and their children. While a great variety of solder alloys have been proposed as alternates to the tin-lead system, none are likely to be in significant general use in the next five (5) years.

Alternatives to hot air solder leveling as a means of preserving PCB solderability will grow in popularity. OSPs and electrolytic tin are likely to be among the most popular alternatives. The switch to "no clean" soldering will continue, but there will be a thrust to develop true "no-residue" fluxes as flux residues begin to affect signal noise, cross talk and impedance in very high clock speed circuitry[19]. In the near-future variable-convection nitrogen reflow ovens and other soldering equipment will be developed to minimize or eliminate the needs for any solder fluxes. Airless flux sprays that minimize the amount of flux used and continually provide fresh flux are already in use.

Increasing use of chip-on-board and large ball grid array packages will be seen. Such products will require guaranteed solderability and either complete cleaning beneath the chip or package, very low residue fluxes, or fluxes fully compatible with epoxy underfills that are likely to be used. Solderability requirements will probably increase in terms of the amount of steam aging that both PCBs and components are going to have to withstand and still maintain their solderability. The use of 16 hours of steam aging as an equivalent to storage may become standard.

Summary

Solderability has been discussed from various points of view, including a definition from process considerations, a review of the development of two quantitative means of measuring solderability, evaluation of the current state of the art of maintaining solderability, and consideration of changes likely to occur in the near future. Emphasis upon ensuring that as-received product has sufficient solderability and upon preserving that solderability will surely increase.

50

References

1. Long, J. B., "A Critical Review of Solderability Testing", Properties of Electrodeposits-Their Measurements, Vol 1, 1975, pp. 102-121.

2. Wild, R. N., "Some Component Lead Soldering Issues - Phase II", IBM Technical Paper 840TP0091, presented at Ninth Annual Soldering Technology Seminar, Naval Weapons Center, China Lake, California, February, 1985.

3. Bowlby, Reed, "Finish First", Circuits Manufacturing, November 1987, pp. 76-86.

4. Novick, D. T., "A Metallurgical Approach to Cracked Solder Joints", Welding Journal, 52 No. 4, April 1973, pp. 154-s - 158-s.

5. Davy, J. Gordon and Randy Skold, "Solderability Testing for Receiving Inspection", Circuits Manufacturing, February 1985, pp. 106-114 and March 1985, pp. 74-82.

6. DeVore, J. A., "Solderability and Surface Mounting", Printed Circuit Assembly, June 1988, pp. 8 - 12.

7. Shipley, J. F., "Influence of Flux, Substrate and Solder Composition on Solder Wetting", Welding Research Supplement, October 1975, pp. 357s-362s.

8. Tench, D. M., et al., "Solderability Assessment Via Sequential Electrochemical Reduction Analysis", J. Applied Electrochemistry, 24, 1994, pp. 18-29.

9. Tench, D. M., et al., "Production Validation of SERA Solderability Test Method", Soldering and Surface Mount Technology, 13, February 1993, pp. 46-50.

10. Tench, D. M., et al., "Reduction of Metallic Surface Oxides Via An Electrochemically-Generated Redox Species", J. Applied Electrochemistry, 25, 1995, pp. 947-952.

11. Hampshire, B. "Steam Aging and Solderability Testing", Circuits Assembly, February 1996, pp. 100-101.

12. DeBiase, J. et al., "Compatibility of PWB Coatings With Assembly Processes and Materials", Electronics Packaging and Production, February 1996, pp. 95 - 100.

13. Ivankovits, J. C. and B. M. Adams, "Effects of Controlled Atmospheres on Solder Wetting", Surface Mount Technology, October 1993, pp. 23 - 26.

14. Zarrow, P. and Debra Kopp, "Organic Solderability Preservatives", Circuits Assembly, February 1996, pp. 32-35.

15. Parquet, D. T., and D. W. Boggs, "Alternatives to HASL: A User's Guide for Surface Finishes", Electronic Packaging and Production, August 1995, pp. 38-42.

16. Boggs, D. W., "Anti-Tarnish: One Alternative to HASL", Electronic Packaging and Production, August 1993, pp. 34-38.

17. Beigle, S. and P. McGrath, "PCB Solderability...with no HASL", Printed Circuit Fabrication, 18 (No 12), December 1995, pp. 14-16.

18. Prasad, Ray R., Effects of Substrate Surface on Solderability", Surface Mount Technology, November 1995, pp. 26-29.

19. Slezak, Eric, "Soldering Materials Trends", Electronics Packaging and Production, December 1994, pp. 11-12.

The At-Temperature Mechanical Properties of
Lead-Tin Based Alloys

W. Kinzy Jones, Y. Q. Liu, Marc A. Zampino, and Gerardo L. Gonzalez
Department of Mechanical Engineering, Florida International University
Miami, Florida 33199, USA
Telephone: (305) 348-2345, Fax: (305) 348-2649
E-Mail: jones@fiu.eng.edu

Abstract

The mechanical properties (elastic modulus, yield point, ultimate tensile strength, and elongation) of five Pb-Sn solder alloys (63Sn/37Pb, 62Sn/36Pb/2Ag, 96Sn/4Ag, 5Sn/95Pb, and 10Sn/90Pb) commonly used in electronic packaging have been determined over the temperature range of -200 to 150 °C using uniaxial tensile test and an acoustic pulse method. It was found that the elastic moduli measured acoustically were 10 to 30 times greater than those measured mechanically, however, it is believed that the viscoelastic response of the materials may make the uniaxial tension test inaccurate for these types of alloys.

Key Words: Mechanical Properties, Pb-Sn Solders, Elastic Modulus

I. Introduction

The recent introduction of new materials for thermal management (AlN, diamond, composites) in the areas of large area dies, flip-chip assembly, and multi-layer, multi-material structures in hybrid and MCM fabrication, have placed new demands on the use and characterization of low temperature joining alloying used for die attach and assembly [1-4]. Although there have been steady advancements in computational modeling tools, such as finite element analysis [5,6], which can accommodate elastic, plastic, fatigue, and creep responses into the models, little is known about the at-temperature properties of the materials to be used to construct the models. In most cases, room temperature or "guestimated" values of the simplest required property (elastic modulus, thermal expansion, tensile strength, Poisson's ratio) are used, resulting in model predictions which are often of little value other than to indicate trends.

The mechanical properties of five Pb-Sn solder alloys were evaluated over the temperature range of -200 to 150 °C. The solder alloys tested can be broken into three groups: 1) Pb-Sn eutectics (63Sn/37Pb and 62Sn/36Pb/2Ag), 2) Pb-Sn solid solutions (5Sn/95Pb and 10Sn/90Pb), and 3) Sn-Ag eutectic (96Sn/4Ag). The elastic

modulus was determined using a mechanical (uniaxial tension) method and an acoustic pulse method. Uniaxial tension testing was also used to evaluate the yield stress (YS), the ultimate tensile strength (UTS), and strain elongation.

II. Experimental Procedures

Uniaxial Tensile Test

Tensile test results provide the response of stress and strain over a range of strain rates. The specimen is deformed, usually to fracture, with a gradually increasing tensile load that is applied uniaxially along the long axis of the specimen. This test gives information on elongation, UTS, yield point and verifies elastic modulus. The ASTM E8 standards cover sample preparation.

The test was carried out on an Instron Model 1011 tensile testing machine. Several strain rates were used $(0.0003 \, s^{-1}, 0.00067 \, s^{-1}, 0.0016 \, s^{-1}, 0.003 \, s^{-1}, 0.016 \, s^{-1})$ so as to find the one which minimized viscoelastic response of the materials. For the data presented in this paper the strain rate was $0.003 \, s^{-1}$.

Tensile samples had a total length of 3.8 cm, a gage length of 1.25 cm, and a thickness of 0.16 cm. Although not conforming to ASTM E-8, this sample size was

G. Harman and P. Mach (eds.), Microelectronic Interconnections and Assembly, 53-58.

selected so as to obtain small grain size in casting which would be more representative of the grain size after soldering. Three samples per test configuration were used and the results were averaged to reduce experimental error.

In order to perform testing over the range of -200°C to 150 °C, the tensile tester was fitted with a thermally insulated container and special small grips made of Inconel. The sample was immersed in a solution of liquid nitrogen/alcohol for low temperatures or heated glycerin for high temperatures. Sample temperature was monitored by a thermocouple. mass of the mixture maintained the temperature uniformly over the testing procedure.

Acoustic Pulse Method

Acoustics describes the phenomenon of mechanical vibrations and their propagation in different materials. Determining the modulus of isotropic solids requires two experimental measurements: 1) the density of the material, ρ, and, 2) the speed of propagation of longitudinal plane wave (V_L). The test specimen must have large lateral dimensions compared with the transducer diameter to eliminate mode conversion due to reflection at a lateral surface and guided wave model effects, since the equations for calculating elastic constant is based on the assumption of linear elastic wave propagation in an unbounded half-space. The modulus of elasticity is calculated by determining the transit time taken for the sound pulse to travel through the specimen in the pulse echo technique, or $E = (V)^2\rho$, where V is the extensional velocity, E is the Young's modulus and ρ is the density of the material. Using this technique, the shear modulus can also be determined, and hence, Poisson's ratio.

The acoustic pulse method was performed using a Panametrics 5055R Pulser-Receiver, a ETV-100 Magnetostrictive Transducer, and a magnetostrictive remendur wire wave guide. The sample was attached to the wave guide wire using a epoxy resin. The dimensions of a typical sample used for this test was 4.0 cm long, 0.40 cm width, and 0.019 cm thickness. The temperature of the sample was controlled by liquid immersion similar to that done for the uniaxial tension tests.

III. Discussion and Results

Yield Strength

The yield strength of the alloys was determined using the uniaxial tension test method over the temperature range of -200 to 150 °C. As shown in Figure 1, the yield strength of all the alloys decrease with increasing temperature, however, the eutectic and high tin alloys show a significant decrease with increasing temperature, while the high lead solders, 5Sn95Pb and 10Sn90Pb, remain fairly constant across the temperature range. The yield strength of the five alloys appears to converge at approximately the same yield strength at 150 °C.

Ultimate Tensile Strength

Uniaxial tension testing was used to obtain the ultimate tensile strength of the alloys over the temperature range of -200 to 150 °C. As shown in Figure 2, the ultimate tensile strength (UTS) decreases with temperature, with all the alloys approaching the same value at high temperature. The silver-bearing alloys, 96Sn/4Ag and 62Sn/36Pb/2Ag, had the highest strength of all the alloys, whereas, the high lead alloys, 5Sn95Pb and 10Sn90Pb, exhibited the lowest strengths.

Strain Elongation

The total elongation δ_T is contains both, the uniform elongation δ_U and the neck elongation dN. The total elongation increases with temperature for all materials. As shown in Figure 3, the elongation of all the alloys is almost identical except for the eutectic alloys, 63Sn/37Pb and 62Sn/36Pb/2Ag, which exhibit superplasticity (denoted by dashed lines) above 25 and 100 °C, respectively.

As shown in Figure 4, the uniform elongation, δ_U, of the eutectic alloys, 63Sn/37Pb and 62Sn/36Pb/2Ag, is almost constant with temperature change. However, for 5Sn/95Pb and 10Sn/90Pb a peak appears at approximately 50 °C which is attributed to both twinning and slip mechanisms. In the case of 96Sn/4Ag, a phase transformation occurs at approximately room temperature causing the decrease in the elongation.

The neck elongation of the alloys is shown in Figure 5, with trends similar to those for the total elongation. Again, the eutectic solders exhibit superplasticity. In contrast to total elongation (Figure 3) the high lead alloys, 10Sn90Pb and 5Sn95Pb, now show a smaller neck elongation than the 96Sn4Ag alloy at all temperatures.

Elastic Modulus

Acoustic measurements of the elastic modulus of the five alloy systems were made over the temperature range of -200 to 100 °C, and are shown in Figure 6. All the alloys show a decrease in the elastic modulus with increasing temperature following a nearly linear trend. As shown in the figure, the elastic moduli break into three groups: 1) Pb-Sn solid solutions (10Sn90Pb and 5Sn95Pb) which have the lowest elastic moduli, 2) Pb-Sn eutectics (63Sn/37Pb and 62Sn/36Pb/2Ag), and 3) Sn-Ag eutectic (96Sn/4Ag) which has the highest moduli. A linear trend-line is given for each of the three groups in Figure 6. It is interesting to note that the temperature dependence of the Pb-Sn alloys is approximately 0.08 Gpa/°C similar, whereas, the Sn-Ag eutectic's dependence is 0.40 Gpa/°C, which is about 5 times greater. The data in Figure also show a convergence of the elastic moduli of all the alloys above 80 °C.

The elastic moduli was also obtained using the uniaxial tensile test over the temperature range of -200 to 150 °C, with the results shown in Figure 7. The data show similar trends to those found acoustically, except for relative magnitude of the moduli. A comparison of the elastic moduli shown in Figures 6 and 7 show that the acoustically measured moduli are 10 to 30 times greater than those obtained mechanically. This trend appears consistent with the results presented by Savage and Getzan (1990) for 63Sn/37Pb and 60Sn/40Pb [7]. Furthermore, the results found in this study for 63Sn/37Pb show excellent agreement with the data in their work.

The discrepancy between the elastic moduli measured mechanically and acoustically may be explained by comparing the test methods. Since the acoustic pulse method is based on the traveling of sound waves through the sample, it is independent of dislocation motion, localized deformations, or grain structure morphology. Hence, it is strain rate independent. In contrast, the uniaxial tension test is very strain rate dependent for viscoelastic materials such as the alloys in this study, which exhibit creep responses at relatively low temperatures.

During uniaxial testing, the elongation of the sample is intermittently stopped for a short period so that applied stress can be determined for the elongation. Stress relaxation was observed to occur during these periods which may have effectively lowered the measured elastic modulus. However, the exact mechanisms involved are not fully understood and are currently under investigation.

IV. Conclusions

The yield strength decreases with temperature increase for eutectic alloys (63Sn/37Pb, 62Sn/36Pb/2Ag,

and 96Sn/4Ag). However, for the high lead alloys (5Sn/95Pb and 10Sn/90Pb), the yield strength changes little below room temperature because of its twinning deformation mechanism.

The UTS decreases with temperature increase the value of the UTS approaching approximately the same value at 150 °C. The data show that the addition of silver to the eutectic alloys produces the strongest materials, and that additional lead reduces the UTS.

The elongation decreases with temperature increase for all solders. The lead-tin eutectic alloys (63Sn/37Pb, 62Sn/36Pb/2Ag) exhibit superplasticity. Also, twinning in the high lead alloys causes a small decrease in the neck elongation and a significant increase in the uniform elongation as compared to the other alloys.

The elastic modulus decreases linearly with increasing temperature over the range of -100 to 100 °C. The data show that elastic modulus is driven mostly by the amount of lead, and to a lesser extent, the addition of silver, in the material.

Due to the viscoelastic nature of these alloys, uniaxial tensile testing may not be an accurate method for measuring the elastic modulus at temperatures above 0.3 times the melting point due to stress relaxation during the testing process.

Acknowledgments

The authors wish to acknowledge Teledyne Electronic Technologies (TRAP program) and the U.S. Air Force Base (FAST Center for Cryo-Electronics for their support of this research.

References

1. W.K. Jones, R.W. Harshbarger, W. Jiang, and R. Zhuge, "Evaluation of Braze Process for AlN Packages," *Int. Journal of Microelectronics*, Vol. 14, No. 1, March, 1991.
2. R. Zhuge, W.K. Jones and S.S. Maganti, "Stress Analysis of Multilayer Multichip Module (MCM-D) Structures Using the Series Solution of Elasticity Theory, *Proc. International. Microelectronics Symposium*, Japan, 1992.
3. G.L. Gonzalez, Y.Q. Liu and S.S. Maganti, "Physical Properties of Low Temperature Solders and Die Attach Materials," Proc. of Southcon 95, Ft. Lauderdale, 1995.
4. H.S. Morgan, "Thermal Stresses in Layered Electrical Assemblies Bonded with Solder., Trans. of ASME J.

56

of Electronic Packaging, Vol.113, pp 350-354, December, 1991.

5. S.S. Maganti and W.K. Jones, "AuSn Brazing for Hermetic Seal Ring Attachment to Post-Processed MCM-D's", Technical Report, 1995.

6. Y. Pao, K.L. Chen, and A.Y. Kuo, "A Nonlinear and Time Dependent Finite Element Analysis of Solder Joints in Surface Mounted Components Under Thermal Cycling, *Proc. Materials Research Society, Symposium*, Vol. 226, pp 23-28, 1991.

7. E.I. Savage and G.D. Getzan, "Mechanical Behavior of 60 Tin/40 Lead Solder at Various Strain Rates and Temperatures, *Proc. Int. Sym. on Microelectronics, ISHM*, Chicago, 1990.

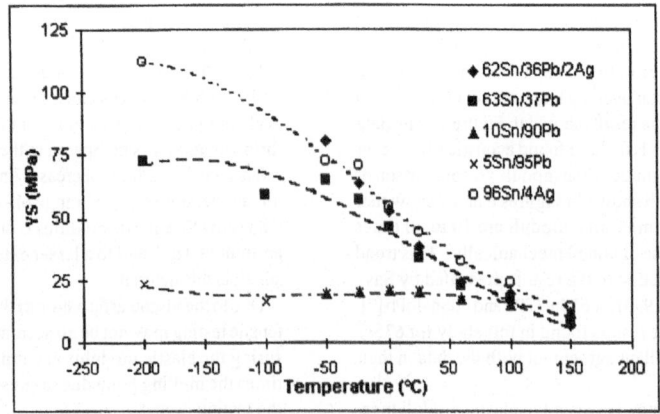

Figure 1. Yield strength obtained using uniaxial test method.

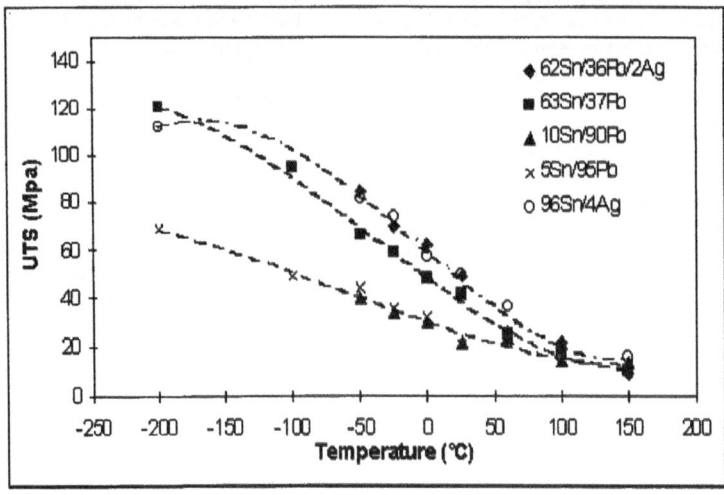

Figure 2. Ultimate tensile strength (UTS) obtained using uniaxial tension test.

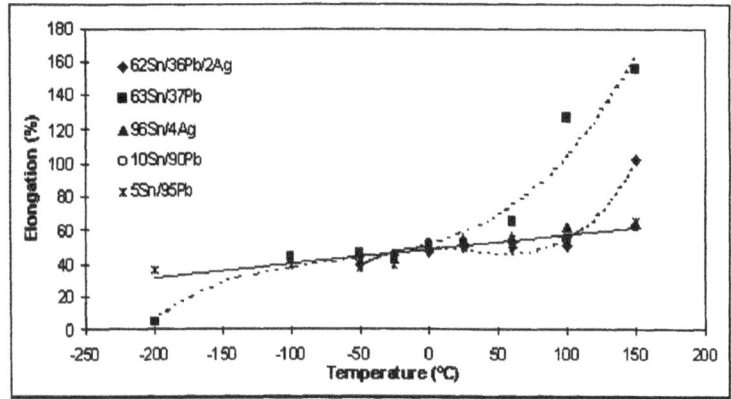

Figure 3. Total elongation obtained using uniaxial tension test.

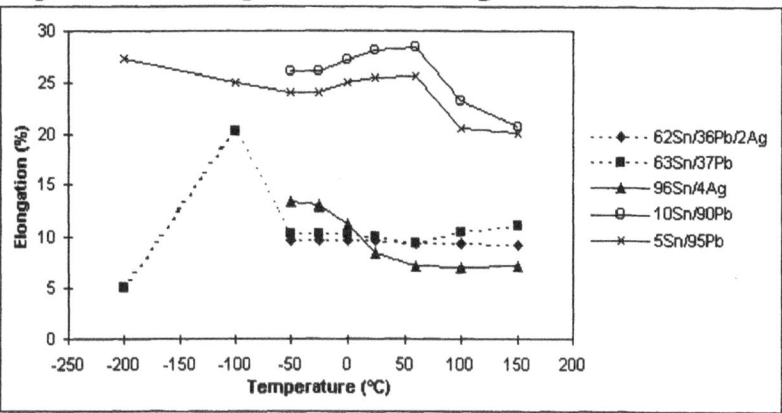

Figure 4. Uniform elongation of tained using uniaxial tension test.

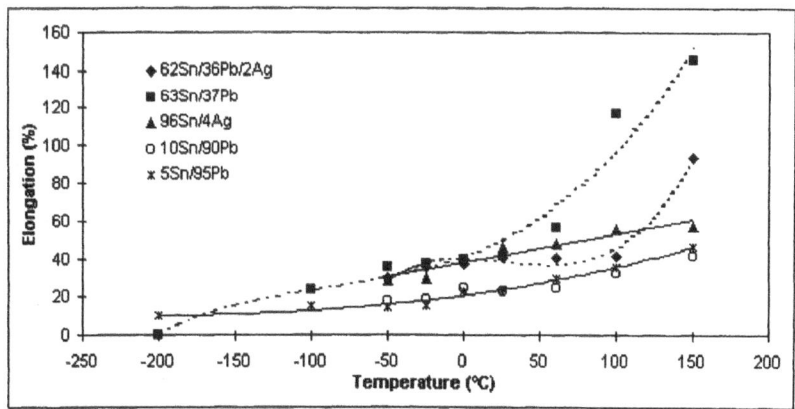

Figure 5. Neck elongation obtained using uniaxial tension test.

58

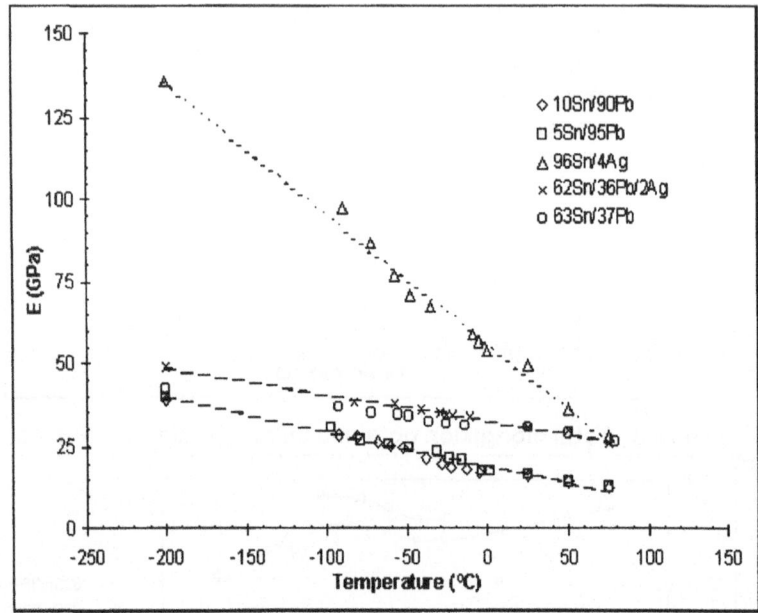

Figure 6. Young's Modulus obtained using acoustic pulse method.

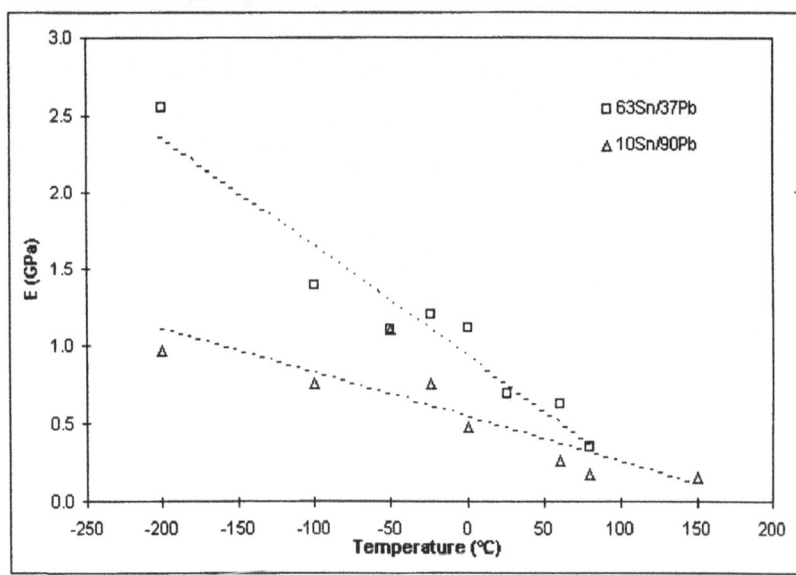

Figure 7. Young's modulus obtained using uniaxial tension test.

Flip Chip Technology: Is It Time of Mass Production?

Carlo Cognetti
SGS THOMSON Microelectronics
Agrate (Milan) - Italy

1.- INTRODUCTION

Development and early use of Flip Chip Technology dates back in the 70s, when it was applied *on ceramic substrates* by a very limited number of Companies in computer by first and than in automotive systems.

Mass productions were essentially based on *laminate substrates* and on plastic first level package with wire bonding interconnection and epoxy encapsulation.

Since then, progress of plastic package was very impressive: *packaging density* (ratio between chip size and package footprint on the board) moved from a few percent of Dual In line Package (DIP) and early Small Outline Package (SOP), to the 30-40% of to-day most aggressive Plastic Quad Flat Package (PQFP) and Thin Small Outline Package(TSOP).

However, requirement of further IC complexity (tab.1) and system miniaturisation is so strong that this evolution of plastic package is presenting limitations in an increasing number of volume applications.

Basically, the industry is demanding smaller dimensions - *size* and *thickness* - of the system and higher performance of the semiconductor chip, with *higher pin count* and operating *speed*.

In principle, flip chip technology - with very high packaging density, minimum profile, minimum signal distortion, and capability of very high I/O count - is *THE ANSWER* to all of above requirements in the same time.

It seems that a convergence of technologies is taking place which - by the first time - may pay back the cost of infrastructure the industry needs for using flip chip in mass productions, with the development of relevant processes on chip and on the next level of interconnection.

The paper will overview the reasons of this convergence presenting the present trends at silicon and system level; alternative solutions - as the Chip Size Package (CSP) - will also be discussed which, *if cost effective*, will become the most serious competitor of flip chip.

G. Harman and P. Mach (eds.), Microelectronic Interconnections and Assembly, 59-75.
© 1998 Kluwer Academic Publishers.

2. - WAFER BUMPING AND FLIP CHIP

Key point of the flip chip technology is the existance of bumps on the die surface, deposited on top of the final metallization, in selected areas opened in the final passivation.

Several methods exist for deposition of bumps: evaporation through a mask, electroplating, chemical plating (electroless).

The most common process is shown in fig.1. Wafer processing is at the final step, with the final passivation opened on the external contacts (fig.1.1) ; a continuous film of suitable metals (TiWCu, TiW/Ni) is sputtered on the wafer and provides both a good adhesion to the aluminium exposed in the contact and to the bump which will be deposited (fig.1.2); after a mask of photoresist is applied (fig.1.3), a layer of Sn or In alloy is grown (fig.1.4). Photoresist is then removed (fig.1.5) and the base metallization etched using the overlaying bump as mask (fig.1.6). As final operation, bumps are melt in order to form a dense and regular sphere of solder alloy.

It must be noticed that other technologies use Ni or Cu or Au as main material of the bump : in this case the final reflow is not needed.

After testing and singulation, die is mounted "face down" on the next level of interconnection which may be a single chip package (BGA and Pin Grid Array - PGA are an ideal carrier) or directly on the board.

The industry is developing mass reflow methods: target is the utilisation of flip chip more or less as an SMD, with similar pick and place, reflow and rework concepts.

3. - FLIP CHIP AND SIZE REDUCTION

Ratio between chip size and the area needed to connect the chip to the board is a parameter which may represent the degree of evolution of a given interconnection technology.

This parameter is known as "packaging density" or "efficiency of interconnection" .In fig.2, three cases are considered:

1] Chip packaged in PQFP

A PQFP of last generation is shown, for example a (10x10) sq mm PQFP, which can accommodate a (7x7) sq mm chip. As the lead length is 1 mm, package footprint is (12x12) sq mm and packaging density Dqfp is:

$$D \; qfp \sim 30\%$$

This is an excellent performance, which can explain the increasing popularity of PQFP in portable electronics; it is one order of magnitude better than old DIP and early SO.

2] Wired Chip on board

In this case, chip is glued of the board and wire bonded. Considering again a 7x7 sq mm die and 1.5 mm long wires, area needed for interconnection is (10x10) sq mm and efficiency Dwb is:

$$Dwb \sim 50\%$$

3] Flip Chip on board

The third case is a bumped chip, flipped on the board. No package is used, as in the previous case, but the additional space is minimum.

In theory, chips can be accommodated very close to each other. In practice, the interconnection density for flip chip Dfc is:

$$Dfc \sim 70\text{-}75\%$$

with tremendous impact on system miniaturisation.

It should be considered that important differences exist between the two last cases:wired chip on board (chip and wire) is a consolidated technology offering the big advantage of flexibility in the board lay-out, as wire bonding process gives some freedom in adapting a certain number of die sizes and bonding pad configurations to the same footprint.

However, wire bonding process is relatively slow (4 to 6 wires per second , which means minutes for a high pin count device) and impractical on large substrates.

On the opposite side, flip chip offers no tolerance in adapting lay-out and board footprint - this has to be considered its most important disadvantage - but, at least in principle, it can become compatible with the same mass placement and reflow process of Surface Mount Technology (SMT) and with the same infrastructure. Indeed, this is the target of intensive development work in many Companies, both equipment suppliers and final users.

4. - FLIP CHIP AND LOW PROFILE

In previous paragraph, size reduction has been considered as a result of shrinkage of footprint. But also the z-dimension has impact on the final product , as it is evident in disc drives, portable phones, flat panel displays, PCMCIA cards.

Profile of first level package has been reduced in the past 5 years from more than 3 mm typical of Plastic Leaded Chip Carrier (PLCC) and early PQFP to leass than 1 mm for the Ultra Thin Small Outline Package (UTSOP) and Ultra Thin Quad Flat Pack (UTQFP) of last generation.

As shown in fig. 3, the evolution under 1.4 mm requires a thickness of the die - and therefore of the wafer the die is coming from - which is lower than the standard production.Wafer thinning is performed by mechanical grinding of wafer backside; requested die thickness for ultra thin packages is <250-300 /um, which becomes very critical with 8" (20 cm) diameter wafers.

This is a first drawback of very thin package; but others exist, as worpage and reliability, which are very sensitive to small variations in the process and in the characteristics of materials.

Flip chip technology offers an effective answer to the request of very thin mountings and profiles consistently lower than 0.5 mm can be obtained, with the possibility of producing 1.5 mm cards, populated in both sides.

5. - FLIP CHIP AND HIGH PIN COUNT

All families of ICs are moving to a denser content of functions, which implies a higher number of connections to the external circuit. This is especially evident in high performance ASICs, which are expected to move up to 1000 I/O in the near future.

But while I/O count is growing, chip size is getting smaller due to the continuous increase of scale integration : in some case, chip dimensions are defined by the number of bonding pads, which are accommodated along the periphery of the die at a constant pitch of 100 /um. For example, a 10 mm die has optimised design if I/O count does not exceed about 100 on each side - with a total of about 400. If more than 400 pads are needed with the same die size, design becomes *pad limited* : this means that the silicon has some empty area and the cost is impacted .

Pad limited dice are often experienced with the 0.5 /um technology and will be very common with the next generations at 0.35 and 0.25 /um.

In order to reduce the problem, wire bonding technology is evolving to smaller pitch, from 100 /um to 80 and than 50 u/m (fig.1) . Up to about 800 I/O will be possible in the previous example of a 10x10 mm die.

This is not bad.

However, in the common practice no active structure is designed under the bonding pad , in order to avoid reliability problems in case the structure is mechanically damaged at the bonding process. This means that the bonding pad and a small surrounding area are "passive" for what refers to both the circuitry and its cost.

Considering that the passive area of one pad is about 10.000 sq microns, in the previous case of 400 pads in a 10x10 mm die , the total passive area is about 4 sq mm , equivalent to the 4% of die size and cost. A real breakthrough is offered by the flip chip technology, which allows to accommodate the bonding pads in a matrix of contacts inside the die area . If their pitch is 250 /um , the matrix has order 40 , with potentially 1600 pads in the 10x10 mm die.

It must be noticed that flip chip soldering does not cause stress in the underlying layers and all chip area is "active" , with some cost saving.

As a consequence of above advantages, BGA and PGA packages will move from wire bonding to flip chip in order to cover very high pin count designs. Indeed, BGA and PGA have a very adequate structure: basically they are multilayer chip carriers with a system of interconnections from the upper side to the lower side. Chip is glued and bonded (fig. 4) onto the upper side, and than protected against external agents.

At this point, in the BGA, ball contacts are applied on the lower side of the package, which is now ready to be surface mounted on the external circuit. Basically, BGA is the surface mount version of PGA, with outer contacts distributed according to an area array configuration and no more along the package periphery.

Early BGA and PGA used a ceramic base plate; chip was accommodated in a sealed cavity. Due to cost constraints, to-day mass production moved to plastic materials, laminate baseplate and epoxy encapsulation

Main characteristics of BGA are the very high pin count capability - up to 1000 -, and the good flexibility in design. Indeed, while the leadframe is a "rigid" system with one layer of interconnections, the BGA / PGA substrate is a flexible system, with more layers of interconnection and capability of much finer lines than a leadframe.

In the last five years Ceramic ball grid arrays were industrialised, whose improved design was able to carry very high pin count flip chips. It is now the moment of developing equivalent solutions in laminate material, in order to serve high volume, cost sensitive markets. A possible structure is shown in fig. 5 : Flip chip is mounted on top of a laminate substrate, which in the cost effective versions has 2-4 copper layers, but can move up to 7-10 layers if needed.

After chip mounting, a liquid coating is dispensed, which fills the gap between chip and substrate, for environmental and mechanical protection. The structure is compatible with the placement of a metal lid on top of the silicon chip, which is a good vehicle for removal of the heat from the IC to the external ambient.

A big amount of work was done and will be done in order to optimise the reliability of this structure, which is basically given by the quality of underfill: its main job is the reduction of mechanical stress from the substrate to the chip when structure is submitted to thermal cycling and shear stress is generated at the solder joints, due to the large mismatch in the thermal expansion coefficients. Joint integrity is quite critical and only with good underfilling material and process control, satisfactory results can be achieved.

It is an important subject, which however is beyond the purposes of this paper.

6. - FLIP CHIP AND IC SPEED

According to tab. 1, operating frequency of ASICs, will move from 150-250 Mhz to 300-600 MHz in the next 5 years.

In the same time, faster memories are designed, and with the improvement of satellite communication, low cost RF modules are needed. In several cases, first level package may become the limiting factor of device speed, due to the parasitic effect of leadframe and wires.

Inductance can be used as parameter for representing the "electrical" performance, of a given interconnection technology.

In case of PQFP , this parameter ranges between 8 and 12 nH, respectively for the short lead located at the centre of each package side and for long leads at the corners.

Inductance can be decreased embedding ground plane in the package body. This is the case of power enhanced PQFP, where the ground plane is a metal heatsink intended to improve thermal dissipation. Shielding offered by this plane can reduce inductance from previous 8-12 nH to 6-8 nH.

A better performance is given by BGA, by taking advantage of its multilayer structure. BGA can be designed in order to minimise parasitic effects and high speed lines can be shielded by surrounding ground and power supply planes.

Best measured performance of wire bonded BGA is 2-2.5 nH, compatible with advanced ASICs, in 0.5 /um technology . It is interesting to notice that , in the optimized package, design gives little contribution to device noise, which is mostly due to the wires, their length and shape.

Espected inductance of flip chip BGA is below 1.0 nH, with the same optimized package considered in the previous example. This means a flip chip BGA can cover the speed requirements considered in tab.1.

7. - *CHIP SCALE PACKAGE vs FLIP CHIP*

In par. 3 it has been considered that most aggressive miniature PQFP have a "packaging density" not far from 30%.

In several companies, developments are underway in order to get new packages whose interconnection density is 50-60%, "custom tailored" on the die.

The name "Chip Scale Package or Chip Size Package (CSP)" , is given to this small interconnection system. In fig.6 , an early member of CSP family is represented, known as microBGA [r] .

It uses a kapton interposer which connects a system of peripheral bonds towards the silicon to an array of balls towards the substrate. On chip connection is capable of very fine pitch (50 μ) typical of Tape Automated Bonding (TAB) technology. A polymer mechanical buffer is placed between chip and interposer, whose function is the mechanical decoupling of chip and substrate, in view of better resistance to thermo-mechanical stress.

It can be noticed that in MicroBGA [R] ball array is smaller than the chip itself, with "fan-in" configuration. Ball array has a pitch of 0.5 mm , which, in principle, can accommodate 300-350 contacts in a 10 x 10 mm^2 die .

This and similar CSPs, offer three possible advantages on the flip chip:

1]   "standard" outline and footprint, in the sense that the industry will choose a few of these CSP and will standardise their dimensions and the position of the contacts to external PCB, which makes "footprint compatible" and therefore interchangeable products coming from different sources;

2]   easier handling and testing, because chip is mechanically protected and fits to a "standard" socket;

3]   easy compatibility with existing SMT infrastructure - because, once packaged and protected, chip can be positioned and soldered on the PCB with the same volume process used for all other SMDs.

On the contrary, flip chip take advantage of wafer level bumping, with associated high productivity, offers extreme miniaturisation and may become a cost effective solution.

MicroBGA is a registered trade mark of Tessera inc. USA

----

8. CONCLUSIONS

Advanced   electronics industry is facing a number of limitations deriving from present interconnections technology, which impacts high performance ICs - in terms of pin count and speed - as well as highly miniaturised systems., in terms of size and tickness.

The flip chip technology   is taking advantage of "bumps" grown in correspondence of electrical contacts in the chip; when mounted on a first level package - like BGA and PGA - or directly on a second level substrate, chips are flipped and bumps are soldered on corresponding areas.

This technology was developed more than 20 years ago for specific computer application, on ceramic substrates. To-day, the industry is devoting significant efforts to transfer the same approach on plastic (laminate) substrates.

In fact, as represented in fig. 7,   there is a convergence of needs whose ultimate solution is considered to be the flip chip technology. Essential condition of this evolution is the development of an industrial infractructure in terms of materials, equipment, processes - compatible with the existing SMT industry. If this will be achieved, movement to flip chip technology will take place in several market segments, due to the intrinsic saving.

There is a serious competitor of this process: a very miniaturised, cost effective package, with packaging density close to   60%.   The industry is developing a number of types, using leadframe or plastic interposer, known as Chip Scale Packages (CSP). Main advantage of CSP when compared to flip chip is the same standard footprint for different devices and manufacturers.It is an enormous advantage, which, however, must stay cheap in order to stay competitive. Cost effectiveness of CSP has not yet been proven.

Captions of the figures

1]   Typical bumping process by electroplating. A thick layer of soft solder alloy is
       deposited on selected areas of the chip and than melted, in order to have spherical
       bumps - Process takes place on wafer .

2]   "Packaging density" (ratio between chip size and PCB area needed to connect the
       chip) in three typical cases : PQFP, Chip on board and flip chip.

3]   Evolution of package profile in the last few years. Thickness was decreased from 3
       mm to 0.7-0.8 mm.

4]   Evolution of BGA and PGA, from wire to flip chip bonding, caused by increased pin
       count and IC speed.

5]   Structure of flip chip BGA, showing the chip soldered on a laminate carrier, the
       underfill and the lid, with the double function of mechanical protection and heat
       removal.

6]   Example of Chip Scale Package (CSP) : microBGA by Tessera Inc. USA.

7]   Evolution towards flip chip interconnection is ruled by four concurrent driving forces:
       pin count and speed from the view point of the IC; miniaturisation in size and profile
       from the view point of the system. For system miniaturisation, Chip Scale Package
       (CSP) is the most serious competitor of flip chip.

IC Technology Roadmap

Tab.1

Semiconductor Industry Association (SIA)

	1995	1998	2001
Chip size logics [mm2]	400	600	800
DRAM [mm2]	200	320	500
metal layers	4-5	5	5-6
* Power Diss [W/die]	15	30	40
* I/O count	750	1500	2000
* Speed off chip [MHz]	100	175	260
on chip	200	350	500

Bumping by electroplating

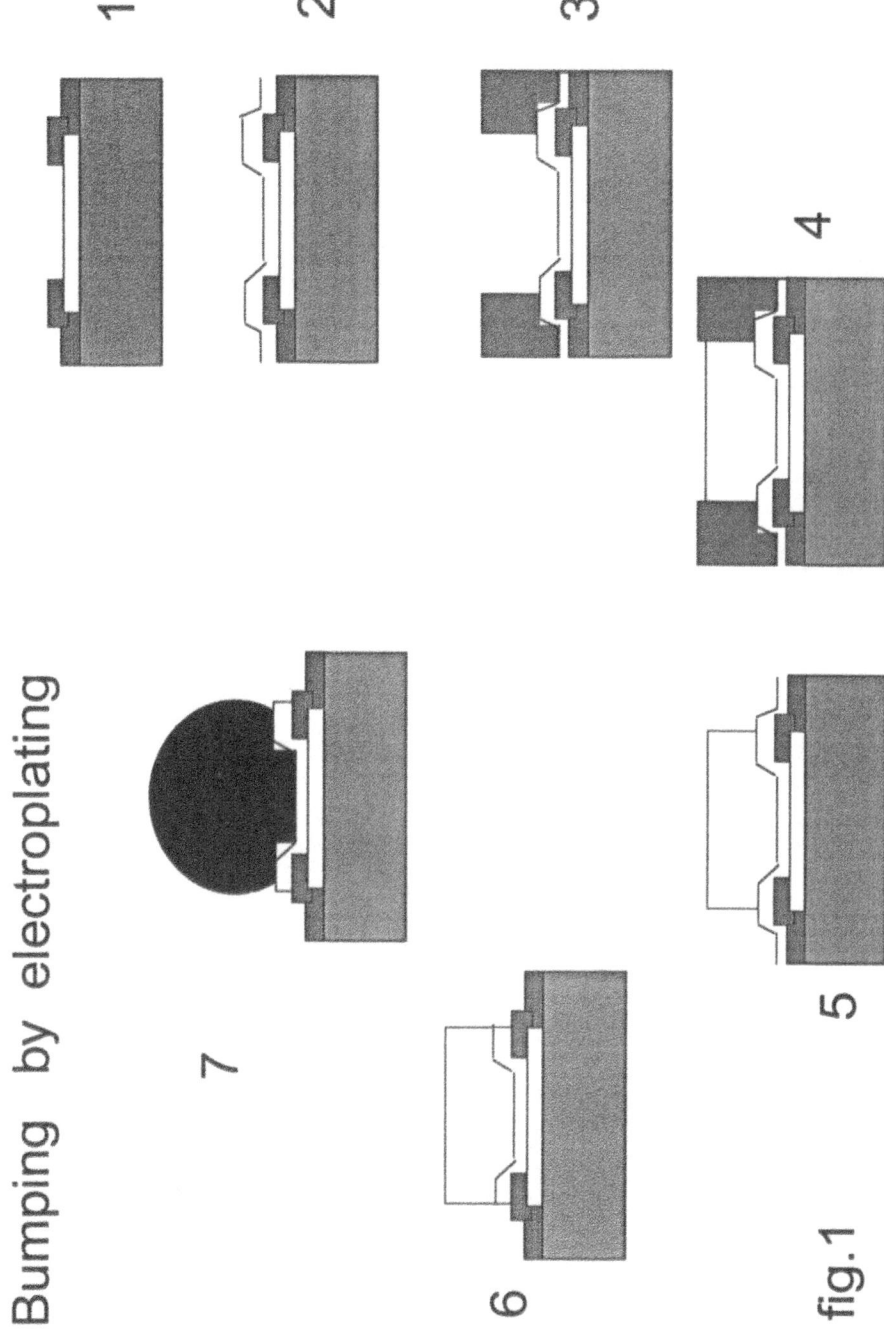

fig.1

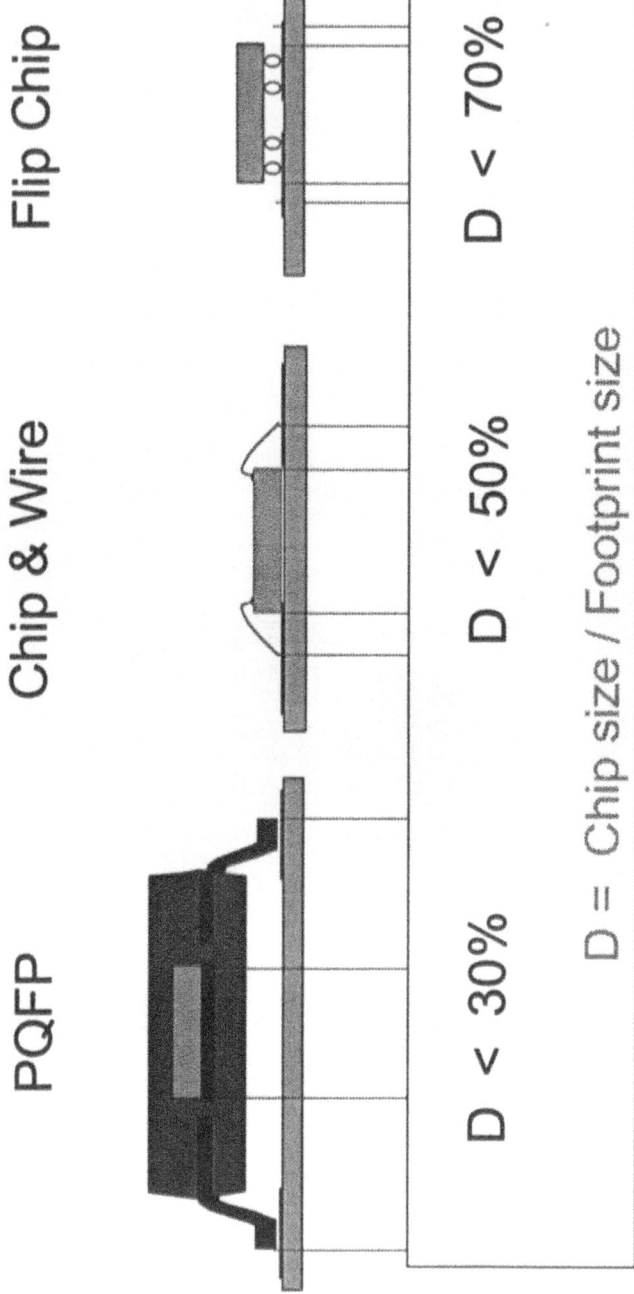

70

fig.2

Packaging Density D

PQFP   Chip & Wire   Flip Chip

D < 30%   D < 50%   D < 70%

D = Chip size / Footprint size

fig.3

Evolution of package profile

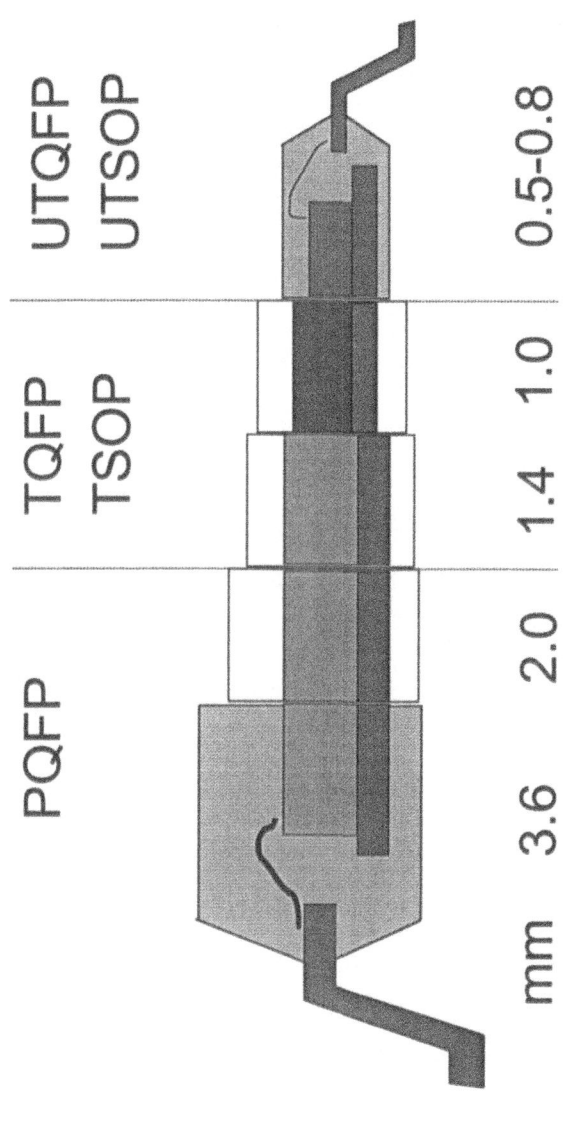

PQFP

TQFP
TSOP

UTQFP
UTSOP

mm   3.6   2.0   1.4   1.0   0.5-0.8

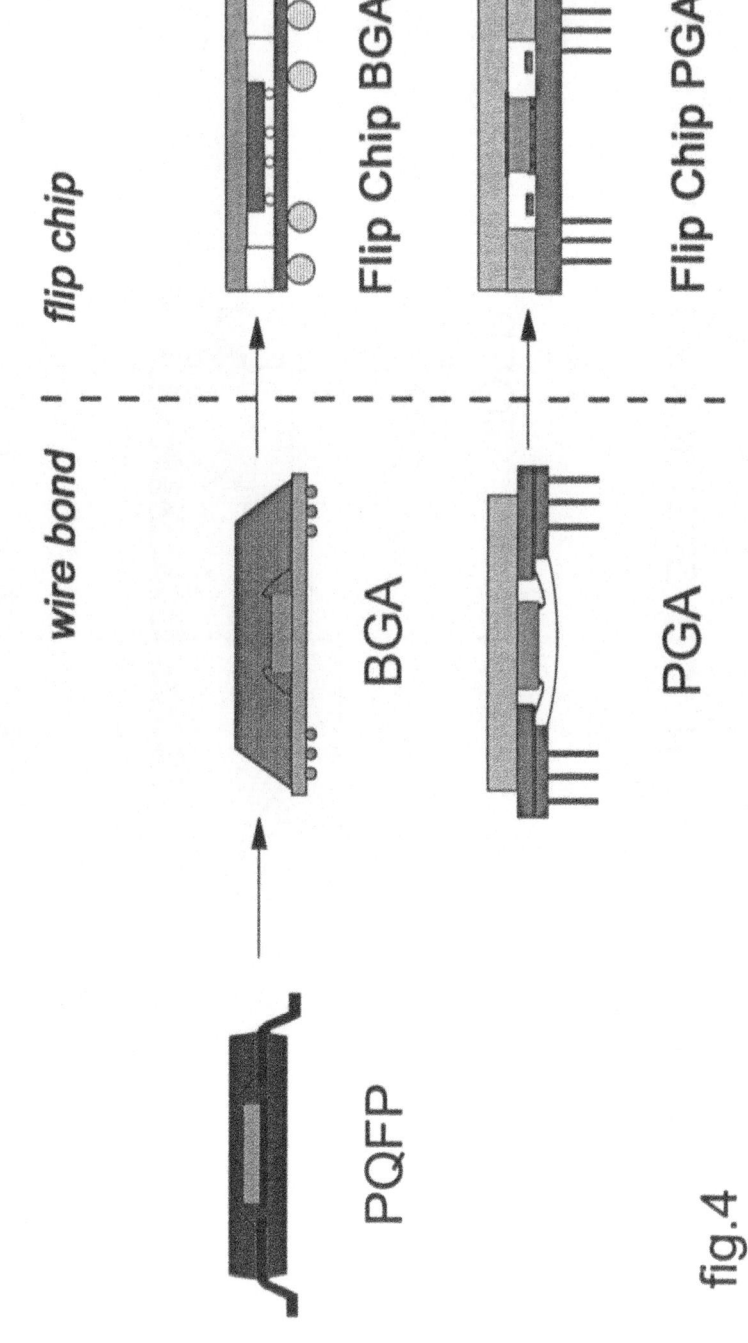

High Performance Package Roadmap

wire bond

flip chip

PQFP

BGA

Flip Chip BGA

PGA

Flip Chip PGA

fig.4

Tape Ball Grid Array (TBGA)

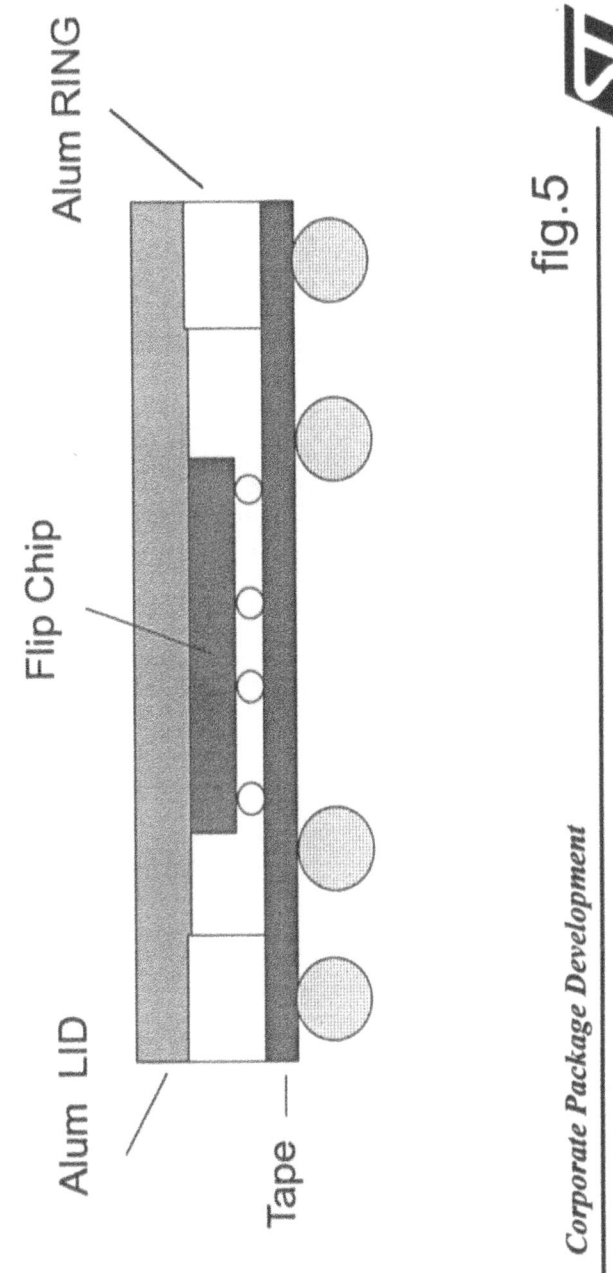

Alum RING

Flip Chip

Alum LID

Tape

fig.5

Corporate Package Development

73

micro BGA

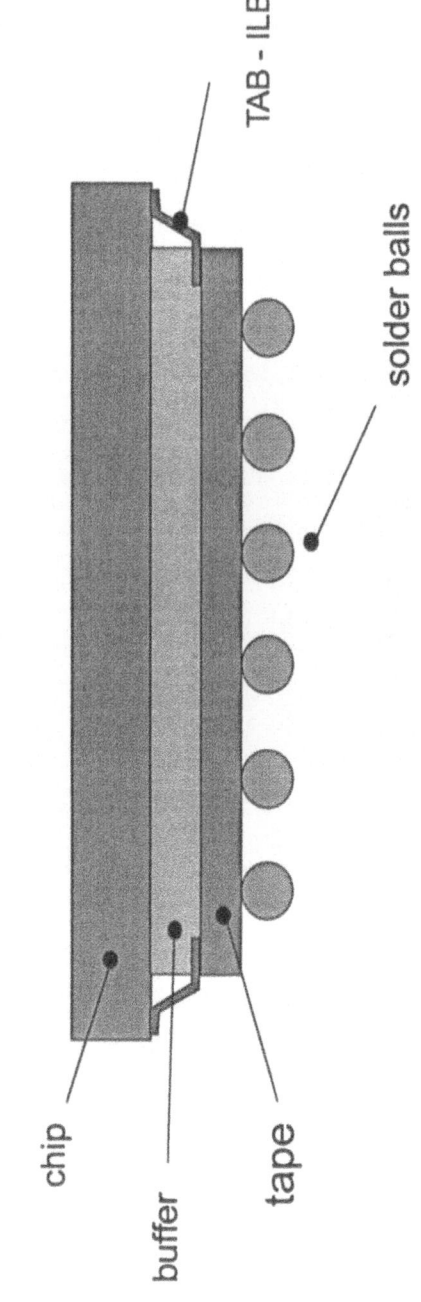

TAB - ILB

solder balls

chip

buffer

tape

✓ Developed by Tessera Inc (USA)

fig.6

fig.7

Convergence
to Flip Chip

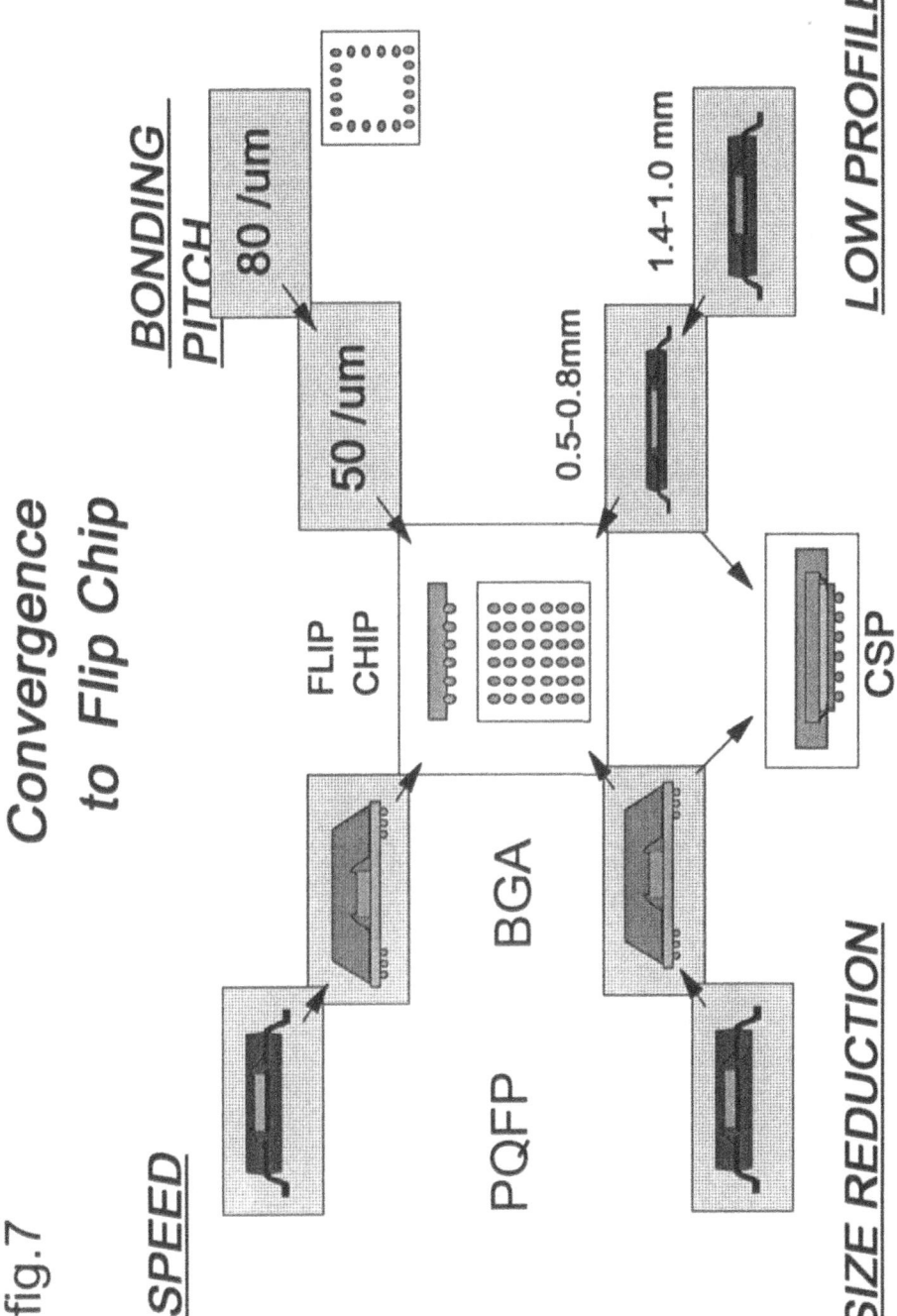

SPEED

BONDING PITCH

LOW PROFILE

SIZE REDUCTION

PQFP   BGA

FLIP CHIP

CSP

80 /um

50 /um

0.5-0.8mm

1.4-1.0 mm

<u>Flip-Chip - The Ultimate Solution</u>
<u>Comparing Flip-Chip and Chip Scale Packaging</u> *

Bill Brox and Katarina Boustedt
IVF
Argongatan 30, S-431 53 Molndal, Sweden
ph: +46 31 706 6141, fax: +46 31 27 6130,
email: kb@ivf.se, homepage: http://www.ivf.se

ABSTRACT:

Flip-chip is becoming increasingly more important and interesting since it has some of the most desired advantages, namely short interconnects for improved electrical performance, and reduction of size and weight leading to cost savings since less silicon and board material is used. Still there are some severe drawbacks, mainly the process time, especially related to underfill and the poor availability of low-cost substrates with the fine lines and spacings required for area array flip-chips.

The remaining challenges with flip-chip may lead potential users to turn to a new package type called Chip Scale Package (CSP). These have the benefits of ordinary surface mount components, being encapsulated, tested, and mountable in high-speed machines, and the benefits of flip-chips in size, weight, and electrical performance. Future users have voiced questions as to the availability of CSPs and the general prospect of a good infrastructure for employing CSPs. The infrastructure is still hardly in place for flip-chips, which have been in production since 1964, so there are reasons to doubt that it will be much easier to receive bare die for packaging in CSP formats. The willingness in a company to adopt either flip-chip or CSP technology is based on the previous experience, the company background. This means that a company assembling bare die today will have less inertia in switching to flip-chip than a company working with component placing. The latter will more readily turn to CSPs.

In some cases, the ruggedness of the CSP is questioned. Is it really as robust as a standard SMT package? The automotive industry have requested that some CSPs be qualified according to bare chip standards. The reliability of the package, and, even more importantly, the reliability of the package soldered onto a board still need further, unbiased investigation.

---

* Paper was presented. Written paper never received.

G. Harman and P. Mach (eds.), Microelectronic Interconnections and Assembly, 77.
© 1998 *Kluwer Academic Publishers.*

Solder and Solderless Flip Chip Assembly on a   MCM-D in a BGA Package *

Peter Bodo, Hans Hentzell, Jan Strandberg, Joacim Haglund and Sima Valizadeh
Industrial Microelectronics Center
Teknikringen 3, S-58330 Linkoping, Sweden

Abstract

With increased demands on system integration, performance and low production cost, flip chip mounting of dies has become an attractive technology. Flip Chip assembly using 60/40 solder is a well established technology but due to environmental concerns non solder based solutions will very soon be required. In this work we will compare assembly using solder bumps with assembly using adhesive bumps both for the interconnection between the chip and the substrate and between the package and the board.

Using specially designed MCM-D test vehicles, flip chip mounting of dies has been evaluated with respect to electrical and mechanical reliability, tested by temperature and power cycling tests. The study includes solder bumping, solderless bumping and adaptive mounting of dies.

Dummy Si dies with a Tin-SiOxNy Al SiOxNy structure were used. The TiN serve as a resistance layer for power dissipation. The Al layer contains temperature sensors and daisy chains for signal continuity control. Pads were deposited with Ti-Cu-Ni for electroplating of Sn/Pb 60/40 solder bumps. The size of dies range from approximately 7-150 mm^2. The pad pitch range from 140-250 μm in structures of circumferential bumps as well as rectangular and circular bump arrays. The largest number of bumps on one die is 3,136.

The Si based MCM-D substrates have and Al-BCB -Al-CB structure with Ti-Cu-Ni-Au deposited on pads. There are temperature sensors and routing for signal continuity control. One type of substrates being 42 x 42 mm is assembled in a custom design PBGA package manufactured by Bull SA. This vehicle is designed for tests in a specific application being a part of the front end read out electronics (FERMI) for the large hadron collider (LHC) detector to be built at CERN.

In this presentation we will report on assembly of dies on the Si-substrate and assembly of the package on the board. Regarding performance and reliability of modules, a comparison between different solder plating processes as well as solderless bumps and their applicability to different pad pitches and structures will be presented.

* Paper was presented. Written paper never received.

G. Harman and P. Mach (eds.), Microelectronic Interconnections and Assembly, 79.
© 1998 *Kluwer Academic Publishers.*

Solder Bumping for Flip Chip Interconnections

P. Annala, J. Kaitila, J. Salonen and I. Suni

VTT Electronics
Otakaari 7 B, 02150 Espoo, Finland

Introduction

Flip chip mounting of bare dice is gaining widespread use in microelectronics packaging. The main drivers for this technology are high packaging density, improved performance at high frequency, low parasitics, potentially high reliability and low cost. Many companies have made significant efforts to develop technology for bump processing, bare die testing and underfill encapsulation to take the benefit of all potential advantages.

For flip chip mounting, the bonding sites have to be prepared for a direct attachment of the chip on the substratrate metallisation. In most cases this is accomplished by building a highly conducting bump on the chip side. Hence, additional steps are required in the back end of the wafer processing or, alternatively, on the level of individual chips. Two major techniques have been used to add bumps on wafer level: physical vapor deposition (either evaporation or sputtering) of the bump metallurgy or electrochemical plating of the solder. The first one requires vacuum processing of relatively thick films. It was pioneered by IBM already in the 60s as a highly reliable high-density packaging method [1]. As a result of the large capital investment and low efficiency this approach has been usually considered too expensive for low-cost volume production. The chemical plating techniques are viable alternatives for vacuum processing with high efficiency and potential savings in equipment cost [2]. Again, two generic processes are available for chemical deposition: electroless plating and electrolytic plating. The electroless method, while extensively used for the deposition of nickel, copper and noble metals, has not emerged as a major deposition method for solder alloys. Although quite straightforward, electroless chemistry does not offer simple control over solder composition and uniformity. The electrolytic plating process of solder alloys has been under extensive study by many groups. It is highly selective and efficient compared to vacuum deposition. It can be scaled to large wafer diameters, small bump size and fine pitch. Binary or ternary solder alloys can be deposited from mixed solutions or by sequential steps of elemental metal deposition.

The present work focusses on the development of a solder bump manufacturing process using electrolytic deposition. Both low temperature ($T_m < 200$ °C) and high temperature $T_m > 300$ °C) solder alloys are discussed. The seed layers required for adhesion and plating are also considered as a crucial part of the structure and as a limiting factor for a self-aligned process.

Process description

The overall process flow of bump fabrication is presented in figure 1. The bump fabrication can be viewed as a sequence of additional steps in the back-end of wafer processing. The starting interface is a fully processed wafer with an insulating passivation layer and openings at the bonding pads. For reconfigurable circuits, additional distribution layers can be included but usually the bumps are built directly on the bonding pads. For electroplating, a continuous seed layer is needed to provide the cathode contact for plating current. The seed layer also provides the adhesion to the bonding pad and a hermetic overlap with the passivation edge.

G. Harman and P. Mach (eds.), Microelectronic Interconnections and Assembly, 81-86.

82

Seed layer deposition. The seed metal performs two functions in the final bump structure. A solderable metal, such as Ni or Cu, must be used beneath the solder to provide the metallurgical contact. It is also used to limit the solder ball in the reflow phase. An adhesion layer is needed between solderable metal and the pad metallization - usually aluminium. It also anchors the bump edges to the passivation layer - usually silicon oxide, silicon nitride or polyimide. An important requirement is the possibility for

selective etching using the solder as an etching mask after bump deposition. Suitable metals for this purpose includes Cr, Ti, TiN, TiW, W and Mo. Cr and Ti are likely to offer the best adhesion whereas TiW, W and Mo are easier to etch. We have used Mo/Cu and Cr/Cu as seed metals. The layers are deposited sequentially by sputtering, first a 150 nm thick Mo or Cr layer and, subsequently a 500 nm thick Cu layer. The thickness of Cu is very sensitively dependent on the reflow processes where a risk of adhesion loss results from the intermetallic reaction of Sn and Cu. The molten tin tends to consume rapidly the copper layer to form Cu_6Sn_5 and Cu_3Sn intermetallic compounds. These two intermetallics exist in Cu-Sn, Cu-Pb-Sn and Cu-Bi-Sn systems already below the eutectic temperature of SnPb alloys. Hence, the copper thickness has to be scaled to match the foreseeable multiple reflow processes or, alternatively, a separate barrier layer has to be incorporated between Cr and Cu. From figure 2 one can conclude that the dissolution rate of Ni is one order of magnitude slower than the corresponding rate for copper. The activation energy of the dissolution rate is approximately 0,5 eV which explains why the consumption of copper rapidly increases with increasing temperature. IBM has succesfully used a graded Cr/Cu layer to prevent the loss of adhesion. We have increased the safety margin against this degradation mechanism by adding electroplated Cu and Ni layers prior to the solder plating.

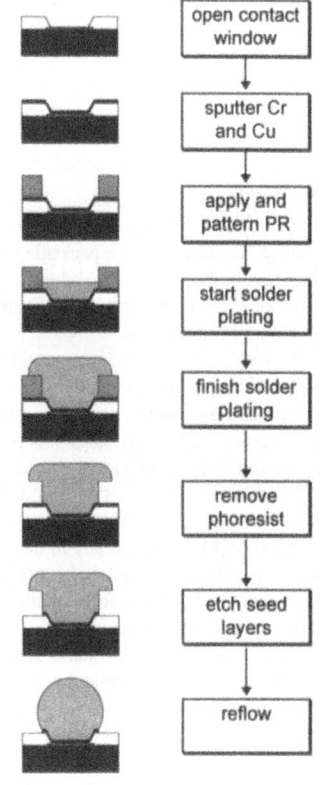

open contact window

↓

sputter Cr and Cu

↓

apply and pattern PR

↓

start solder plating

↓

finish solder plating

↓

remove phoresist

↓

etch seed layers

↓

reflow

Figure 1. Schematic process flow of the solder bump fabrication.

Pattern definition. The bump sites are delineated in a thick photoresist layer. The Shipley AZ4620 photoresist is spin coated on the wafer in a single step to a thickness of 20 μm. The openings in the photoresist are circular to avoid the high local mechanical stress at corners of square holes. A typical solder volume in the bump is 1.3×10^{-12} m^3. As a consequence, the bump assumes a shape of a mushroom with only about 12 % of the volume in the stem. Other investigators have pointed out that this plating geometry results in volume non-uniformities and incomplete photoresist stripping [2]. Hence, alternative photoresist schemes have been suggested including the use of 75-100 μm thick dry-film photoresists [2,3]. It appears, however, that the useful range of bump size and pitch is limited to approximately 100 μm and 200 μm respectively when using dry-film resists. In our experience, mushroom shaped bumps associated with the relatively thin spin coated phoresist can be controlled to the required uniformity without significant problems in photoresit removal. This process can also be scaled down to much smaller bump diameters and pitches.

Solder deposition. The simultaneous deposition of the metals can be achieved by electrolysis from a solution of metal salts. For tin-lead alloys, commercial electrolytes based on fluoborates and alkyl sulfonic acid have been developed. Recently, the use of alkyl sulfonic acid baths has increased due to the toxicity and corrosiveness of fluoborate baths. The metal composition of the deposit depends mainly on the composition of the plating bath and the operational variables, such as the current density used [6]. In our laboratory, the tin-lead bumps have been electroplated using LeaRonal's commercial Solderon M-54 reagents. The singly baked AZ4620 photoresist withstands this acid well at room temperature but not at elevated temperatures. For the Pb-10%Sn alloy, we have used a commercial recipe, but since there is no formula for eutectic Sn-40 %Pb composition, we have developed our own bath recipe. For both binary baths the plated composition is a strong function of the current density (see fig. 3.). As the electroplating proceeds to the cap stage, the area of the bump in contact with the bath increases with time. To achieve a constant composition throughout the plated bump the current density has to be maintained at a constant level. A computer simulation program is used to calculate the contact area and adjust the current so that the current density remains constant.

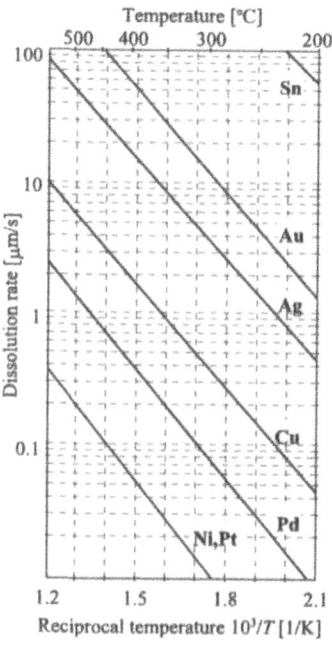

Figure 2. Dissolution rates of various metals in solder as a function of temperature [4]

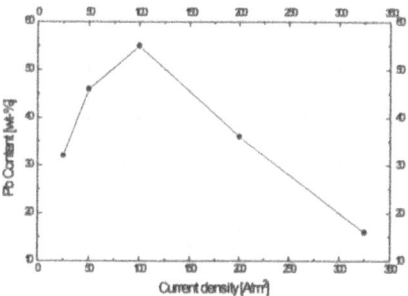

Figure 3. Dependence of the solder alloy composition on the current density for nominally eutectic Sn-Pb plating bath. The solder composition has been determined by RBS.

There is a significant interest in developing low-melting point solder alloys as a replacement for the standard Pb-Sn. Indium containing alloys are useful for lowering the melting point but their application to volume production is hampered by the high cost. Eutectic tin-lead-bismuth alloy is one of the low-melting alloys, the liquidus temperature being only about 95 °C. The ternary phase diagram of tin, lead and bismuth is shown in fig. 4. There is no mutual solubility of Bi and Cu at 250 °C where the Cu_3Sn phase is formed at a rate of approximately 100 nm/min [6]. The low soldering temperature can be made use of to reduce the dissolution rate of Cu in Sn. The development of a ternary Sn-Pb-Bi plating bath is complicated because of several variables (bath composition, current density, total metal concentration in the bath, etc.) After a series of experiments, a bath producing a Sn-Pb-Bi alloy with a melting point of 155°C was achieved.

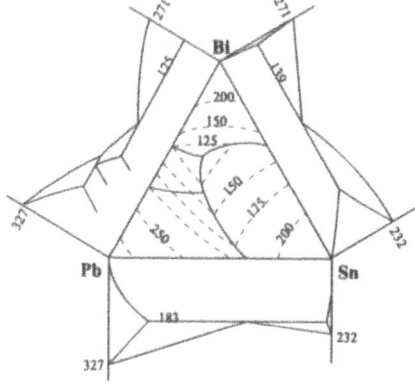

Figure 4. Ternary phase diagram of the Pb-Sn-Bi system showing the equal T_m composition lines.

A more reliable and easier method is to use several elemental baths, each containing only one metal, and to deposit the metals layer by layer. Using elemental baths of tin, lead and bismuth, solder bumps with a near-eutectic composition have been succesfully deposited. The reflow temperature of these tin-lead-bismuth solder bumps was 105°C.

Etching of the seed layer. After solder deposition and photoresist stripping the seed layer is selectively removed. During this stage the solder bumps will be exposed to the same etchants. Hence, these have to be selected to minimize the attack on the solder alloy and the underetching of the seed layers. Our process has gone through a long development to solve all these chemistry related problems. The copper layer is etched in a commercial alkaline Bondenex A copper etchant. This etchant had no visual effect on the bumps. After the reflow the bumps had an appearance of the unetched reference samples. The underetching of Cu was controlled to about 1 µm. The initial choice of using Mo as adhesion layer was primarily based on the simple etching recipe. A standard phosphoric acid etchant including 5.5 % each of acetic acid, nitric acid and deionized water was used to remove Mo [4]. This etchant does not degrade the solder bump appearance but leaves flaky residues if not carefully rinsed. To improve the bump adhesion Cr has been substituted for Mo. The usual acidic etchants severely attack the solder alloys and therefore they are not relevant for the bump fabrication process [2]. An alkaline potassium permanganate solution has proved an efficient etchant for the Cr adhesion layer.

Reflow. The reflow temperatures depend on the solder composition. In our process the reflow was originally carried out in an inert/reducing gas ambient. A simple and reliable alternative to this procedure was developed by transfering the process from gas ambient to a liquid. This brings about a much faster melting cycle because of the rapid heat transport from the liquid to the solder. The reflow medium for the eutectic Sn-Pb and the ternary Sn-Pb-Bi solder alloys is glycerol and for the high lead content Pb-Sn alloy erythritol. The melting points observed by slowly ramping the temperature of the reflow medium are 184 °C for the low temperature alloy and 307 °C for the high temperature alloy, respectively. It appears that these organic liquids have a mildly reducing effect themselves without any additional fluxing agent. In the thermal ramping experiments, the melting occured without overheating in the absence of a thermal shock. Because of the efficient thermal contact between the wafer and the liquid, a relatively small safety margin is required between the melting point and the actual reflow temperature. Reflow temperatures of +220 °C and +320 °C have been used for eutectic and lead rich alloys respectively. Successful reflow in both alloys can be accomplished in 10 s. A bump of a near-eutectic Pb-Sn alloy is shown in fig. 5 as plated, photoresist removed and seed layer etched (a) and after reflow for 10 s in glycerol at 190 °C (b).

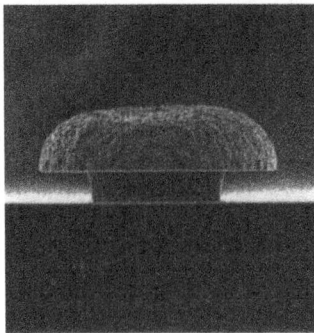

Figure 5a. A near-eutectic Pb-Sn solder bump as plated and photoresist removed.

Figure 5b. The solder bump of Fig 5a after reflow in glycerol at 190 °C.

Plating cup

Many commercial plating cup systems are obviously designed for gold baths in which the flow geometry of the plating solution is not of great significance for plating uniformity. This is because the reducing process of gold at the cathode is so slow that free Au ions are always abundantly available. In lead-tin baths, on the other hand, the reaction is so fast that the growth becomes limited by the supply of new ions to the cathode. To achieve uniform plating results when using lead-tin baths, one has to take care that - in one way or another - the whole wafer surface is exposed to the same average flow during the process. Normally, the geometry is such that the wafer is facing down against the flow escaping at the edge of the wafer. The flow velocity varies across the surface and, as a consequence, the plating uniformity is poor. To improve the plating uniformity we have designed and built a plating cup with a new flow geometry. In this design, the single central inlet and peripheral outlet of the cup are replaced by a regular array of alternating sources and drains of the plating fluid. The structure of the plating cup is shown in figure 6.

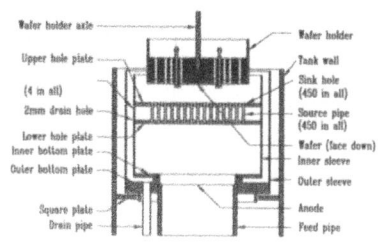

Figure 6. A plating cup designed for high flow uniformity.

Test structures

The electroplated bumps have been tested for mechanical strength and electrical performance using the test structure shown in figure 7. This is a dummy daisy chain chip, 5 mm a side, with 71 solder bumps arranged peripherally. The chip is mounted on a passivated silicon test substrate with metal lands and aluminium interconnections to close the daisy chain loop. Two individual bumps can be probed via four point contacts. The bump strength was checked mounting the soldered test structure into a home made pull tester. For high temperature solder bumps with the electroplated Cu and Ni buffer layers, the measured strength is about 10-15 g/bump with a pad diameter of 80 μm. This should be compared to the value of 30-50 g/bump with a pad diameter of 125 μm reported by a group at IBM [7]. When scaled with the bump size our values are lower by 20 %. The mechanical strength depends on the bump geometry and it is sensitive to eventual underetching of the seed layers [2]. In an additional reflow process at 250 °C for 11 min. after bonding the mechanical strength slightly decreases. This can possibly be attributed to the consumption of the seed metal in the periphery of the bonding pad, thus reducing the effective contact area under the bump.

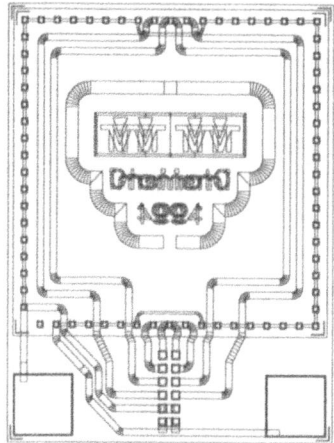

Figure 7. A daisy chain test structure for electroplated solder bumps. The bumped chip is shown transparent and flipped over the test substrate.

86

The bump resistance was determined from the daisy chain structure and the Kelvin test bumps. A typical value for our solder bumps is 8 mΩ. This is close to the bump resistance obtained from the empirical data at IBM where a design target value is 5 mΩ/bump [8].

Bumping cost

To answer the frequently raised question of the bumping cost we carried out an analysis on the various cost factors affecting the final cost of a bumped wafer. The analysis was based on the assumption that commercial production was to be set up using the technology presented in this work. The total cost for running a facility with a full capacity of 150 000 wafers (150 mm diameter) annually was estimated at 3,2 million USD/year or 21,3 USD/wafer. The capital cost is based on 7 % interest rate and a payback time of 10 years. Testing and flip chip assembly were not included in the cost breakdown because they are strongly affected by the application. Figure 8. shows that the total cost per wafer consists of three almost equal factors: investments, labour cost and materials.

Summary

A process based on electroplating of solder alloys can be effectively used to fabricate bumps for flip chip assembly. Compared to vacuum deposition, electroplating offers a selective and potentially cost effective method to deposit thick solder layers. The process can be scaled to large diameter silicon wafers with sufficient plating uniformity. The mechanical and electrical performance of these electroplated solder bumps is closely similar to that reported by other groups. The processing steps described in this work can be incorporated to the back-end of a standard IC process using a single additional mask layer.

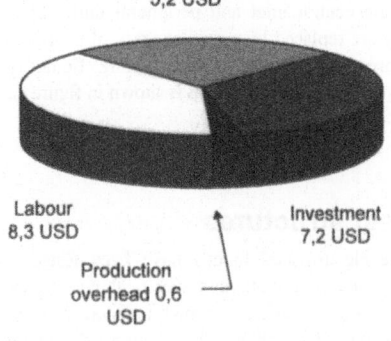

Total cost 21,3 USD/wafer
Materials and services
5,2 USD

Labour
8,3 USD

Production
overhead 0,6
USD

Investment
7,2 USD

Figure 8. Cost breakdown of the wafer bumping process based on a production capacity of 150000 wafers per year. The diagram gives the costs per wafer in USD.

References

1. P.A. Totta and R.P. Sopher, *IBM J. Res.Dev.*, **16**, 226 (1969).
2. M. Datta, R.V. Shenoy, C. Jahnes, P.C. Andricacos, J. Horkans, J.O. Dukovic, L.T. Romankiw, J. Roeder, H. Deligianni, H. Nye, B. Agarwal, H. M. Tong. and P. Totta, *J. Electrochem. Soc.* **142**, 3779 (1995).
3. E.K. Yung and I. Turlik, *IEEE Comp., Hybrids, Manuf. Technol.* **14**, 549 (1991).
4. J. Salonen and J. Salmi, *Physica Scripta* **T 54**, 230 (1994).
5. R.J. Klein Wassink, *Soldering in Electronics* (Electrochemical Publications Ltd, Scotland 1984).
6. J.K. Kivilahti, *IEEE Comp., Hybrids, Manuf. Technol. B*, **18**, 326 (1995).
7. L.F. Miller, *IBM J. Res. Develop.* **13**,239 (1969).
8. In *IBM C4 Product Design Manual, Volume I: Chip and Wafer Design*, DOC#: SPIC4IDM1-001, (IBM - Technology Products, East Fishkill, Hopewell Junction, New York), p. 4-4.

Reliability of Solder Joint Interconnections in Thermally Matched Assemblies

E. Suhir

Bell Laboratories, Lucent Technologies
Physical Sciences and Engineering Division,
Murray Hill, New Jersey 07974, U.S.A.

We report on the results of theoretical and experimental investigations of the thermally induced stresses in, and mechanical reliability of, flip chip solder joint interconnections in a thermally matched assembly ("Si-on-Si"). The mechanical behavior of encapsulants and their effect on the stresses in solder joints is also discussed. We conclude that matched thermal expansion between the chip and the substrate leads indeed to a more reliable interconnection. However, application of thermally matched materials should not be regarded as an omnipotent "panacea" which puts right all the mechanical "troubles".

Introduction

Typically, an appreciable thermal expansion mismatch exists between the soldered components. This results in shear deformations of the joints (see, for instance, [4,5,8]). There is an obvious incentive to employ, if possible, similar materials for the soldered components to minimize the adverse effect of the expansion mismatch. Such an approach has been implemented, particularly, in the advanced Si-on-Si packaging technology [1,13]. When similar materials are employed as adherends, and the temperature change is the same throughout the structure (thermal cycling conditions), the stresses and strains in the solder joints are due exclusively to the mismatch between the solder and the material of the soldered components. The effect of such a "local" mismatch was addressed by the author in application to Si-on-Si flip-chip assemblies [9,12], and by J.-P. Clech et at [2], in application to electronic components surface mounted on printed circuit boards.

In this paper we report on the results of theoretical and experimental investigations of the mechanical reliability of flip-chip solder joint interconnections used in Si-on-Si technology. We address also the role of the encapsulation material in such interconnections. Encapsulation of integrated circuit devices, which is done to protect them from moisture, ionic contaminants, radiation, and hostile environment [14], can result in additional thermal stresses caused by the thermal expansion mismatch of the encapsulant with other materials in the package. This can cause serious problems in the device reliability [11]. Such stresses are believed to be especially large in a

situation, when the expansion of the encapsulant in the throughthickness direction is constrained, and, because of that, the coefficient of expansion in this direction can be substantially larger than the specified bulk value [3]. This study has been carried out primarily in application to the Bell Laboratories advanced VLSI MCM package design [1].

The theoretical analyses were aimed at the assessment of the magnitude and the distribution of the thermal stresses in the solder joints, and to evaluate the effect of different encapsulation materials and technologies on these stresses. We assessed also the interaction between the "local" and the "global" stresses in soldered components in a situation, when the thermal expansion (contraction) mismatch of the component materials is not small, and therefore the effect of the mismatch of the solder with the component materials need not be accounted for. In such a situation the "local" mismatch is due to the interaction of the soldered materials within the soldered region. Only the case when this region occupies the peripheral portion of the assembly is considered. The detailed derivations are given in the Appendix.

The objective of the experimental investigations was to identify potential reliability problems and accumulate reliability statistics.

Theoretical Analyses

Stresses and Strains in Unencapsulated Interconnections

The analysis of stresses and strains in unencapsulated flip-chip solder joints due to their thermal contraction mismatch with the soldered components was performed on the basis of the theory-of-elasticity approach [9]. The solution to the elastic problem was obtained, using modified Bessel functions. Some of the calculated data are given in Table 1. The parameter of the solder joint aspect ratio

$$\xi = \pi \frac{d}{h} \qquad (1)$$

was chosen equal to 3, 6, and 10. These correspond to the ratios of the solder joint diameter, d, to its height, h, equal to 0.995, 1.910, and 3.183, respectively. In all the calculations, Poisson's ratio was taken equal to $\nu = 1/3$. The distributions of stresses and strains along the joint's

87

G. Harman and P. Mach (eds.), Microelectronic Interconnections and Assembly, 81-86

Table 1. The calculated strains and stresses in solder joints of different diameter-to-height ratios.

r/d	0	0.2	0.5	0.7	0.9	0.95	1.0
			$\xi = 3.0$				
$\varepsilon_r/\Delta\alpha\Delta t$	2.377	2.345	1.930	0.818	−2.114	−3.365	−4.885
$\varepsilon_t/\Delta\alpha\Delta t$	2.377	2.367	2.268	2.038	1.490	1.271	1.000
$\varepsilon_z/\Delta\alpha\Delta t$	−4.402	−4.252	−3.090	−0.815	4.296	6.334	8.769
$\gamma_{rz}/\Delta\alpha\Delta t$	0	2.034	3.271	2.147	−2.628	−4.828	−7.598
$\sigma_r/E\Delta\alpha\Delta t$	2.048	2.104	2.279	2.145	1.168	0.658	0
$\sigma_t/E\Delta\alpha\Delta t$	2.048	2.121	2.532	3.060	3.871	4.134	4.413
$\sigma_z/E\Delta\alpha\Delta t$	−3.047	−2.844	−1.486	−0.920	5.975	7.931	10.240
$\tau_{rz}/E\Delta\alpha\Delta t$	0	0.763	1.227	0.805	−0.985	−1.811	−2.849
			$\xi = 6.0$				
$\varepsilon_r/\Delta\alpha\Delta t$	0.927	1.005	1.968	2.499	−0.824	−3.757	−8.444
$\varepsilon_t/\Delta\alpha\Delta t$	0.927	0.992	1.269	1.569	1.561	1.367	1.000
$\varepsilon_z/\Delta\alpha\Delta t$	−2.168	−2.268	−3.036	−2.735	3.978	8.716	15.892
$\gamma_{rz}/\Delta\alpha\Delta t$	0	1.225	6.624	6.121	1.695	−3.298	−11.770
$\sigma_r/E\Delta\alpha\Delta t$	0.460	0.551	1.626	2.874	2.917	1.927	0
$\sigma_t/E\Delta\alpha\Delta t$	0.460	0.542	0.801	2.176	4.706	5.770	7.083
$\sigma_z/E\Delta\alpha\Delta t$	−1.861	−1.904	−2.127	−1.052	6.519	11.281	18.252
$\tau_{rz}/E\Delta\alpha\Delta t$	0	0.459	2.484	2.295	0.636	−1.237	−4.414
			$\xi = 10.0$				
$\varepsilon_r/\Delta\alpha\Delta t$	0.420	0.464	1.053	2.709	2.267	−2.085	−12.490
$\varepsilon_t/\Delta\alpha\Delta t$	0.420	0.434	0.570	0.908	1.456	1.401	1.000
$\varepsilon_z/\Delta\alpha\Delta t$	−1.226	−1.277	−1.884	−3.240	1.035	8.544	24.353
$\gamma_{rz}/\Delta\alpha\Delta t$	0	0.169	1.556	5.575	7.786	0.839	−17.469
$\sigma_r/E\Delta\alpha\Delta t$	0.0269	0.0645	0.594	2.315	5.268	4.332	0
$\sigma_t/E\Delta\alpha\Delta t$	0.0269	0.0417	0.232	0.964	4.660	6.947	10.406
$\sigma_z/E\Delta\alpha\Delta t$	−1.208	−1.242	−1.609	−2.147	4.345	12.304	27.921
$\tau_{rz}/E\Delta\alpha\Delta t$	0	0.0634	0.583	2.091	2.920	0.315	−6.551

ε_r, σ_r = radial strain and stress
ε_t, σ_t = tangential (circumferential) strain and stress
ε_z, σ_z = axial strain and stress
γ_{rz}, τ_{rz} = shear strain and stresses

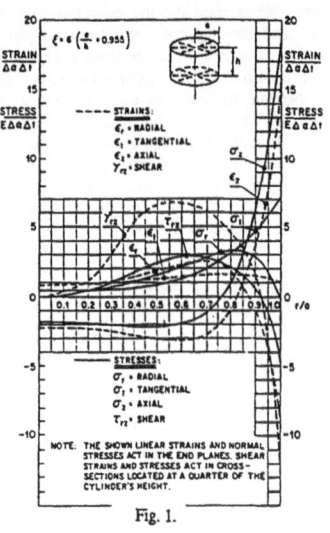

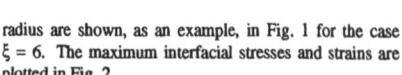

Fig. 1.

The distribution of strains and stresses along the joint's radius.

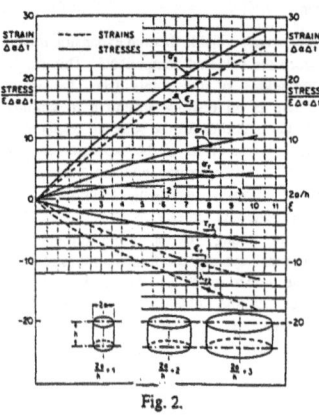

Fig. 2.

The maximum strains and stresses along the joint vs aspect ratio parameter $\xi = \pi \dfrac{d}{h}$.

radius are shown, as an example, in Fig. 1 for the case $\xi = 6$. The maximum interfacial stresses and strains are plotted in Fig. 2.

Examine, for instance, a 5% Sn/95% Pb solder joint ($E \approx 2000$ kg/mm^2, $\alpha = 2.8 \times 10^{-6}$ 1°C) interconnecting a silicon chip to a silicon substrate ($\alpha = 3.0 \times 10^{-6}$ 1/°C). The soldering ("zero stress") temperature is 320°C. If the lowest testing temperature is, say, −50°C,

then the nominal thermal mismatch strain is $\Delta\alpha\Delta t = 0.925\%$. For a solder joint with a diameter $d = 115$ μm and a height $h = 60$ μm, the calculated ξ value is $\xi = 6$. The distributions of the stresses and strains are those shown in Table 1 and Fig. 2. The maximum calculated stresses are very high, and, if the strains were not restricted, would inevitably result in a failure of the joint. The maximum normal strains occur on the contours of the end planes and are tensile.

According to the Table 1 data, these strains are as high as

$$15.892 \times 0.00925 = 0.147 = 14.7\% \; .$$

Since, however, the static elongations at failure for the solder material in question are even higher and vary from 30% to 60% [4], the margin of safety seems, nevertheless, sufficient, at least if the number of periodic loadings is not very large.

The ultimate shear strain can be assessed by the formula

$$\gamma_u \approx \frac{2(1+\nu)}{\sqrt{3}} \, \varepsilon_u \; . \qquad (2)$$

Here ε_u is the ultimate tensile strain and ν is Poisson's ratio. The formula (2) reflects an assumption that the well-known von-Mises relationship for the yield stresses in tension and in shear holds for the corresponding strains as well. With the ε_u value varying in the range from 30% to 60%, the shear strain varies, with $\nu = 1/3$, from 40% to 93%. The maximum shear strain in the example in question is

$$- 11.77 \times 0.00925 = -0.1089 = -10.89\% \; ,$$

i.e. considerably smaller than the ultimate strain of the solder material.

We would like to point out, however, that although the safety margins seem to be large enough, as far as the ultimate ("static") strains are concerned, they may turn out to be insufficient from the viewpoint of the fatigue strength of the joint. If this is the case, the stresses and strains can be brought down by reducing the diameter-to-height ratio of the solder joint. If, for instance, this ratio is reduced by a factor of two, the maximum axial strain in the examined example becomes about 8.1%. Then, if the Manson-Coffin relationship is used to assess the number of cycles till failure, the expected increase in the fatigue life can be evaluated by the formula

$$\frac{N_2}{N_1} = \left(\frac{\varepsilon_1}{\varepsilon_2} \right)^m , \quad m = 2.5 \to 3.0 \; . \qquad (3)$$

As is evident from this formula, a twofold reduction in the maximum strain results in an about $5.7 \to 8.0$ times longer fatigue life of the interconnection. It should be emphasized, however, that the relationship (3) was obtained for thermally mismatched assemblies, i.e. for a "global", rather than a "local", mismatch, and therefore may not be accurate enough in the case of thermally matched materials.

In addition to analytical investigations, finite element calculations were carried out using standard ANSYS programs. The calculations were performed for 2 mil, 3 mil, and 4 mil diameter joints. In all the calculations, the height of the joint was taken h = 2 mils (50.8 μm). The obtained distribution of the axial stresses is shown, as an illustration, in Fig. 3 for a 4 mil diameter joint (d/h = 2). The calculated maximum axial stresses for a 2 mil and 4 mil diameter joints are 34.5 kg/mm² and 49.3 kg/mm², respectively. These stresses, computed using Table 1 data, are, however, 192.0 kg/mm² and 350.2 kg/mm², i.e. significantly higher. This should be attributed primarily to the fact that the finite-element method "automatically" averages the stresses within the given element. Both analytical and numerical data indicate that a twofold reduction in the diameter-to-height ratio results in an about $1.4 \to 1.8$ fold reduction in the maximum axial stress.

Effect of the Encapsulation Material

The following encapsulation technologies and materials are or can be used in flip-chip structures: 1)

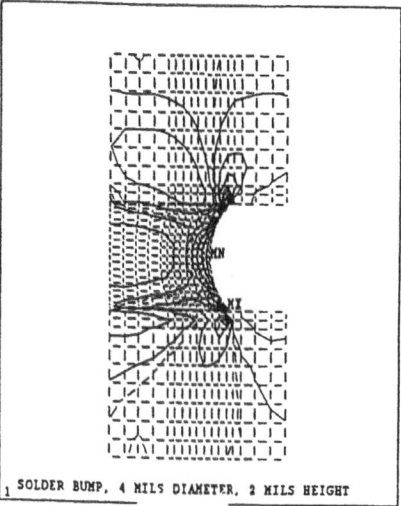

1 SOLDER BUMP, 4 MILS DIAMETER, 2 MILS HEIGHT

Fig. 3.

The distribution of the axial normal stresses in a joint, based on finite-element calculations (d/h = 2).

encapsulant fills in the entire underchip space (silicone gel, epoxy); 2) encapsulant covers the underchip surfaces with thin conformal coating (polyimide, polyxylylene).

The technology with conformally coating encapsulants looks especially attractive for advanced VLSI modules with very large chips. Additional advantage of such a technology is due to the fact that the stress in solder joints is independent from their number. Therefore solder joints in the assemblies with different number of conformally coated joints are supposed to have the same reliability. It should be pointed out also that polyxylylene, unlike other encapsulants, is applied at room temperature and because of this results in an almost symmetric thermal stress cycle. This is deemed to be a favorable factor from the viewpoint of the fatigue strength of the joints.

The formulas for the stresses in the solder joints and in the encapsulation material itself were obtained, using strength-of-materials approaches [11]. The stresses in the joints were calculated for 5 μm thick coatings, assuming that the change in temperature is 200°C, in the case of the silicone gel and a low expansion epoxy, 75°C in the case of polyxylylene, and 250°C in the case of polyimide coating. The calculations showed that the stresses in solder joints encapsulated with low expansion epoxy, exceeded the yield stress $\sigma_y = 5500$ psi (3.9 kg/mm²) of the solder material, while all the other encapsulants resulted in substantially lower stresses. The lowest stresses were found when high mix ratio silicone gel was used. Although this material has a very high coefficient of expansion, its Young modulus is so low that the calculated stresses in the solder joints never exceeded 2 kg/mm². It should be emphasized that epoxy encapsulants with moderate or even high coefficients of expansion (contraction) could turn out to be more feasible than widely applied low expansion epoxies. This is due to the fact that high expansion epoxies will result in elevated compressive stresses in solder joints. This is supposed to reduce somewhat the large tensile axial stresses in the joints.

It is clear that if an encapsulant fills in the entire underchip cavity, the encapsulation material itself is understressed. The situation might be different, however, in the case of conformally coating encapsulants. These can be viewed, in effect, as thin films "fabricated" on solder joints, that play the role of "thick substrates". Therefore the stresses in conformally coating encapsulants can be quite high. In the above example, however, the calculated stresses in the polyxylylene and in the polyimide are only about 0.51 kg/mm^2 and 1.10 kg/mm^2, respectively. The yield stress of the polyxylylene, according to the manufacturer's data, is 8000 psi (5.6 kg/mm^2), and the yield strength of the polyimide materials is in the range between 5000 psi (3.5 kg/mm^2) and 8000 psi (5.6 kg/mm^2). Hence, the strength of both encapsulants is thought to be sufficient. The stresses in conformally coating encapsulants are expected to remain more or less constant with the change in their thickness, as long as this thickness is small in comparison with the cross-sectional dimensions of the solder joints. The stresses in the joints, however, increase with an increase in the thickness of the coating. The final selection of an encapsulation material should be done, of course, with consideration of all the factors affecting the performance of the joints, including electrical, chemical, and technological requirements.

Although the study has been aimed at the evaluation of the performance of solder joints in thermally matched structures, we assessed also the interaction of the "local" and "global" shearing stresses in a situation when the mismatch between the soldered materials is significantly, i.e. in another extreme case. In Si-on-Si technology such a situation is possible in power cycling conditions, when the temperature of the chip can be substantially higher than the temperature of the substrate. In such a situation, the coefficient of expansion of the solder material is of a relatively small importance, and the "local" mismatch is due to the interaction of the soldered components within the soldered region. The case when this region occupies only the peripheral portion of the assembly is considered. The analysis has indicated (see Appendix) that one should try to avoid very narrow soldered areas, unless a very compliant joint is employed. This is deemed, however, impractical.

Experiments

The experimental evaluations of the mechanical behavior and reliability of flip-chip solder joint interconnections included shear-off tests, static thermal loading, liquid-to-liquid thermal shock cycling, real-time X-ray inspection, and cross-sectional microscopy [12].

The main purpose of the *shear-off tests* was to make sure that the adhesive strength provided by the interconnection is sufficiently large, and that the joints fail cohesively, rather than adhesively. The performed tests confirmed that in almost all the cases failures occurred in the bulk of the material. The tests have shown also that the actual shear strength of the solder material was consistently larger than the shear strength of the bulk 5% Sn/95% Pb solder reported in the literature [4]. In addition to shear-off tests, *twist-off tests* [10] were performed, and good agreement between the ultimate shear force and the ultimate torque was observed.

In order to assess the ultimate strength of the interconnections, when subjected to *extensive static thermal loading*, two samples of the multichip module were exposed sequentially to –50°C in a refrigerator, to –80°C in dry ice, and to –180°C in liquid hydrogen. No failures were detected. Electrical testing was carried out after each temperature level. Measurements were made,

using two-probe tester. It was found that heating up of the samples prior to electrical measurements was important: in many cases "opens" could be detected only at elevated temperatures. For this reason all the measurements were made at an elevated temperature of about 85°C.

Liquid-to-liquid thermal shock cycling tests were carried out for nine multichip module samples. The tests were conducted until about half of the entire solder joint population failed. The results are shown in Table 2. Percentage of the failed joints versus number of cycles is shown in Fig. 4. The calculated failure rates are plotted in Fig. 5. For the steady-state portion of the reliability

Table 2. Solder joint population failed vs number of cycles.

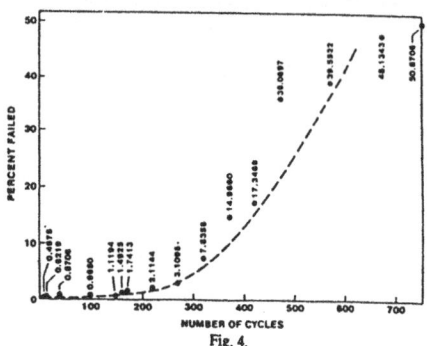

Fig. 4.
Percentage of the failed solder joints vs number-of-cycles.

("bath-tub") curve, the computed failure rate is within the range between 0.25×10^{-4} and 1.0×10^{-4} 1/cycle. Such a rate level is thought to be low enough compared to the existing data for electronic equipment of general purpose [6]. Based on the obtained reliability data, it can be concluded that the technology in question has proven to be reliable. All the samples (excluding, probably, just one, 200-6) showed very good dependability. One of the samples (201-3) survived 750 cycles without a single failure at all.

The experimental data give a reason to believe that liquid-to-liquid shock cycling can be effectively used for screening purposes as well: some joints in a "weak" sample 200-6 failed after the first $10 \rightarrow 40$ cycles, while the joints in "strong" samples performed very well for

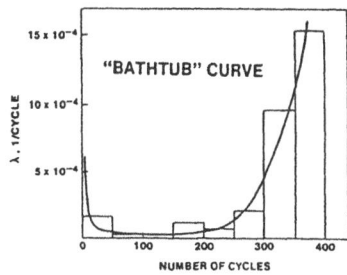

Fig. 5.
Calculated failure rates (the reliability curve).

quite a long time. Note, that the relatively early failures observed in the samples 200-3 and 200-4 could be due to the fact that these samples were subjected to high level static thermal loading prior to liquid-to-liquid cycling.

The results of thermal shock cycling have indicated also that voids, as long as they do not result in actual or potential "shorts" (joints with large voids are usually bigger than void-free joints, and because of that gaps between the joints with large voids and the adjacent joints are smaller) seem to be a lesser reliability problem, than usually perceived. Two samples, one heavily "loaded" with voids (202-4) and one completely void-free (202-1), showed more or less equal and rather high reliability. No failures were detected in either of them after 320 cycles. The fact that the chip fell off somewhat earlier in the sample with voids (after 670 cycles) should not be necessarily attributed to the presence of voids. Additional investigations are needed, of course, to evaluate the size, number, location, and method of detection of permissible voids in the given design. Obviously, voids which can result in "swelling" of the bumps and/or in potential "bridges" ("shorts") should not be permitted.

The obtained data can be used for a tentative assessment of the expected fatigue life of a solder joint interconnection in actual operation conditions. Since the accelerated testing was carried out in the temperature range between −65°C and 150°C, the change in temperature was $\Delta t_a = 215$°C. In actual operation conditions, the temperature of the chip changes from room temperature of about 20°C to the operation temperature of, say, 80°C. Assuming that heat transfer between the chip and the substrate is so good, that the power cycling conditions need not be considered, we conclude that the temperature change in operation conditions is 60°C. Assuming further that the maximum strain in a solder joint is proportional to the change in temperature, we find that the accelerated maximum strain ε_a is larger than the operational strain ε_0 by a factor of $\frac{215}{60} = 3.58$. The failure rate curve in Fig. 5 indicates that the accelerated number-of-cycles-till-failure can be chosen, rather conservatively, as $N_a = 150$. Then, using the relationship (3), with $m = 2.5$, we obtain the following expected number of cycles till failure in actual operating conditions:

$$N = N_a \left[\frac{\varepsilon_a}{\varepsilon} \right]^m = 150 \times 3.58^{2.5} = 3637.$$

With one cycle per day, the expected fatigue life of the joint is about 10 years. The calculated probability-of-failure is shown in Table 5 for different numbers of cycles and failure rates.

Real-time X-ray inspection can produce a useful X-ray image which clearly exhibits voids, "bridges", "swallen" joints, etc. X-ray inspection seems to be especially useful for screening purposes. Although it cannot provide, of course, any quantitative information on the mechanical reliability of the interconnection, this nondestructive technique provides valuable qualitative information about the configuration and the geometry of the joints. The following major failure-mode-analysis (FMA) observations, obtained using real-time X-ray inspection, were reported: poor reflow due to the lack of self-alignment at reflow of the "halves" of the joint on the chip and the substrate; poorly shaped joints; "bridges" between the joints; large voids.

Cross-sectioning microscopy was applied to evaluate the actual shapes of solder joints and to detect possible imperfections which, for one reason or another, were not revealed by nondestructive methods. This experimental technique was used also to determine, whether liquid-to-liquid testing can lead to an appreciable distortion of the joints' shape. We found that the diameter-to-height ratios has indeed increased substantially for many joints after liquid-to-liquid cycling tests. Therefore we conclude that although such testing can be regarded as a feasible experimental technique for the evaluation of the solder technology and, possibly, also for screening purposes, it cannot be recommended as a suitable testing technique in manufacturing.

Table 3. Calculated probability-of-failure for different number of cycles and failure rates.

FAILURE RATE / NUMBER OF CYCLES	0.25 x 10⁻⁴	0.50 x 10⁻⁴	0.75 x 10⁻⁴	1.0 x 10⁻⁴	1.25 x 10⁻⁴	1.50 x 10⁻⁴
100	0.25 %	0.50 %	0.75 %	1.00 %	1.24 %	1.49 %
200	0.50 %	1.00 %	1.50 %	1.98 %	2.47 %	2.96 %
500	1.24 %	2.47 %	3.88 %	4.88 %	6.06 %	7.23 %

Conclusions

The following major conclusions can be drawn from the performed analyses:

- All the strains and stresses in flip-chip solder joints are the largest either on the lateral surface itself or in the vicinity of this surface. Hence, it is the peripheral part of the joint which is primarily responsible for its mechanical performance.

- The maximum strains and stresses in the solder joints act in the axial direction and concentrate at the joint's "corners". These strains and stresses are significantly larger than all the other strain and stress categories.

- The maximum axial stresses at the lateral surface are tensile and can possibly result in an ultimate ("static") failure or in a crack initiation during thermal cycling. The adverse effect of the concentrated axial stresses is aggravated by tensile tangential stresses, which, unlike tangential strains, are rather large, and, in combination with large axial stresses, result in a three-dimensional stress condition. This increases the likelihood of fracture initiation at the "corners" of the joint, especially at low temperatures, when the solder material is more prone to fracture formation.

- In the range of diameter-to-height ratios of practical interest ($d/h = 0.5 - 3.0$), the strains and stresses increase with an increase in this ratio. As a "rule of thumb", one can assume that they are approximately proportional to the diameter-to-height ratio.

- In addition to the thermally induced stresses due to the mismatch of the solder material with the chip and the substrate, the flip-chip solder joints experience thermal stresses caused by their mismatch with the encapsulants. Low modulus and conformally coating materials results in smaller stresses than high modulus epoxies. On the other hand, application of moderately or even high expansion epoxies (with relatively high Young's moduli), resulting in elevated compressive axial stresses in the joints, can be an effective means to bring down elevated tensile axial stresses at the joint's "corners".

- Based on the obtained data, we conclude that, from the point of view of mechanical reliability of solder joint interconnections, the evaluated technology is quite reliable, and can be successfully used in advanced VLSI package designs. Application of thermally matched materials should not be regarded, however, as a "panacea" which would put right all the reliability "troubles".

Acknowledgement

The author acknowledges, with thanks, the contributions made by R. Zappulla, J. Liu, S. M. Huang, G. N. Poli, P. J. McCord, R. J. Thompson, M. J. Neubelt, R. M. Richman, and A. W.-C. Lin, as well as valuable discussions with C. J. Bartlett, P. M. Hall, K.-L. Tai, T. D. Dudderar, J. M. Segelken, E. N. Fuls, and P. A. Heimann.

REFERENCES

1. C. J. Bartlett, J. M. Segelken, and N. A. Tenekedges, "Multichip Packaging Design for VLSI-Based Systems", *37-th ECC Proceedings*, May 1987, Boston, Mass., pp. 518-525.

2. J.-P. Clech, F. M. Langerman, and J. A. Augis, "Local CTE Mismatch in SM Leaded Packages: A Potential Reliability Concern", *40-th ECTC Proceedings*, May 1990, Las Vegas, Nevada, pp. 368-376.

3. P. M. Hall, "The Effect of Expansion Mismatch on Temperature Coefficient of Resistance of Thin Film", *Applied Physics Letters*, vol. 12, No. 6, March 1968.

4. P. M. Hall and W. H. Sherry, "Materials, Structures, and Mechanics of Solder Joints for Surface Mount Microelectronics Technology", *Proc. of the 3-rd Int. Conf. on Interconnection Technology in Electronics*, Fellbach, West Germany, 1986, pp. 47-61.

5. J. H. Lau and D. W. Price, "Solder Joint Failure in Surface Mount Technology: State-of-the-Art", *Solid State Technology*, vol. 28, No. 10, October 1985, pp. 91-104.

6. Military Handbook - "Reliability Prediction of Electronic Equipment, MIL-HDBK-217D, Dept. of Defense, Washington, DC, 1982.

7. E. Suhir, "Calculated Thermally Induced Stresses in Adhesively Bonded and Soldered Assemblies", *ISHM Int. Conference*, Atlanta, GA, 1986.

8. E. Suhir and Y.-C. Lee, "Thermal, Mechanical, and Environmental Durability Design Methodologies in Electronic Packaging", in Electronic Packaging, *Electronic Materials Handbook*, vol. 1, "Packaging", ASM International, 1989, pp. 45-75.

9. E. Suhir, "Axisymmetric Elastic Deformations of a Finite Circular Cylinder, With Application to Low Temperature Strains and Stresses in Solder Joints", *ASME J. Appl. Mechanics*, vol. 56, No. 2, 1989, pp. 328-333.

10. E. Suhir, "Twist-off Testing of Solder Joint Interconnections", *ASME J. Electr. Pack.*, vol. 111, No. 3, 1989, pp. 165-171.

11. E. Suhir and J. M. Segelken, "Mechanical Behavior of Flip-Chip Encapsulants", *ASME J. Electr. Pack.*, vol. 112, No. 4, 1990, pp. 327-332.

12. E. Suhir, "Mechanical Behavior of Flip Chip Solder Joint Interconnections in Thermally Matched Assemblies", *42-th ECTC Proceedings*, May 1992, San-Diego, Calif., pp. 563-572.

13. T. Yamada, et al., "Low Stress Design of Flip-Chip Technology for Si-on-Si Multichip Modules", *Proc. of the Int. Symp. on Electronic Packaging*, Orlando, Fla., 1986.

14. C. P. Wong, "Integrated Circuit Encapsulants", in *Polymers in Electronics*, Encyclopedia of Polymer Science and Engineering, 2-nd ed., vol. 5, New York, John Wiley, 1986.

APPENDIX

"Global" and "Local" Thermal Mismatch Stresses in an Elongated Bi-Material Assembly Soldered at the Ends

An elongated bi-material assembly soldered at the ends is a useful analytical stress model which can provide valuable insight into the interaction of "global" and "local" thermal mismatch stresses in some microelectronic structures. The "global" and "local" shearing stresses acting on the adherends at low temperature conditions are schematically shown in Fig. A-1 (the component #1 has a higher coefficient of expansion than the component #2). The "global" mismatch of the adherends is counted with respect to the midcross-section of the assembly as a whole, while the "local" mismatch is counted with respect to the midcross-sections of the bonded regions. In the analysis which follows we develop engineering formulas for the evaluation of the "global", "local", and total shearing stresses in the solder layer (joint). The major assumption underlying the analysis is that there is appreciable thermal expansion (contraction) mismatch between the soldered components, and that the solder layer (joint) is compliant enough (in the x-y plane), so that the mismatch between the solder and the soldered materials need not be considered. In other words, the term "local mismatch" used in this Appendix has to do with the stresses developed within the region occupied by the solder layer (joint) due to the thermal expansion (contraction) mismatch of the soldered components. A situation when, for one reason or another, the mismatch between the solder material and the soldered components should be considered is not addressed in this analysis.

Let a bi-material assembly (Fig. A-1) manufactured at an elevated temperature be subsequently cooled down to a low (room or testing) temperature. If the interfacial shearing stress, $\tau(x)$, caused by the thermal contraction mismatch of the materials, were known, then the longitudinal interfacial displacements of the assembly components within the soldered region could be evaluated by the formulas (Suhir, 1986):

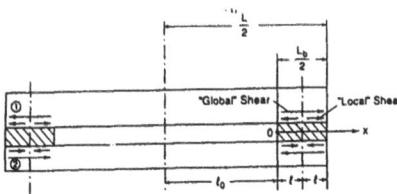

Fig. 1

Bi-material Assembly Soldered at the Ends

$$
\left.
\begin{aligned}
u_1(x) &= -\alpha_1 \Delta t\, x + \lambda_1 \int_0^x T(\xi)\,d\xi - \\
&\quad - \kappa_1 \tau(x) - \frac{h_1}{2} w_1'(x) \\
u_2(x) &= -\alpha_2 \Delta t\, x - \lambda_2 \int_0^x T(\xi)\,d\xi + \\
&\quad + \kappa_2 \tau(x) + \frac{h_2}{2} w_2'(x)
\end{aligned}
\right\} . \quad \text{(A-1)}
$$

Here α_1 and α_2 ($\alpha_1 < \alpha_2$) are the coefficients of thermal expansion (contraction) of the materials, $u_1(x)$ and $u_2(x)$ are the interfacial longitudinal displacements of the low and the high expansion components, respectively,

$$
\lambda_1 = \frac{1-\nu_1^2}{E_1 h_1}, \quad \lambda_2 = \frac{1-\nu_2^2}{E_2 h_2} \quad \text{(A-2)}
$$

are the longitudinal compliances of the components, E_1, ν_1 and E_2, ν_2 are the elastic constants of the materials, h_1 and h_2 are the thicknesses of the components,

$$
\kappa_1 = \frac{h_1}{3G_1}, \quad \kappa_2 = \frac{h_2}{3G_2} \quad \text{(A-3)}
$$

are their interfacial compliances, $G_1 = \dfrac{E_1}{2(1+\nu_1)}$ and $G_2 = \dfrac{E_2}{2(1+\nu_2)}$ are the shear moduli of the component materials,

$$
T(x) = T_0 + \int_0^x \tau(\xi)\,d\xi \quad \text{(A-4)}
$$

is the thermally induced force in the components cross-sections, T_0 is the magnitude of this force in the midportion of the assembly (outside the soldered regions), $w_1'(x)$ and $w_2'(x)$ are the angles of rotation, and Δt is the change in temperature. The origin of the coordinate x is at the inner edge of the bonded region (Fig. A-1).

The first terms in the formulas (1) are the unrestricted ("stress-free") thermal contractions. The second terms are the displacements caused by the thermally induced forces $T(x)$. These displacements are evaluated on the basis of Hooke's law under an assumption that they are evenly distributed over the cross-section x. Clearly, this assumption does not adequately reflect the actual situation: the component's displacements are the largest at its interface, where the induced forces are applied. This is accounted for by the third terms in the formulas (1). These terms provide corrections for the non-uniform distribution of the longitudinal displacements over the given cross-section. These terms reflect an assumption that such corrections are proportional to the magnitude of the shearing stress in this cross-section and are not affected by the stresses in other cross-sections.

Assuming that the solder layer (joint) is subjected to shear only, and is, in addition, thin (compliant) enough, so that its own thermal contraction need not be considered (Suhir, 1986), one can write the condition of the compatibility of the interfacial displacements as follows:

$$u_1(x) = u_2(x) - \kappa_a \tau(x) , \qquad (A-5)$$

where

$$\kappa_a = \frac{2h_a}{3G_a} \qquad (A-6)$$

is the interfacial compliance of the attachment, $G_a = \dfrac{E_a}{2(1+v_a)}$ is the shear modulus of its material, and h_a is the attachment thickness. Substituting the formulas (A-1) into the condition (A-5), we obtain:

$$\kappa \tau(x) - \lambda_{12} \int_0^x T(\xi)\,d\xi + \frac{h_1}{2}\, w_1'(x) +$$

$$+ \frac{h_2}{2}\, w_2'(x) = \Delta\alpha\Delta t x . \qquad (A-7)$$

Here

$$\lambda_{12} = \lambda_1 + \lambda_2 , \quad \kappa = \kappa_1 + \kappa_2 + \kappa_a \quad (A-8)$$

are the total axial and the total interfacial compliances of the assembly, and $\Delta\alpha = \alpha_2 - \alpha_1$ is the difference in the coefficients of expansion.

We assume that the difference in the deflections $w_1(x)$ and $w_2(x)$ of the assembly components has an insignificant effect on the shearing stress (clearly, this difference cannot be ignored when the "peeling" stresses are evaluated). Then we have:

$$\kappa \tau(x) - \lambda_{12} \int_0^x T(\xi)\,d\xi + \frac{h}{2}\, w'(x) = \Delta\alpha\Delta t x . \quad (A-9)$$

In this equation, $w(x) = w_1(x) \approx w_2(x)$, is the deflection function of the assembly, and $h = h_1 + h_2$ is its total thickness. By differentiation, we obtain:

$$\kappa \tau'(x) - \lambda_{12} T(x) + \frac{h}{2}\, w''(x) = \Delta\alpha\Delta t . (A-10)$$

No axial external forces, nor bending moments are applied at the edge $x = 2l$. Therefore $T(2l) = 0$, and $w''(2l) = 0$. These conditions can be translated in the following boundary condition for the shearing stress function $\tau(x)$:

$$\tau'(2l) = \frac{\Delta\alpha\Delta t}{\kappa} . \qquad (A-11)$$

The bending moment acting over the assembly cross-sections can be evaluated as

$$M(x) = -T(x)a , \qquad (A-12)$$

where a is the distance between the midplanes of the assembly components. Then the average curvature $w''(x)$ of the assembly is

$$w''(x) = -\frac{a}{D}\, T(x) , \qquad (A-13)$$

where D is its flexural rigidity. This can be calculated as

$$D = \frac{\bar{E}_1 h_1^4 + \bar{E}_2 h_2^4 + 2\bar{E}_1\bar{E}_2 h_1 h_2(2h_1^2 + 2h_2^2 + 3h_1 h_2)}{12(\bar{E}_1 h_1 + \bar{E}_2 h_2)} , (A-14)$$

where $\bar{E}_1 = \dfrac{E_1}{1-v_1^2}$ and $\bar{E}_2 = \dfrac{E_2}{1-v_2^2}$ are the "effective"

Young's moduli of the materials. If both components were of the same thickness ($h_1 = h_2$), then the formula (A-14) would yield:

$$D = \frac{\bar{E}_1^2 + \bar{E}_2^2 + 14\bar{E}_1\bar{E}_2}{96(\bar{E}_1 + \bar{E}_2)}\, h^3 , \qquad (A-15)$$

where $h = h_1 + h_2$ is the total thickness of the assembly. If, in addition, $\bar{E}_1 = \bar{E}_2 = \bar{E}$, then the formula (A-15) would result in the following well-known relationship: $D = \dfrac{\bar{E}h^3}{12}$. If one of the components is significantly thinner than the other (say, $h_2 \ll h_1$), the formula (A-14) results in a similar relationship: $D = \dfrac{\bar{E}_1 h_1^3}{12}$.

Introducing (A-13) into (A-10), we obtain:

$$\kappa \tau'(x) - \lambda T(x) = \Delta\alpha\Delta t , \qquad (A-16)$$

where

$$\lambda = \lambda_{12} + \frac{ah}{2D} \qquad (A-17)$$

is the total axial compliance of the assembly with consideration of the effect of bending.

Differentiating the equation (A-16), and taking into account the notation (A-4), we obtain the following differential equation for the shearing stress function $\tau(x)$:

$$\tau''(x) - k^2\tau(x) = 0 , \qquad (A-18)$$

where the eigenvalue k is

$$k = \sqrt{\frac{\lambda}{\kappa}} . \qquad (A-19)$$

The equation (A-18) has the following solution:

$$\tau(x) = C_1 \sinh kx + C_2 \cosh kx , \qquad (A-20)$$

where C_1 and C_2 are the constants of integration. Then the equation (4) can be written as

$$T(x) = T_0 + \frac{C_1}{k}(\cosh kx - 1) + \frac{C_2}{k}\sinh kx .(A-21)$$

In the case of "global" mismatch only the solution (21) can be written, putting $C_1 = kT_0$, as follows:

$$T(x) = T_0 \cosh kx + \frac{C_2}{k}\sinh kx .$$

The boundary condition $T(2l) = 0$ results in the following relationship between the constants of integration:

$$C_2 = -C_1 \cotanh 2kl = -kT_0 \cotanh 2kl ,$$

Hence,

$$\tau(x) = kT_0(\sinh kx - \cotanh 2kl \cosh kx) , \qquad (A-22)$$

$$T(x) = T_0(\cosh kx - \cotanh 2kl \sinh kx) . \qquad (A-23)$$

At the origin ($x = 0$),

$$\tau(0) = -kT_0 \cotanh 2kl , \quad T(0) = T_0 .(A-24)$$

In the middle of the soldered region ($x = l$),

$$\tau(l) = \frac{kT_0}{2 \sinh kl} , \quad T(l) = \frac{T_0}{2 \cosh kl} . (A-25)$$

At the assembly end ($x = 2l$),

$$\tau(2l) = -\frac{kT_0}{\sinh 2kl} , \quad T(2kl) = 0 . \quad (A-26)$$

For long enough bonded areas (large l values) and/or sufficiently stiff attachments (large k values), the formulas (A-24)-(A-26) yield:

$$\tau(0) = -kT_0, \quad T(0) = T_0$$
$$\tau(l) = \tau(2l) = 0, \quad T(l) = T(2l) = 0 \left.\right\} \quad \text{(A-27)}$$

The obtained results indicate that, in the case of long bonded areas and/or stiff attachments, the "global" interfacial shearing stresses concentrate at the inner ends of the bonded regions. In another extreme case, i.e. for very short and/or very compliant attachments, the "global" shearing stress is uniformly distributed along the interface, and the resulting force in the assembly components decreases linearly from T_0 to zero:

$$\tau(0) = \tau(l) = \tau(2l) = -\frac{T_0}{2l},$$
$$T(0) = T_0, \quad T(l) = \frac{T_0}{2}, \quad T(2l) = 0 \left.\right\} \quad \text{(A-28)}$$

In the case of a "local" mismatch, the shearing stress $\tau(x)$ is distributed antisymmetrically with respect to the midcross-section $x = l$ of the bonded region, and therefore the following condition must be fulfilled:

$$\tau(l) = 0. \qquad \text{(A-29)}$$

Then the solution (20) yields:

$$C_2 = -C_1 \tanh kl, \qquad \text{(A-30)}$$

and can be written as follows:

$$\tau(x) = C_1 (\sinh kx - \tanh kl \cosh kx). \qquad \text{(A-31)}$$

The constant C_1 can be determined from the boundary condition (11):

$$C_1 = \frac{\Delta\alpha\Delta t}{k\kappa}. \qquad \text{(A-32)}$$

Thus, the interfacial shearing stress due to the "local" mismatch is

$$\tau(x) = \frac{\Delta\alpha\Delta t}{k\kappa} (\sinh kx - \tanh kl \cosh kx)$$
$$= \frac{k\Delta\alpha\Delta t}{\lambda} \frac{\sinh[k(x-l)]}{\cosh kl}. \qquad \text{(A-33)}$$

The resulting axial force, acting in the assembly components, is

$$T(x) = -\frac{\Delta\alpha\Delta t}{\lambda} (1 - \cosh kx + \tanh kl \sinh kx)$$
$$= -\frac{\Delta\alpha\Delta t}{\lambda} \left[1 - \frac{\cosh[k(x-l)]}{\cosh kl} \right]. \qquad \text{(A-34)}$$

At the origin ($x = 0$),

$$\tau(0) = -\frac{k\Delta\alpha\Delta t}{\lambda} \tanh kl, \quad T(0) = 0. \quad \text{(A-35)}$$

In the middle of the bonded region ($x = l$),

$$\tau(l) = 0, \quad T(l) = -\frac{\Delta\alpha\Delta t}{\lambda} \left[1 - \frac{1}{\cosh kl} \right]. \quad \text{(A-36)}$$

At the outer end ($x = 2l$):

$$\tau(2l) = -\tau_{(0)} = \frac{k\Delta\alpha\Delta\tau}{\lambda} \tanh kl, \quad T(2l) = 0. \quad \text{(A-37)}$$

For long bonded areas and/or stiff attachments (large kl values),

$$\tau(0) = -\tau(2l) = -\frac{k\Delta\alpha\Delta t}{\lambda},$$
$$T(0) = T(2l) = 0, \quad T(l) = -\frac{\Delta\alpha\Delta t}{\lambda} \left.\right\} \quad \text{(A-38)}$$

In the case of short bonded areas and/or compliant attachments (small kl values),

$$\tau(0) = -\tau(2l) = -\frac{\Delta\alpha\Delta t}{\kappa} l,$$
$$T(0) = T(2l) \approx 0, T(l) = -\frac{\Delta\alpha\Delta t}{2\kappa} l^2 = \frac{l}{2}\tau(0) \left.\right\} \quad \text{(A-39)}$$

These formulas are not different from those obtained for continuously bonded assemblies (Suhir, 1986).

The *compressive force* T_0 in the midportion of the assembly can be determined using a compatibility condition similar to the condition (A-5) and written for the cross-section $x = 0$:

$$-\alpha_1\Delta u_0 + \lambda_1 T_0 l_0 - \kappa_1\tau_0 - \frac{h_1}{2} w'(0)$$
$$= -\alpha_2\Delta u_0 - \lambda_2 T_0 l_0 + \kappa_2\tau_0 + \frac{h_2}{2} w'(0). \qquad \text{(A-40)}$$

Here l_0 is half the length of the unbonded midportion of the assembly, and τ_0 is the total shearing stress at the origin ($x = 0$). This stress can be expressed, using (A-22) and (A-35), as

$$\tau_0 = -kT_0 \coth 2kl - \frac{k\Delta\alpha\Delta t}{\lambda} \tanh kl. \qquad \text{(A-41)}$$

Introducing this formula into the condition (A-40) and solving the obtained equation for the force T_0, we obtain:

$$T_0 = -\frac{\Delta\alpha\Delta t}{\lambda} \frac{kl_0 + \tanh kl}{kl_0 + \coth 2kl}. \qquad \text{(A-42)}$$

In the case of very large kl and/or kl_0 values, the formula (A-42) yields:

$$T_0 = -\frac{\Delta\alpha\Delta t}{\lambda}. \qquad \text{(A-43)}$$

For small kl values, when one can assume

$$\tanh kl \approx kl, \quad \coth 2kl \approx \frac{1}{2kl}, \qquad \text{(A-44)}$$

and the formula (A-42) yields:

$$T_0 = -2\frac{\Delta\alpha\Delta t}{\lambda} k^2 l l_0 = -2\frac{\Delta\alpha\Delta t}{\kappa} l l_0. \text{(A-45)}$$

This formula indicates that, as long as the conditions (A-44) are fulfilled, the compressive force T_0 can be reduced by employing an attachment material with low curing temperature (for lower Δt) and high interfacial compliance κ. This compliance can be increased, as evident from the formula (A-6), by increasing the thickness of the adhesive layer (the height of a solder joint), and by employing low modulus adhesives (say, adhesives with low glass transition temperature) or very compliant ("soft") solders. The formula (A-45) indicates also that the length $2l$ of the bonded area should be kept, for lower thermally induced forces, as small as possible. It should be emphasized that all these measures can be effective only provided that the equations (A-44) are fulfilled. Since the second condition in (A-44) is stronger than the first one, we conclude that the increase in the attachment compliance can lead to a lower compressive force T_0, if the kl value is appreciably smaller than unity.

The "global" shearing stress can be evaluated, using the formulas (A-22) and (A-42), by the formula

$$\tau(x) = -\frac{k\Delta\alpha\Delta t}{\lambda} \frac{kl_0 + \tanh kl}{kl_0 + \coth 2kl} \times$$
$$(\sinh kx - \coth 2kl \cosh kx). \qquad \text{(A-46)}$$

Then, considering the formula (A-33) for the shearing stress due to the "local" mismatch, we obtain the

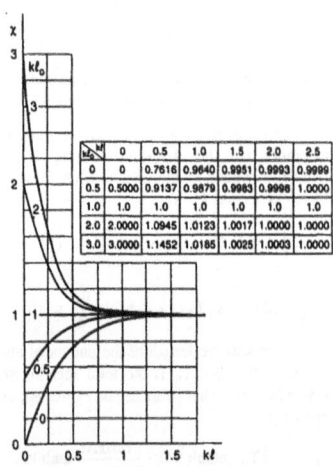

kl_0 \ kl	0	0.5	1.0	1.5	2.0	2.5
0	0	0.7616	0.9640	0.9951	0.9993	0.9999
0.5	0.5000	0.9137	0.9879	0.9983	0.9998	1.0000
1.0	1.0	1.0	1.0	1.0	1.0	1.0
2.0	2.0000	1.0945	1.0123	1.0017	1.0000	1.0000
3.0	3.0000	1.1452	1.0185	1.0025	1.0003	1.0000

Fig. 2

Ratio of the Maximum Shearing Stress in an Assembly Bonded at its Ends to the Maximum Stress in a Continuously Bonded Assembly

following formula for the *total stress*, caused by the combined effect of the "global" and "local" mismatches:

$$\tau(x) = \frac{k \Delta \alpha \Delta t}{\lambda} \frac{\sinh kx + kl_0 \cosh kx}{kl_0 \sinh 2kl + \cosh 2kl} \quad \text{(A-47)}$$

The values of the shearing stress at the ends of the bonded (soldered) region are:

$$\left. \begin{aligned} \tau(0) &= \frac{\Delta \alpha \Delta t}{\kappa} \, l_0 \, \frac{1}{kl_0 \sinh 2kl + \cosh 2kl} \\ \tau(2l) &= \frac{k \Delta \alpha \Delta t}{\lambda} \, \frac{\tanh 2kl + kl_0}{1 + kl_0 \tanh 2kl} \end{aligned} \right\} . \quad \text{(A-48)}$$

For large kl values we obtain:

$$\tau(0) = 0 , \quad \tau(2l) = \frac{k \Delta \alpha \Delta t}{\lambda} . \quad \text{(A-49)}$$

In the case of small k values, the formulas (A-48) yield:

$$\tau(0) = \frac{\Delta \alpha \Delta t}{\kappa} \, l_0 \, \tau(2l) = \frac{\Delta \alpha \Delta t}{\kappa} \, (l_0 + 2l) . \quad \text{(A-50)}$$

Thus, as follows from (A-49) and (A-50), the maximum shearing stress occurs at the outer end of the bonded region and is expressed by the second formula in (A-48). This formula can be written as

$$\tau_{max} = \tau_{\infty} \chi(kl_0 , kl) , \quad \text{(A-51)}$$

where

$$\tau_{\infty} = \frac{k \Delta \alpha \Delta t}{\lambda} \quad \text{(A-52)}$$

is the maximum shearing stress in the case of a continuous bond (i.e. in the case when only "local" mismatch stresses exist), and the factor

$$\chi = \frac{kl_0 + \tanh 2kl}{1 + kl_0 \tanh 2kl} \quad \text{(A-53)}$$

considers the effect of the combined action of the "global" and "local" stresses, as well as the effect of the attachment compliance on the total maximum shearing stress. This factor is equal to $\chi = 1$, in the case of a long soldered area and/or stiff attachment, and is equal to

$$\chi = k(l_0 + 2l) , \quad \text{(A-54)}$$

in the case of a small k value. If, in addition, the length $2l$ of the bonded area is small compared to the length $2l_0$ of the midportion of the assembly, then

$$\chi \approx kl_0 . \quad \text{(A-55)}$$

In the case of stiff adherends and compliant adhesive (i.e. low λ values and large κ values, leading to low k values) the formulas (A-51), (A-52) and (A-54) yield:

$$\tau_{max} = \frac{\Delta \alpha \Delta t}{\kappa} \, (l_0 + 2l) = \frac{\Delta \alpha \Delta t}{2\kappa} \, L_a , \quad \text{(A-56)}$$

where L_a is the assembly length. This formula indicates that when a sufficiently compliant attachment is used, the maximum shearing stress becomes independent from the stiffness of the soldered components. In practical design, the adherends should be simply made sufficiently stiff to retain planarity.

The factor χ is plotted in Fig. 2 versus kl values for different kl_0 values. As evident from this figure, one should try to avoid very short lengths of the soldered areas, unless a very compliant solder is used. The latter case seems, however, impractical.

REFERENCE

E. Suhir, "Stresses in Bi-Metal Thermostats", ASME J. of Applied Mechanics, vol. 53, No. 3, Sept. 1986.

MECHANICAL STRESS IN MICROELECTRONIC INTERCONNECTS

PETER J. GIELISSE, MEIRONG TU,
DONGMING Y. WHITE, AND YANG XU
FAMU-FSU College of Engineering
Florida State University
Tallahassee, Florida 32310
U.S.A

1. Abstract

This article examines the origin and nature of mechanical stress and possible failure of interconnects in microelectronics, involving both the global interconnect and the intermetal dielectric.

Results from thin film processing illustrate the importance of thin film microstructure (topography) and its relation to that of the underlayer in determining the interconnect properties.

Materials characterization, stress measurements and modeling of stress effects in interconnect underlayer structures, have been accomplished with an Optical Surface Scattering (OSS) technique, the application of a custom windows-based software package and the use of holographic interferometry. Results from each category are presented.

2. Introduction

With the advance of large scale integration technologies, thousands to tens of thousands of devices are fabricated on a silicon chip. This ever increasing density results in high heat fluxes and stress levels in microelectronic interconnects. Comprehensive knowledge of material properties, thermal dissipation patterns and stress distributions is, therefore, of critical importance.

The internal thermal performance of various assembly and interconnect variations is mostly associated with the paths "through the substrates" under each chip in a module. The thermal resistance value depends on the type of interconnection between film and substrate which, in turn, is dependent on the topography and nature of the substrate surface. The imperfect nature of the contact across surface asperities causes the actual heat transfer area to be only a part of that of the actual area. A knowledge of the precise surface character resulting from materials processing, will directly impact on the proficiency of methods to reduce stress levels, to attain product

97

G. Harman and P. Mach (eds.), Microelectronic Interconnections and Assembly, 97-108.
© 1998 Kluwer Academic Publishers.

uniformity, and on the ability to institute true quality control and process improvements. An optical scattering system (OSS) has been developed in which the angular distribution of scattered light sensitively characterizes a sample in the spatial frequency domain.

The stresses in interconnects, resulting from a mismatch in thermal properties of different materials, are distributed asymptotically. The peeling stress at the free edge, the location of the singularity point, is one of the most prevalent failure modes in multilayer structures. The singularity stress can be expressed by two parameters: the order of stress singularity and the intensity of stress singularity. The dominant factors influencing both the order of the stress singularity and the intensity of the stress singularity are the properties of the materials that participate in the package [1], the thickness of each of the layers [2] and the complications that arise from certain geometries. A hybrid element FEM package "MULTILAYER STRUCTURES" has been developed, in which hybrid elements cover the region of singularity behavior. It has been designed to specifically treat heat transfer and stress analysis in multilayer structures.

3. Technical Approaches

3.1 Hybrid finite element method (HFEM).

Since analytical solutions for the residual stress levels along the substrate-film and film-film interfaces do not cover the region close to the free edge (Fig. 1), a hybrid element FEM approach was introduced.

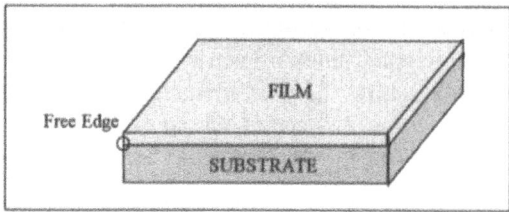

Figure 1. Layered Structure Displaying Free Edge and Singularity Point.

Stress at the free-edge along the interface of a multilayer structure behaves in a singular way. It has been shown [3] that the stress at this location can be expressed as,

$$\sigma_{ij} = \kappa_{ij} r^{-\lambda} \tag{1}$$

where κ_{ij} is the intensity of the stress singularity, λ is the order of the stress singularity and r is the distance from the free edge. The two singularity parameters κ_{ij} and λ, have been used as evaluation parameters in judging the reliability of multilayer structures [3].

A complex variable technique has been used to calculate κ_{ij} and λ, based on a complex function representing stress and displacement. This method was initially introduced in reference [3] and provides basically the same solution as introduced by other methods such as in the Mellin Transformation Method [4] and the Superposition Method [5].

As in conventional finite element approaches, the compatibility equations and boundary conditions are satisfied in our hybrid element method as well. In addition, the hybrid element satisfied the equilibrium equation $\sigma_{ij,j} = 0$ in yielding the solution provided in equation (1). The variation function for the problem is then given by,

$$\Pi = \sum \Pi_h + \sum \Pi_r = \sum_1^{M_h} [\int_{\partial A_-} (\tilde{u} - \frac{1}{2}T_i u_i) dh - \int_{A_-} \frac{1}{2}\varepsilon_{ij}^0 \sigma_{ij} dA] + \sum_1^{M_r} [\int_{A_-} (\frac{1}{2}u_{i,j}S_{ijkl}^{-1}u_{k,l} - \varepsilon_{kl}^0 S_{ijkl}^{-1}u_{i,j}) dA] \quad (2)$$

in which h refers to the hybrid elements and r to the regular elements.

The stiffness matrix and force vector can subsequently be calculated after the proper interpolation functions for the inter-element boundaries have been chosen [6].

3.1.1 Windows software development

The modeling of stress effects in multilayer structures has been implemented through the development of an appropriate computer program titled "MULTILAYER STRUCTURES". The computer program is user-friendly as well as menu-driven. The popup menus are for data input (Fig. 2 gives an example), for analysis choices and to compose output serving specific application needs.

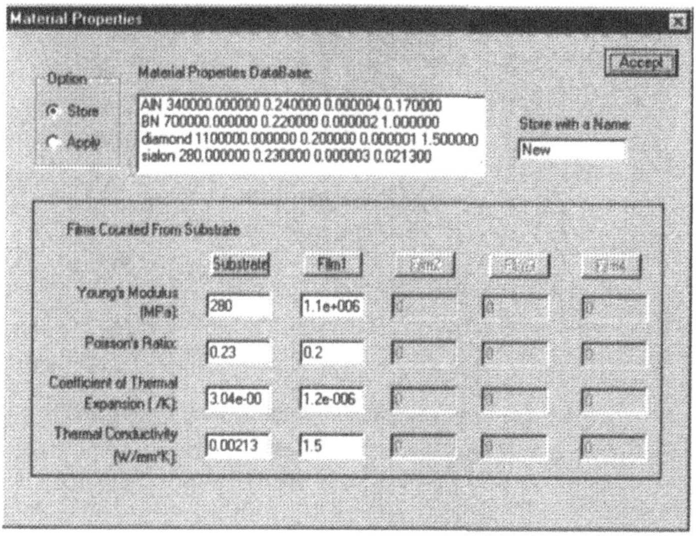

Figure 2. Material Properties Menu of the MULTILAYER STRUCTURES Software.

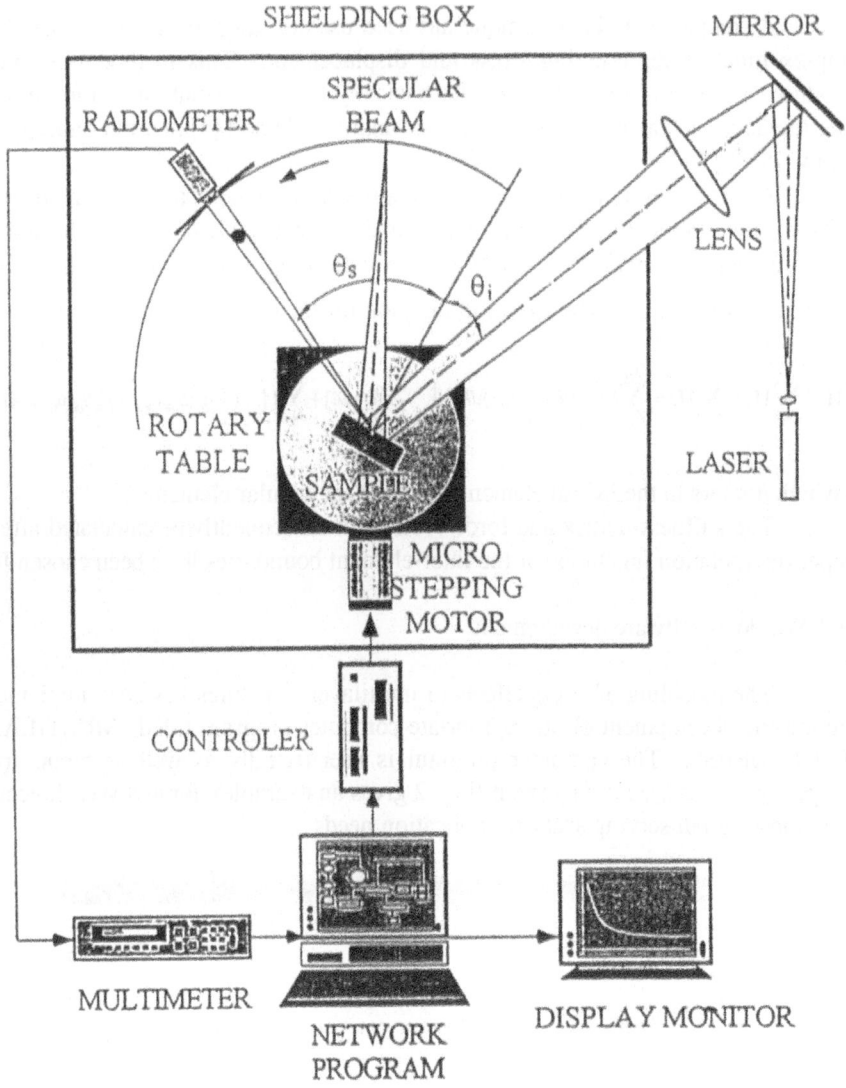

Figure 3. Schematic of the Optical Scattering System (OSS).

The program is easy to use for anybody familiar with the Windows operating system. The input calls for choice of geometry, thermal conditions and material properties. Depending on application needs, one of three analysis methods (direct FEM, analytical solution, or hybrid FEM) can be chosen. Output menus list stress, strain and displacement respectively.

3.2 Optical Scattering System (OSS)

The OSS technique has been applied to surface characterization of substrates and of multilayer films. The principle of the OSS technique is based on scattering resulting from the interaction of electromagnetic radiation (light) with matter. The far-field scattered light distribution from a rough surface under full optical illumination depends on the properties of the surface material and on the configuration of the microirregularities. By detecting the intensity of the scattered light over the angular distribution, surface character can be determined and quantified in the spatial frequency domain. The commonly accepted functions that are used to quantify surface features are: BRDF (Bidirectional Reflectance Distribution Function) and PSD (Power Spectral Density) [7]. Integration of the volume under the PSD function yields, for example, the RMS roughness value. Other parameters can be derived as well.

The main components of the optical scattering system are light source, microstepping motor, rotary table, sample holder, photo sensor, and a computer based network program for motion control and data analysis (Fig. 3). The laser source is polarized (s-polarized), spatially filtered, expanded and finally brought to focus on the detector path. The scattered light refracted by the surface of the sample is collected by a low-power detector-receiver. The detector-receiver is mounted on a rotatable optical rail. The acceptance aperture for the receiver must be well defined, since the solid angle, subtended by the receiver spot diaphragm for the sample, is used in the calculation of the bidirectional scatter distribution (BRDF) [8]. A diaphragm is placed at a position on the observation circle in front of the receiver. A multimeter has been used to measure the intensity of the scattered light. Readings were taken at 1/60 degree intervals over the range specular to five (5) degree, and every 1/12 degree for the 5 to 80 degree range. A network program performed the task sequence and was created with the aid of a Hewlett-Packard Visual Engineering Environment (HP VEE) software package. The program controls all the microstepping motor functions, reads the data from the multimeter, and analyzes the data. The final results are displayed in spreadsheet with line plot format. It takes about 30 seconds to complete a measurement.

4. Applications and Results

4.1 Stress Level Modeling

The "MULTILAYER STRUCTURES" program has already proven its applicability in various areas. Examples have been given in reference [1], which treat the impact of film thickness in the stress singularity levels, and in reference [2], in which the application of both hybrid element FEM analysis and the analytical solutions approach were used to determine the "optimal" choice of a substrate material for diamond thin film deposition. Reference [2] also provides an example of the use of transition (functionally gradient) layers to reduce stress levels in thin film deposition of materials with thermal properties distinctly different from those of the substrate.

Figure 4 shows a typical failure encountered in thin film processing. An 0.8 μm aluminum nitride thin film deposited onto a 0.7 mm thick aluminum nitride substrate has spalled off the surface. Note the extreme plastic distortion of the otherwise brittle material. The film was sputter deposited onto a room temperature substrate. Even though the film and the substrate have the same chemical composition, different microstructures can apparently cause a mismatch of material properties leading to total failure.

Figure 4. Typical Failure Mode in Thin Film Deposition.
(AlN on AlN Substrate)

Yet another aspect of importance in multilayer stress modeling relates to complications introduced by specific system geometries, such as the dependence of the stress level on the angles at which participating materials join at the singularity point.

We have noticed that cracks are often located near edges at steps and at or near bends. These edge effects are important areas of concern especially in the multilayer package area in which such discontinuous geometries abound. As an example, the residual thermal stress levels for various interface geometries, in the case of diamond film deposition on SIALON substrates, are given in Fig. 5. The (90°-90°) geometry clearly displays considerably lower stress levels than the (180°-90°) or the (180°-45°) type interfaces. The later types should be avoided by design, if at all possible.

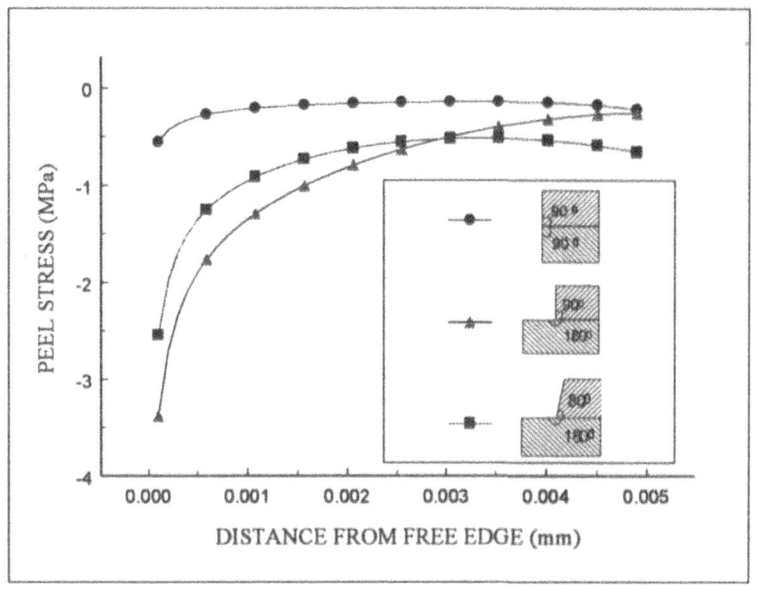

Figure 5 (a). Peel Stress Behavior for Three Different Singularity Geometries.

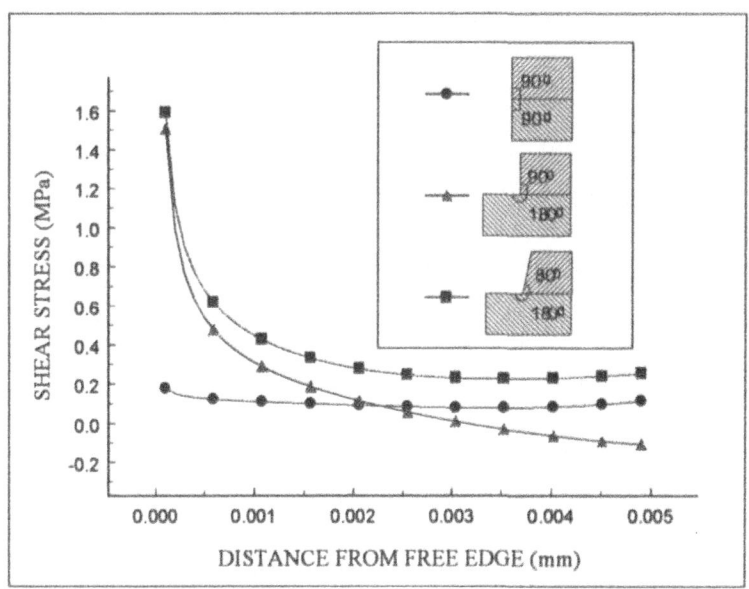

Figure 5 (b). Shear Stress Behavior for Three Different Singularity Geometries.

4.2 Surface characterization

SIALON thin films deposited on three different types of silicon (100) substrates, one superpolished with a roughness of about 50 A, one etched (2900A) and a ground specimen (3450A), were characterized by the OSS method.

Figures 6(a), (b) and (c) are SEM micrographs of the thin films on these substrates. The film deposited on the superpolished substrate reveals a surface character comparable to that of the substrate itself. The film on the ground silicon substrate displays a coarse columnar growth structure, while the etched substrate shows an in-between character.

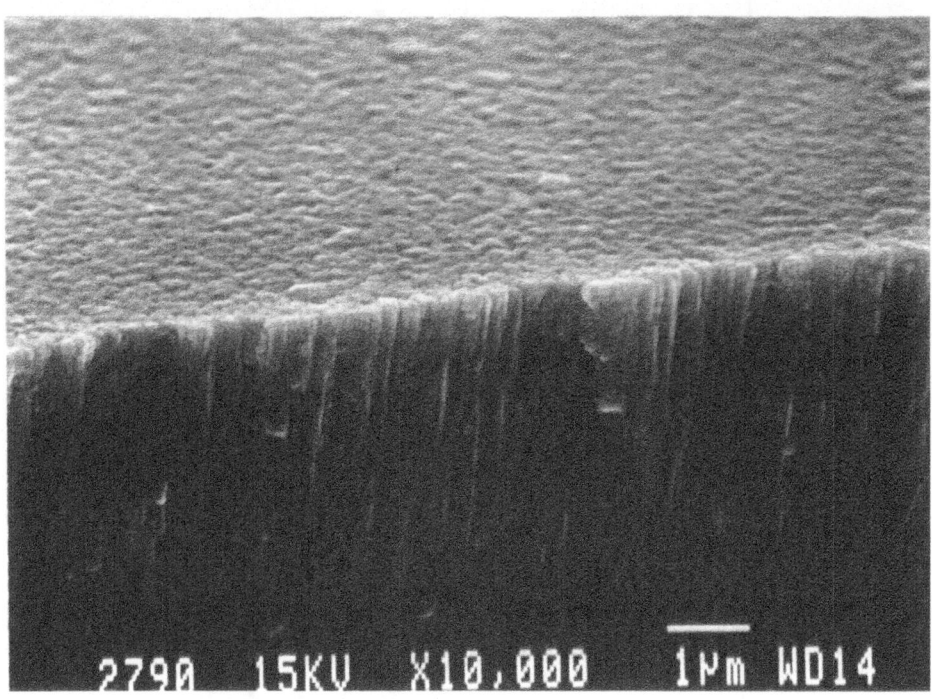

Figure 6 (a). SEM Micrographs of SIALON Thin Film Deposited on
Superpolished Silicon.

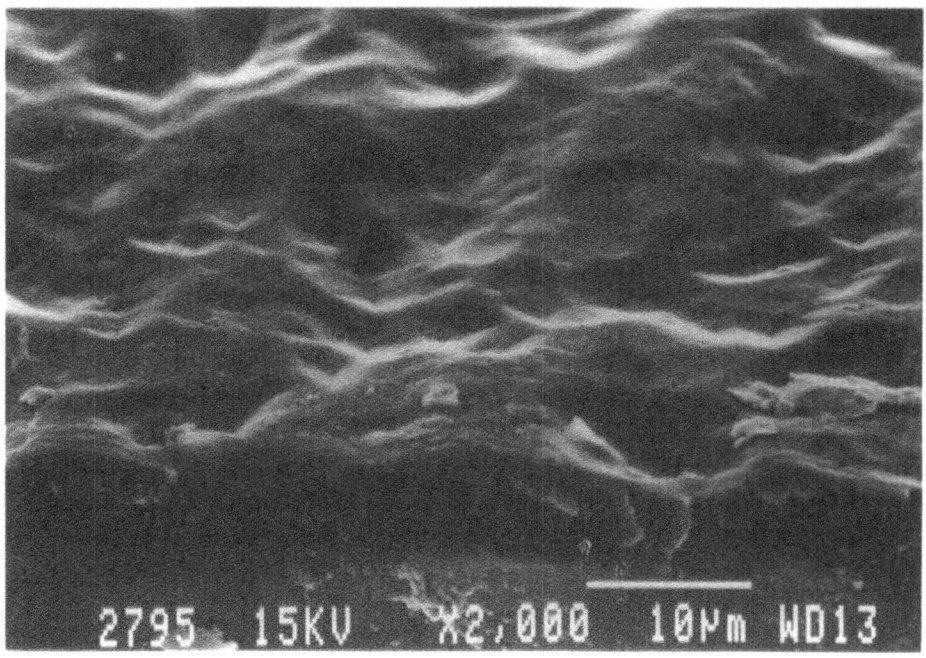

Figure 6 (b). SEM Micrographs if SIALON Thin Film Deposited on Etched Silicon.

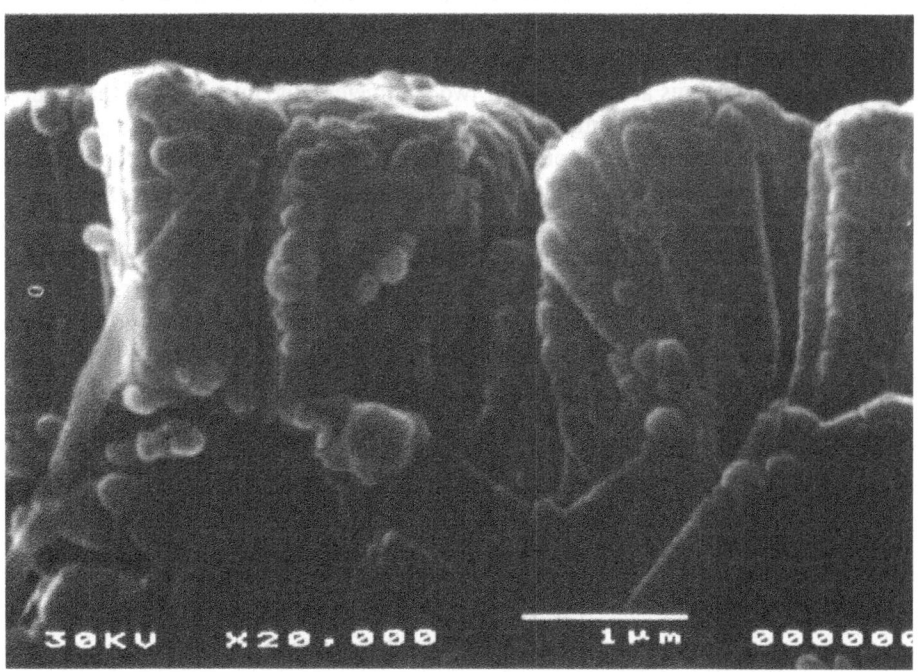

Figure 6 (c). SEM Micrographs if SIALON Thin Film Deposited on Ground Silicon.

Figure 7. displays the power spectral density (PSD) functions of the as-deposited film surfaces as a function of the spatial frequency. The PSD curves obtained from the superpolished silicon substrate surface and from the SIALON films deposited thereon, appear virtually identical. The films deposited on the ground and etched silicon surfaces are similar in character through the entire frequency range. The roughness values are, however, considerably different from those deposited on the superpolished silicon surface. The etched substrate surface shows the higher reflectance near the specular direction ($v=0$) and less (light) scattering out of the reflection direction than that of the film deposited on it , which is also the case for the ground substrate and film. The PSD versus frequency plots for the films deposited on the ground and etched substrates are typical of randomly scattering surfaces and can be considered "rough". There is no sharp drop in PSD intensity with frequency as is the case for surfaces that can be considered "superpolished", i.e., surfaces with an overall roughness less than 100 A. The important finding here is that the film surface appears to "mimmick" or "copy" the character of the substrate surface, displaying a roughness and structure that is not much different from that of the substrate itself. The film structure is affected not only by the roughness but also by the specific microstructure of the substrate.

5. Summary

It could be shown that the magnitude of peel and shear stress levels in thin film interconnects can be modeled using a combination of traditional analytical solutions and a hybrid finite element (HFEM) approach. The latter is required to accurately treat stresses at the free edges, the so called singularity points. The stress levels at these locations are governed by two parameters, the intensity and the order of the stress singularity. The dominant factors influencing the values of these parameters are the mechanical and thermal properties of the participating materials, the thickness of each (film) layer and the way in which materials join at the singularity points. A "MULTILAYER STRUCTURES" program allows, among others, evaluation of the applicability of functionally gradient material structures, the choice of "best substrate" and the determination of the influence of geometric parameters on film stress level.

An optical scattering system made possible high resolution (50A) surface analysis, which clarified the dependence of film type and surface character on substrate properties.

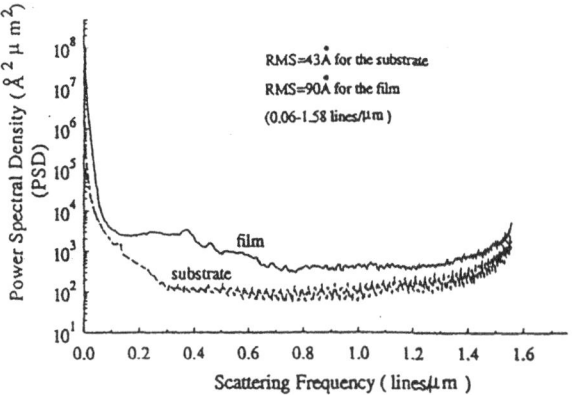

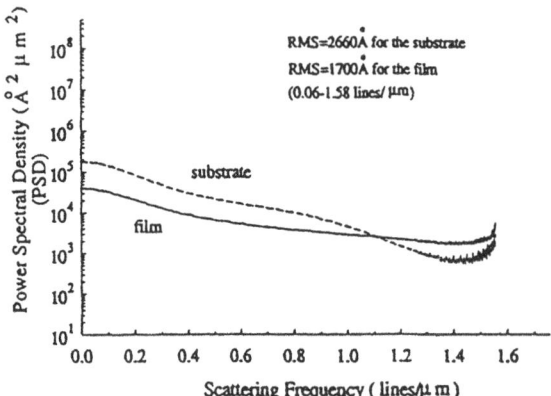

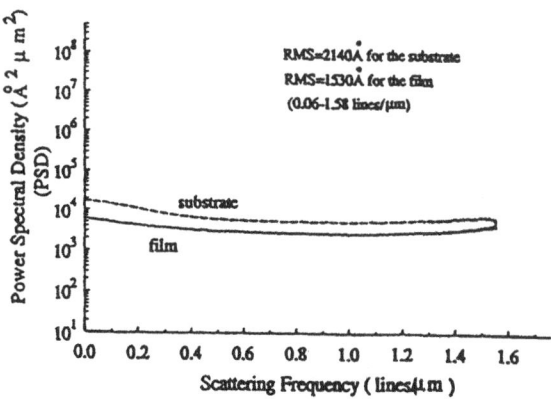

Figure 7. The PSD Function vs. Frequency for a SIALON Film Deposited on a Silicon Superpolished Substrate (top), a Film Deposited on an Etched Substrate (center), and a Film Deposited on a Ground Substrate (bottom).

6. References

[1] Yuan, D. and Gielisse P.J., (1996) Software "Design Tool" for Multilayer Structures, *Proceedings, Annual Meeting ISHM*.

[2] Gielisse, P.J., Niculescu, H., Yuan, D., Schmidmeier, R. and Chen, K.L., (1994) Modeling Stress Distributions in Multichip Module (MCM) Structures, *Proceedings International Conference on Electronic Technologies*, Windsor, England, pp227-232.

[3] Theocaris, P.S., (1974) The Order of Singularity at a Multi-Wedge Corner of a Composite Plate, *Int. J. Eng. Sci.*, 12, pp107-120.

[4] Bogy, D.B., (1968), Edge-bonded Dissimilar Orthogonal Elastic Wedges Under Normal and Shear Loading, *J. Appl. Mech.* 35, pp460.

[5] Hetenyi, M., (1960) A Method of Solution for the Elastic Quarter-Plane, *J. Appl. Mech.* 27, pp189.

[6] Chen, K.L. and Kuo, A.Y., (1991) Edge Stresses of a Multi-layered Device Under Transient Thermal Loading, *Proceedings ASME Winter Annual Meeting*.

[7] Stover, J.C., (1990) Optical Scattering Measurement and Analysis, *McGraw-Hill New York*.

[8] Bennett, J.M. and Mattsson, L. (1989) "Introduction to Surface Roughness and Scattering, *Optical Society of America*, Washington D.C..

Low Cost Interconnection Technology
for Fast Prototyping of Multichip Modules

Zsolt Illyefalvi-Vitéz; János Pinkola; László Gál; Endre Tóth

Department of Electronics Technology, Technical University of Budapest
Goldman t. 3., Budapest, H-1521, Hungary
illye@ett.bme.hu; pinkola@ett.bme.hu; gal@ett.bme.hu; tothe@ett.bme.hu

ABSTRACT: Recognising the immediate need for improvement of the technological capabilities of low cost, high performance electronic circuit modules in countries of Central Europe, the leading research institutions and small/medium-size enterprises of Hungary, Romania and Slovenia with the support of relevant institutions of the United Kingdom and Belgium have applied for the assistance of the European Union to establish fast prototyping low cost multichip module (MCM) technology facilities. In this Project the Department of Electronics Technology, Technical University of Budapest (TUB) will improve its laminate based (MCM-L) technology and will establish mounting and bonding facility for the assembling of the modules. The Project will realise research and technology development targeted on industry, improving the research capabilities of the Hungarian industry in the filed. The manufacturing system will integrate new processing technologies with well-established ones in all development phases, including design, manufacturing, assembling and test; its application will result in faster prototyping and shorter production period. In addition, problem-oriented knowledge of MCM technology can be given to students, who are interested in modern realisation of circuit modules.

1. INTRODUCTION

The success of the electronics industry of a region of smaller co-operating countries or a larger country mainly depends on the ability to use the most cost-effective and possibly advanced interconnection and packaging technology. The reason for this is that the performance of today's electronic equipment is primarily limited by the interconnections between components and subsystems and not by the high-speed, very large scale integrated (VLSI) circuits from which the systems are constructed. Thus, **to achieve high performance at a reasonably low cost, it is imperative to safeguard and exploit the existing design, manufacturing, assembly, testing technology and intellectual potential in the field of electronics interconnection and packaging.**

The idea of the establishment of fast prototyping low cost multichip module technology facilities can be reasoned by the fact that the research institutions and industrial companies of Hungary, Romania and Slovenia (the involved countries of Central/Eastern Europe) had rather good research and manufacturing facilities and gathered fairly good experience in the field of development and manufacturing of electronic circuit modules. During the last one or one-and-half decade, however, because of the steadily deteriorating economic conditions and the changing political/economical systems, these facilities have partly become obsolete and the experience also needs updating. Meanwhile in the highly developed countries new technologies providing products of higher quality and precision, better performance, increased reliability, and faster operation have emerged. Multichip module technology, utilising and/or uniting the advantages of ceramic (C), depositing (D) and laminating (L) techniques, is a promising solution of the requirements.

The Project will enable the participating countries to solve the specific problems concerning their electronics assembling industry, and to achieve economic development in the field. From the running of the established CHEAP-MULTI-CHIP facilities the European Industry will also get much benefit.

G. Harman and P. Mach (eds.), Microelectronic Interconnections and Assembly, 109-117.
© 1998 *Kluwer Academic Publishers.*

2. TYPES OF MULTICHIP MODULES

In the design and manufacturing of each level of interconnection of any electronic system, the driving force is to lower cost and improve performance. From these points of view the followings are of great importance:

- The *decrease of the distance between chips* in order to achieve faster operation. This requirement can be fulfilled by increasing the functional density and integration level of the applied very large scale integrated (VLSI) circuits and/or increasing the density of the interconnection system.
- The *reduction of the cost per connection* in order to reduce cost. The most cost-effective solution can be achieved by increasing the integration levels of both the chips and the interconnection systems, thus making more connections simultaneously, reducing the use of materials and processing, which results in the reduction of the cost per connection. From this a general rule can be formulated, that the cost per connection is the higher the farther the connection from the centre of the chip is.
- The *use of catalogue devices*. The customisation on the chip level is very expensive unless large quantities are to be used. The use of less application specific integrated circuits together with custom designed interconnection system can minimise the duration of the design-manufacturing-test cycle, and can result in a cost-effective complex solution.
- The *optimum partitioning of the circuitry* in order to improve thermal performance by using materials with high thermal conductivity, instead of using more complicated and expensive heat-transfer mechanisms.
- The *exploitation of the existing design, manufacturing, assembly, and testing technology* in the field of interconnection and packaging.

Taking into account all these requirements, **multichip module (MCM) technology seems to be a promising solution**. The position in the hierarchy of this modern electronics interconnection and packaging technology is between the application-specific integrated circuits (ASICs) and application-specific electronic assemblies, i.e. printed board assemblies, surface mounted assemblies (SMAs) and hybrids. MCMs can be subdivided into three general types, viz. MCM-C(eramic) uses screen-printed-and-fired materials layers as part of multilayer ceramic substrates; MCM-D(eposited) has a base substrate with deposited dielectric and metal (conductive/resistive) circuit layers; and, MCM-L(aminated) uses clad/plated copper conductors as part of multilayer, reinforced, organic laminate dielectric substrates.

MCM-C is an extension of thick-film hybrid interconnection and packaging technology, as circuit chips are mounted on insulating ceramic substrates and are interconnected by means of screened and fired conductor pastes. Vertically the conductor patterns are also separated by screened and fired insulating (dielectric) layers, and connected by conductive paste filled vias. Current state of the art for this screen printing and firing technology in high-volume production is 250 μm diameter vias, and 150 μm wide lines and spaces. From the assembling point of view, MCM-C modules are mostly designed using fixed chip-and-wire parts and surface mounted devices (SMDs) connected with bond wires to signal (conductor) layers.

The *MCM-D* type module is an electronic assembly using a substrate with one or more sequentially deposited dielectric and conductive (including resistive) layers. The conductive thin films are deposited by evaporation or sputtering, and patterned by photolithography and etching. Spin or spray coated polyimide and/or vacuum deposited silica layers are used to separate the conductive layers, while vias are dry or wet etched in the course of photolithographic processing prior to the consecutive metallization in order to prepare the necessary vertical interconnections. Since MCM-D technology offers the finest pattern resolution, usually adhesive attached and wire-bonded bare chips are used as active elements. Interconnect density characterised by 25 μm signal traces, 75 μm chip-to-chip spacings, and vias as small as 15 μm diameter can be achieved.

The current state of the art process for the manufacture of *MCM-Ls* is essentially identical with that used for highly dense multilayer printed boards, but includes higher requirements of fine-line pattern definition, etching, selective plating, additive plating, special materials etc. The five standard types of processes, which are characteristic of MCM-L technology, are as follows: photoresist application for imaging the different layers; chemical etching of clad and/or plated copper foils/layers;

lamination and curing of precured adhesive dielectric layers to insulate and fix the conductive patterns of the multilayer structure; drilling of holes through single layers for buried and blind vias and through the entire laminate for plated through holes; and, plating of drilled holes to accomplish electrical connections between the conductive patterns. In high-volume production this processing currently results in 300 μm diameter holes (vias), and 100 μm wide lines and spaces. The usual technique for assembling MCM-L modules is to fix and wire bond the dies and SMDs to the printed boards, although other methods, such as flip chip and tape automated bonding (TAB), are also in use.

3. DEVELOPMENT STRATEGY OF MCM-L TECHNOLOGY AT TUB

The research and development efforts made by the Department of Electronics Technology at TUB are to upgrade processing technologies and to establish new facilities in order to **produce low cost, high performance laminated based multichip modules**. In accordance with the phases of MCM-L technology the tasks are as follows:

1. *select and apply a suitable CAD system*;
2. *upgrade manufacturing facilities and processing technology of laminates* up to a level to fulfil the low cost, fast prototyping requirements (quantitatively, the aim is to realise 0.3 mm diameter plated-through holes/vias with 0.1 mm pattern resolution, suitable for 0.25 mm pitch);
3. *establish mounting and bonding facilities* for the assembling of MCMs (the aim is to handle 0.25 mm pitch medium-size VLSI chips);
4. *upgrade final test and performance evaluation* hardware/software tools and *production control* facilities.

These four tasks require four different strategies, mainly determined by the existing facilities and possibilities, and much less being the matter of decision. The Department will concentrate on the second task, and will continue upgrading the facilities and improving processing technologies of printed circuit laminates. In order to be able to produce not only laminates but also modules, it is imperative to establish mounting and bonding facilities (task No.3.), which needs an efficient support (a "capital injection") from a foundation, a multinational company, or an international development program, like INCO/COPERNICUS. The first and fourth tasks are in closer connection with application, as a consequence it could and should be solved in co-operation with the users, and the aim of the Department is to upgrade its possibilities to a level and its capacity to an extent which are necessary for education and for the analysis and control of the manufacturing.

It is the **Laborcomplex for the Technology of Printed Wiring Boards** at the Department where the laminates are fabricated and the processes are investigated. The basic equipment of the PWB Laborcomplex is as follows:

- LGS-2 Lasergraph (plotter) for mask film direct exposition (max. 480x570 mm, 0.25 μm accuracy);
- SAQURA automated developer for mask film preparation;
- Fabrication line for dry film photoresist technology (Du Pont, Germany);
- Etching apparatus for printed circuit boards (DEA-180/B, USA);
- CNC drilling machine (POSALUX MULTIFOR 10, Switzerland);
- Electroplating line for the fabrication of plated-through-hole printed circuit boards;
- Electroplating line for gold plating of direct connectors;
- Press for the fabrication of multilayer printed boards (Bradley and Thurton, England).

The equipment of the PWB Laborcomplex provides good possibilities for the fabrication and technological development of multilayer plated-through-hole printed wiring boards with max. 8 layers, min. 0.3 mm hole diameter, 0.2 mm line width and spacing. The fabrication of various other components made by chemical and/or electrochemical processing, such as chemically etched metal parts, is also possible. Considering the higher requirements of the laminates of MCM-Ls, these processing technologies need improvements.

As it was mentioned in Chapter 2., there are five standard types of processes, which are characteristic of the technology of multilayer printed circuit laminates. Accordingly, the development strategies of these processes and the processing sequences at the Department are as follows:

1. *Dry film photoresist application* will remain the characteristic process for imaging the different layers, although liquid photoresists offer better resolution and, as a consequence, it is also within the scope. Dry film photoresist imaging applied in an additive or a semiadditive processing sequence (Figure 1.), however, provides fine resolution, and a proper solution for MCM-Ls. As part of the continuing reconstruction, aqueous dry film photoresist technology, including the suitable stripping equipment and process, were recently introduced at the Department [1], primarily in order to meet the requirements of environmental protection.

2. *Wet chemical etching* of copper layers after a due reconstruction of the equipment will meet the requirements. Switching to the more advanced dry ion etching is beyond the possibilities, since it is less compatible with the other processes.

3. The equipment for the *lamination and curing of precured adhesive dielectric layers* of the multilayer structure is compatible with the applied materials, and although presently fulfils the requirements, it needs improvement in the near future.

4. *Mechanical drilling of holes* through single layers for buried and blind vias and through the entire laminate for plated through holes will certainly remain the main process in this field. Since the possibility of laser drilling is steadily developing at the Department, its application is also in scope.

5. Plating of drilled holes to accomplish electrical connections between the conductive patterns is a key process in any MCM-L processing sequence. During the last two years the main parts of the equipment of the electroplating line for *panel and pattern plating* [2], and the gold plating line of contact fingers have been changed, developed or modified, and recently *direct plating technology* has been introduced [3], thus the plating process can be considered as the most advanced and reliable part of the sequence.

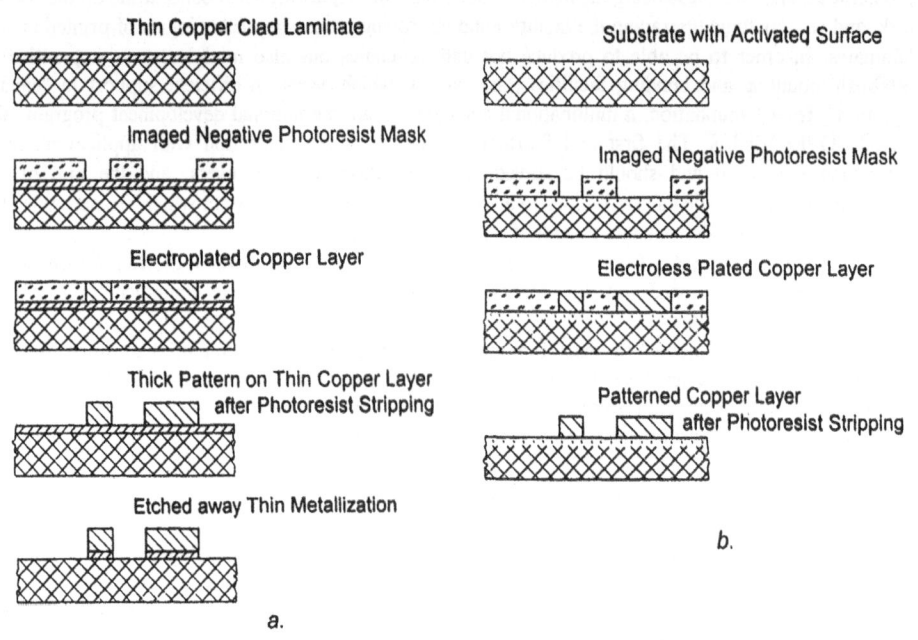

Fig. 1. Semiadditive (a) and additive (b) processing sequences of printed wiring laminates

For the last few years one of the most significant research activity of the Department of Electronics Technology has been laser development and laser material processing. As the main result, the *GRAVILASER* micromachining systems has been developed [4] for structuring surface layers, engraving decorative pictures, marking mechanical or electronic parts, trimming film type components, soldering, drilling etc. The pattern is engraved on the surface applying focused laser beam delivered by computer controlled rotating mirror galvanomotor scanning system [5-9].

The *GRAVILASER* system is capable to accept computer generated figures, especially the imaging and/or trimming patterns of computer simulated and designed hybrid circuits [10-12]. Moreover, this laser system can be a successful tool in a novel technology, in the *combined processing sequence of electroplating, laser surface imaging and wet chemical etching*, which offers a promising solution of fine resolution laminate fabrication (Figure 2.), providing fast prototyping possibilities for multichip modules.

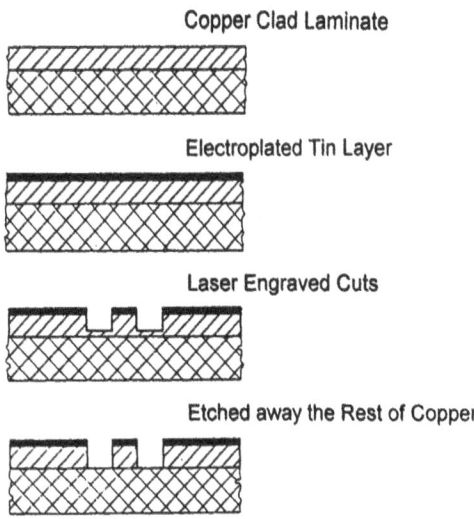

Fig. 2. The combination of electroplating, laser engraving and etching for the fabrication of fine resolution printed wiring laminates

4. INITIAL RESULTS OF LASER IMAGED TIN MASK TECHNOLOGY

Utilising the existing possibilities of the Department initial investigations were carried out. Applying the standard electroplating technology, single-sided copper clad epoxy laminates with nominally 18 μm thick Cu layer were covered by approximately 3 μm thick tin (Sn) layer.

This structures were imaged by laser. The Nd:YAG laser worked in the multi-mode, the Q-switched pulses, as the function of time and the pumping current, are shown in Figure 3. Depending on the different parameters of the laser, the beam intensity in the focal spot was in the 1 to 20 GW/m² range. The samples were moved under the laser spot with a velocity of 10 mm/s. The imaging was carried out with different pumping currents, and multiplied laser processing was also tried. The optimum was found at about 21 A pumping current and single processing.

The laser pulses opened line-shape windows in the tin layer on the surface, while they had only a slight effect on the copper underlayer. Figure 4. shows the Alpha-Step surface profile of the sample across the processed line. The depth is about 5.5 μm, which means that the 3 μm thick tin layer is cut through, the width of the cut is about 33 μm. Large amount of material, mainly tin, can be observed along the edges.

Etching was carried out in a conveyance spraying equipment, using Shipley Neutraetch solution of copper-tetramin-complex. The temperature of the solution was 50 °C, and the etching time was 1 min. The resulted cut profile is presented in Figure 5. The width of the cut is much larger with about 60 μm, and the etching went through the approximately 14 μm thick copper layer. The undercutting of the etching is significant.

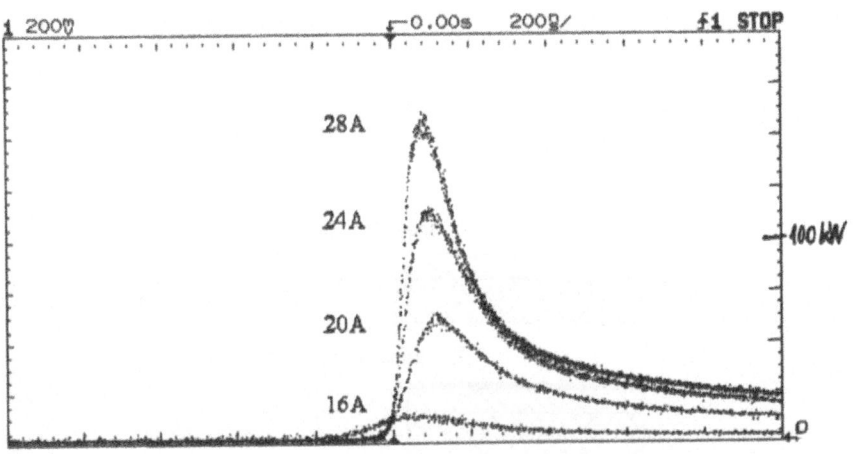

Fig. 3. The shape and power of the applied laser pulses

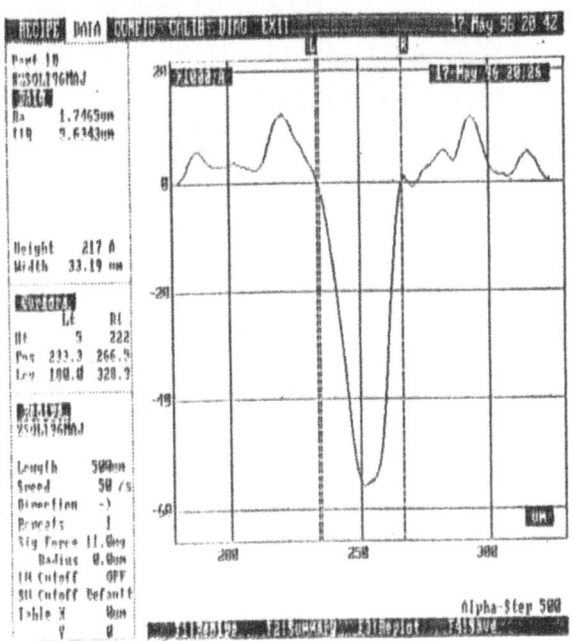

Fig. 4. The profile of the laser processed cut in a tin layer on copper

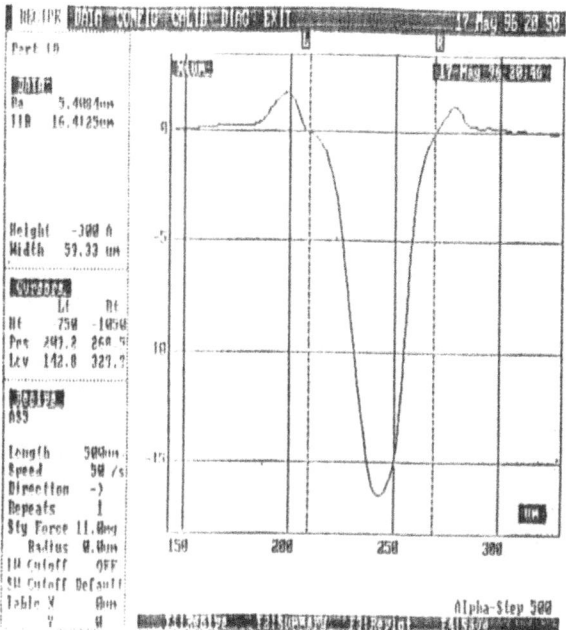

Fig. 5. The shape of the cut after etching

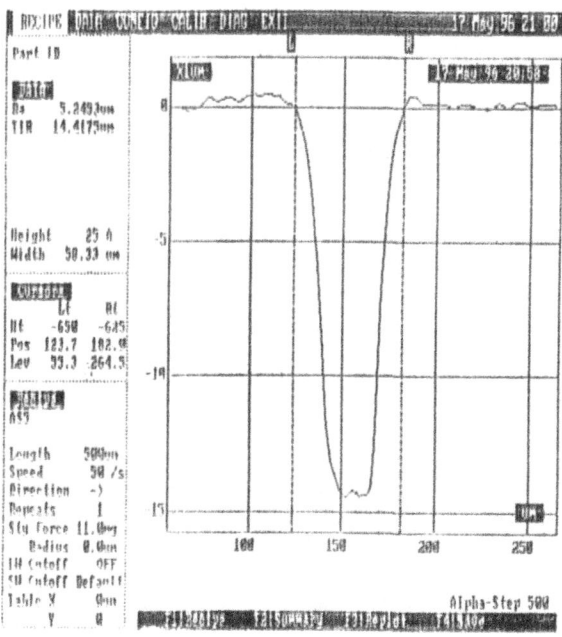

Fig. 6. The shape of the cut after tin mask removal

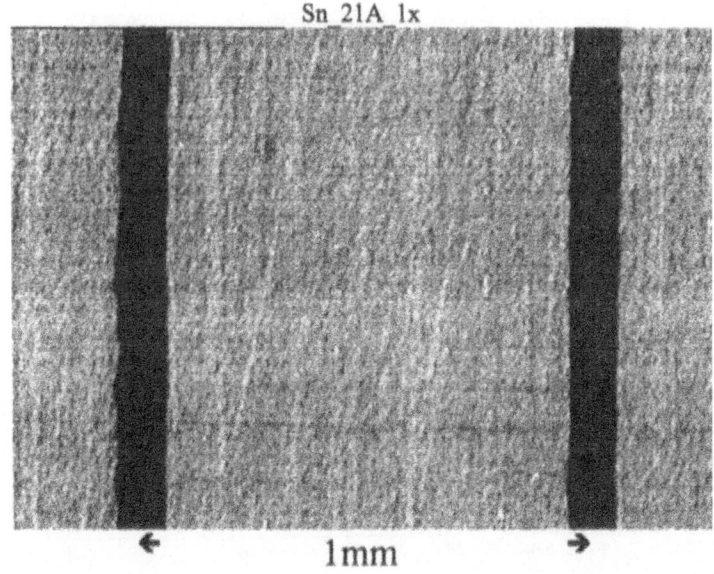

Sn_21A_1x

← 1mm →

Fig. 7. The top view of laser processed and etched cut after tin mask removal

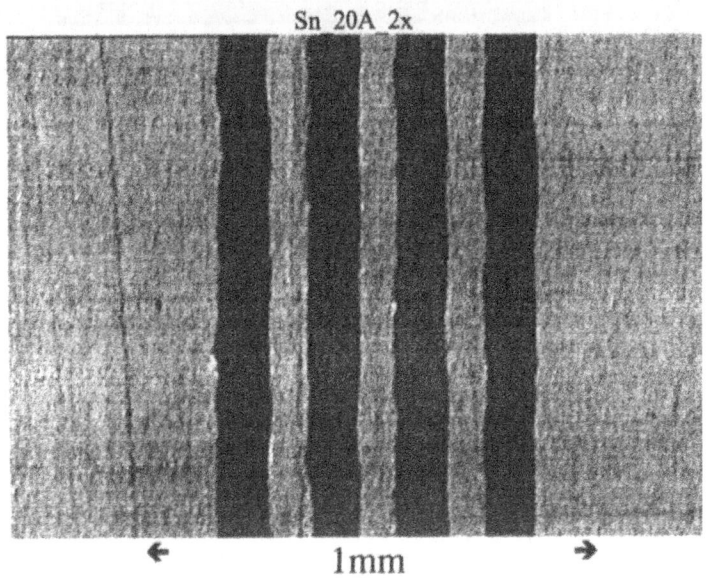

Sn_20A_2x

← 1mm →

Fig. 8. An illustration of the achievable line width and spacing

Figure 6. shows the cross-section of the cut after removing the tin mask. The profile in the copper layer practically did not changed, the width of the cut remained 60 µm, the accumulated materials of tin, however, disappeared from the edges. The microscopic views of the cuts are shown in Figure 7. and 8. The latter Figure is to demonstrate how fine pattern with 40 µm line width and 60 µm wide spacing was easily achieved. It should be mentioned, that with the application of the multi-mode laser beam and a relatively obsolete etching equipment, the possibilities of the technology were not exploited, thus the initial stage of the experimental result should be emphasised.

REFERENCES

1. Tóth,E.; Gál,L.; Pinkola,J.; Hajdu,I.: *Environmental Problems of Masking Technologies used in Printed Wiring Board Production.* Proceedings of the 17th International Spring Seminar on Electronics Technology, May 31-June 3, 1994, Weissig (Germany), pp.333-338.
2. Gál,L.; Hajdu,I.; Tóth,E.; Pinkola,J.: *Examination the Metal Layers of Through-Plated Holes.* Proceedings of the 16th International Spring Seminar on Electronics Technology, April 25-28, 1993, Szklarska Poreba (Poland), pp.119-124.
3. Gál,L.; Hajdu,I.; Pinkola,J.; Tóth,E.: *New Metal Plating Technology of Nonconductive Surfaces.* Proceedings of the 18th International Spring Seminar on Electronics Technology, June 26-30, 1995, Temesvar (Czech Republic), pp.254-258.
4. Illyefalvi-Vitéz,Zs.; Enyedi,L.; Hoffmann,Á.; Kovács,P.; Mann,Gy.; Pinkola,J.; Ruszinkó,M.; Szigethy,D.: *A GRAVILASER mikrogravírozó rendszer (The GRAVILASER microengraving system).* ecMARKinfo, 1995/4, pp.48-49, (in Hungarian).
5. Enyedi,L.; Illyefalvi-Vitéz,Zs.; Ruszinkó,M.: *Analysis of a Laser Beam Deflecting Mechatronics System.* Mechatronics The Basis for New Industrial Development, Computational Mechanics Publications, Southampton (ISBN: 1853123676), Boston (ISBN: 1562522914), (Proceedings of the Joint Hungarian-British International Mechatronics Conference, September 21-23, 1994, Budapest), pp.247-252.
6. Illyefalvi-Vitéz,Zs.: *Mechatronics Technology Improves the Reliability of Motion Systems.* Proceedings of the 18th International Spring Seminar on Electronics Technology, June 26-30, 1995, Temesvar (Czech Republic), pp.47-53.
7. Ruszinkó,M.; Enyedi,L.; Illyefalvi-Vitéz,Zs.: *Sensors in a Complex Laser Processing System.* Proceedings of the 17th International Spring Seminar on Electronics Technology, Weissig, May 31-June 3, 1994, pp.157-160.
8. Enyedi,L.; Illyefalvi-Vitéz,Zs.; Ruszinkó,M.: *Sensor Applications in a Laser Micromachining System.* Proceedings of the Workshop on Thick and Thin Film Sensors and their Application in Ecology, Szklarska Poreba (Poland), June 12-15, 1994, pp.162-165.
9. Enyedi,L.; Fülöp,S.; Illyefalvi-Vitéz,Zs.; Ruszinkó,M.: *Simulations and Measurements in a Laser Beam Delivery Mechatronics System.* Mechatronics'96, Sep 18-20, 1996, Guimaraes (Portugal), (accepted paper).
10. Kovács,P.: *Image Processing Methods for Improving the Quality of Laser Engraved Pictures.* Proceedings of the 18th Spring Seminar on Electronics Technology, June 27-30, 1995, Temesvar (Czech Republic), pp.72-76.
11. Gombás,G.: *Hybrid Topology Design using Artificial Intelligence Technology.* Periodica Polytechnica Ser. El. Eng. Vol.37, No.1, pp.73-78, (1993).
12. Papp,G.: *Simulation of Laser Trimming of Film Resistors.* Periodica Polytechnica Ser. El. Eng. Vol.37, No.1, pp.43-52, (1993).

Opto-Electronic Multi-Chip Modules (OE-MCMs) : Current R&D and Applications to Microelectronic Interconnections

Sayan D. Mukherjee, Professor of Photonics
Department of Physical Electronics, Norwegian University of Science and Technology - NTNU
N-7034 Trondheim, Norway.
Phone: +47 7359 4260   Fax: +47 7359 1441
e-mail: Sayan.Mukherjee@fysel.unit.no

Abstract

Optoelectronic interconnects with its many advantages over electrical connections suffer from its high cost of implementation due to problems associated with optical packaging, especially the coupling of optical components to the outside world. Recent developments in the area of optical packaging and OE-MCM are addressing these issues. New methods of incorporating large numbers of optoelectronic components in board and MCM-level interconnects promise to break the technology- and cost-barriers currently undermining it usefulness. This paper summarizes the background and describes several existing and past projects aimed at bringing optical interconnects to finer grain applications.

1. Introduction

Advancements in computing and signal processing systems require an interconnect medium through which high-speed and highly parallel signals can be routed without sacrificing processor speed/ power and system performance. Recent projection shows that the state-of-the-art processors commercially available for personal computer at year 2000 will be able to deliver clock speed at 500 MHz. The electrical interconnect faces its fundamental physical limits as the performance of these processors and their affiliated memories grow: Figure 1 [1]. Optical or optoelectronic interconnections (OI) offer solutions not possible with electrical interconnects. Some of the main advantages of optical paths utilized in OI are illustrated in Figure 2. Most of these advantages are not utilized in long-haul optical communication.

Some of the earliest OI implementations date back to the late 1980s. An internal project at IBM - Yorktown Heights called NEXUS (1984-1988) demonstrated 10:1 multiplexed, one and four optical fibre links for inter-processor interconnection [2]. Concurrently, Honeywell in the US demonstrated gigabit links using off-the-shelf components in 1988 [3,4]. Further systems analysis suggested that the latency problems associated with time-division-multiplexing (TDM) is best eliminated with the use of bit-parallel (also called direct replacement or space-division-multiplexing (SDM)) approach, in which 8-, 16- or 32-bit parallel interconnection is achieved by using equivalent number of parallel optical paths. To accomplish this, specialized optical transmitter/driver array circuits and optical receiver arrays are being generated continuously [5-8].

G. Harman and P. Mach (eds.), Microelectronic Interconnections and Assembly, 119-140.
© 1998 *Kluwer Academic Publishers.*

OI is now widely accepted as a way to alleviate such bottleneck in platform-to-platform, machine-to-machine, and, inter- and intra-module interconnects which are some of the most pivotal scenarios for the realization of information super highway [9, 10]. Parallel optical links, using silica-based fibre ribbons and plastic fibres, with electronic interfaces, are currently commercially available in small volumes. Several companies are engaged in proprietary developments for larger volume markets such as fibre-in-the-loop and fibre-to-the-curb type applications of telecommunication distribution and access networks. As a result of the relative maturity exhibited by these technologies, R&D efforts in OI have shifted to solving probably the most formidable show-stoppers for OI implementation into the large volume market. It is the high cost of incorporation for optical emitters (semiconductor laser diodes, for example) and optical receivers into finer grain OI sub-systems.

In particular, co-operative R&D among various consortia are on the way to incorporate either higher bit-rate per channel of fibre-ribbon links connecting sub-systems [11], or, larger number of optical components within a sub-system. It is in the latter that optical back-planes, optical daughter-boards (plug-in boards) and optoelectronic multi-chip-modules (MCMs) belong.

2. Acceptance of technological diversity in OI

What distinguishes the optical wavelengths and media currently used for optoelectronic interconnects from those used in long-haul optical communication, is the acceptance of a variety of wavelengths (i.e., optical sources emitting at 650 nm, 800-850 nm, 980-1050 nm, as well as 1300 and 1550 nm) and media not suitable for long-haul. The optical transmitting media are glass slabs, plastic fibres, multi-moded silica- and polymer-based fibres, multi- and single-moded polymeric waveguides on both planar (rigid, flat) and flexible substrates, and free-space (air). To service massive parallelism, the most important parameters associated with the passage of optical signal through these media are the ability to

(1) create large number of parallel channels, with several cross-overs, with the minimum of cross-talk and electro-magnetic emission/absorption,

(2) route these paths to-and-from MCMs, mother- and daughter-boards, and,

(3) connect them quickly and inexpensively to the optical emitters and receivers necessary for electronic-to-optical interfacing and vice-versa.

Items (1) describes some of the advantages of optics summarized in Figure 2. Items (2) and (3) require technological solutions. In addition to these issues, two important aspects distinguish optical interconnects from long-haul communication: monetary and power cost, and system definition. In OI, cost per link per unit bandwidth must compete with, and exceed, that of electrical equivalents, while providing significantly higher benefit in the long run. In optical communication, cost efficiency is not as important. This is because, in optical communication the link is the system, while in OI the link must interface easily into the digital system that it is trying to improve.

Substantial R&D in Europe, the US and Japan have resulted in advancements in the technologies of optical back-planes and direct optical connections to electronic MCMs with optical interfaces. Some revolutionary, but mostly evolutionary, approaches are now pushing these technologies towards the realization of MCMs that will accommodate optical emitters and receivers relatively easily, with either no or minimal changes in the manufacturing processes currently used in electronic packaging and electronic MCMs.

3. Parallel fibre-ribbon links to OE-MCMs

3.1 Parallel fibre-ribbon links

Work in the area of fibre-ribbon interconnects were initiated in the US with a strong commercial effort currently in both US and Japan. An early US-ARPA (Advanced Research Project Agency) funded project OETC - Opto-Electronic Technology Consortium - amongst Lockheed-Marietta (originally the Aerospace and Military Avionics part of General Electric), IBM, AT&T and Honeywell, aimed to develop 32-parallel channel fibre-ribbon link with a total throughput of 16 Gb/s [8]. The VCSEL (vertical cavity surface emitting laser) emitting at ~830 nm was chosen as the optical source [12, 13].

Another independent program supported by US-ARPA is POLO - Parallel Optical Link Organization. This is a consortium of HP, Du Pont, AMP, Univ. of Southern California and SDL (Spectra Diode Laboratories, San Jose) [14]. The objective is to develop 10-20 Gb/s total throughput with 32- or 64-bit parallel channels at $10 per channel cost. The duration of the project is 1994-1997 and the schematic of their demonstrator is in Figure 3. Here again, VCSELs are used. Also incorporated are flexible polymeric optical waveguide arrays using Du Pont Polyguide (trade mark) which is a series of products that are mass produced in sheets with optical waveguides embedded in thin transparent films. These films can be laminated to other circuit-board materials or taken out of the board to other destinations not too far away [15].

In Japan, almost all high-tech companies have built their own parallel link demonstrators, see Table 1.

3.2 Optical backplanes and use of optical polymers

The parallel fibre-ribbon links with industry-standard fibre-connectors, such as the MT connectors, provide box-to-box or rack-to-rack optical interconnections. As mentioned earlier, this item is entering the mass-manufacturing phase. So, the next finer-grain aspect of optical interconnects, the optical backplane and optical interconnections to and from boards attached to the back plane, is a strong current R&D topic. The objectives here are to create system and sub-system demonstrators to convince user industries.

One of the simplest optical solutions for rack-mounted system is called the Apollo demonstrator and is from Ericsson in Sweden [16]. In this demonstrator, silica optical fibres are laminated between flexible polymer sheets, with MT connectors at ends. Different planar or curved sheets of these "flexfoils" are either attached to the slide-in boards or serve to connect between daughter-boards and mother-boards. A European Commission funded program called SPIBOC (Standardized Packaging and Interconnect for Intra- and Inter-Board Optical Communications), with twelve participants, goes further in replacing individual fibres with 8x to 12x fibre-ribbons attached to MT-type optical connectors: Figure 4. Optical data rates are 2.4 Gb/s, and submultiples of it.

A US-ARPA and Airforce funded project, OIT (Optical Interconnect Technology), involving a collaborative effort between Honeywell and GE, attempts to solve the back-plane to board connections using expanded beam technique using half GRIN-rod (here GRIN stands for graded-index) lenses and miniature prisms: Figure 5. Here a set of parallel optical waveguides can be mapped (focused) onto another set of parallel waveguides using only one pair of lenses and one prism, alleviating the need for individual fibre-to-fibre connections used in, e.g., SPIBOC. The complete backplane, board and OE-MCM solution is shown schematically in Figure 6. Here, transparent optical polymers are used for creating the optical waveguides on circuit-board materials.

Optical waveguides have been considered for a while as an alternative to fibres for high performance packaging [18, 19], and passive optical polymers fit the bill perfectly. They can be easily applied, for example by spinning, spraying and by film-transfer. The cure temperatures are not too high, less than 250 degrees C. There are polymers that planarize rough surfaces, others that carry light and lastly, some that achieve (almost) hermetic seals [20, 21]. Some of them also can be defined using photolithographic processes and in some the optical properties, such as refractive index, can be changed by lithographic UV exposure. Therefore, optical polymers with many other complementary properties are finding increased use in solving backplane and OE-MCM issues. The POLO project and OIT are but two of many examples. In addition, optical fibres using plastics and polymers are finding increasing use in short-haul links [22, 23].

Polymeric optical waveguides can be fabricated and patterned both on rigid and flexible substrates (e.g., Polyguides of DuPont). The softness and plasticity of polymers allow them to be shaped using molding techniques, both by compressing and embossing [24], and by injection [25]. Polymers can be dry-etched in reactive plasma environments for structuring over large areas or laser ablated to create localized structures with different vertical angles. Polymers, and in particular polyimides of different sorts, are routinely used in electronic IC and packaging industries. Its uses in traditional fields, therefore, are well documented. Its introduction into optical waveguiding is also no longer a new science [26-29]. In addition, there are active optical polymers in which the influence of electrical field can cause changes in refractive index (electro-optic, or EO, effect) and other non-linear optical properties (see last few chapters in ref. 26). Very simple 1x2 guided wave optical switches have been made by simply changing the refractive indices of polymers with change in temperature brought about by localized heating (thermo-optic, or TO, effect), with performances far in excess of equivalent devices using planar, silica waveguides [30, 31]. So, increasingly, optical polymers are applied to solving the next generation of finer-grain interconnection problems for distributed processing, or, localized information-transfer and information-sharing by optical means, that is, in the incorporation of optical components and their electronic interfaces within multi-chip modules, or MCMs.

However, before we embark upon our discussion on OE-MCMs, credit should be given to European (and some US) efforts in achieving backplane-to-board optical interconnections using glass-slabs, binary optics (gratings) and free-space optical interconnects. Notable are an earlier EU-project called OLIVES which gave rise to a newer project called HOLICS which is almost coming to its end in 1996 [32]. In HOLICS, a glass slab, with internally reflecting light beams carrying some of the fan-out signals like clock, is used as part of the back-plane, with metal lines drawn on it that carry conventional electronic information. The board-to-board communication occurs through free-space, either using lensed LEDs or un-lensed VCSELs (future plan), and photodetectors/optical receivers. Free-space OI and packaging are gaining acceptance in research communities [33], especially for large volume data transfer and in neural-net type applications. At the other end of the spectrum, silicon wafers are used, in a variety of ways, to reduce the necessity of active alignment of laser diodes or waveguides to optical fibres and ribbons [34]. But these are not central to the theme of this paper and so will not be discussed here.

3.3 Optoelectronic multi-chip-modules (OE-MCMs)

Three examples will be given, one each from the US, Europe and Japan.

The US example is OIT, the collaborative effort between Honeywell and GE, described briefly earlier. In this solution to OE-MCM, the optical devices are added to the electronic MCM at the end of the manufacturing of the electronic MCM. Optical layers (that is, those required for optical waveguiding) are created at the end, with adaptive lithography and laser ablation with automatic video-monitoring taking care of the actual positions of the optical emitters relative to the waveguides

[17, 27, 35]. Adaptive lithography is also used for creating the final layer of metal lines to accommodate relatively slackened positional accuracy of electronic dies. Another significant innovation is the use of truncated flexible polymeric waveguides that optically connect the MCM to board-level waveguides by passive positioning of the flexible optical ribbon within a groove: Figure 7. In essence, therefore, it enables the use of standard electronic MCM technologies together with optical packaging technique that is adapted to the former. The innovations and technological demonstrations under OIT have been combined with AMP and DuPont's foundry-made polymer splitters and low-loss polymers from Allied Signal to create a new program called POINT (Polymer Optical Interconnection Technology) in which five US companies participate: AMP, DuPont, Allied Signal, GE and Honeywell. In this current program three kinds of OE-MCMs will be developed and evaluated for performance: (1) Connectorized modules for serial/parallel fibre and polymer waveguide interconnects, (2) Polymer-interconnected modules for intra-cabinet applications (see Figure 6 again), and, (3) Parallel fibre-interconnected modules for cabinet-to-cabinet applications.

The European example is an international project called DONDOMCM (Development Of New Dielectric and Optical materials, testing and comparison with existing polymers and applications in Multi-Chip Modules) with eight participants. A cross-sectional schematic of the intended demonstrator is shown in Figure 8. Here again, standard, or improved electronic MCM technology is merged with the process of attaching edge-emitting (in-plane) semiconductor laser diodes that emits into the final (top-most) polymeric waveguide layer.

The final example is from NTT, Japan [36]. As in the two previous examples, polymeric optical waveguides are patterned after the conclusion of the fabrication of the electronic MCM substrates: Figure 9. The electronic MCM has standard multilayer ceramic-metal substrate followed with multilayer copper-polyimide. The optical waveguides are fabricated atop the Cu-PI multilayer structure. This OE-MCM does not yet have a direct-optical-connection-to-board solution. Instead, it relies on independent, and precisely made, fibre ribbon connectors that couple light to and from optical waveguides easily: Figure 10. This approach, therefore, is a combination of traditional methods of optical connection to the outside and new methods of creating optical connections to optical transmitters and receivers. In a way it is an earlier, and a more established, form of connectorizing than what item number (1) of the US program POINT plans to achieve.

4. Conclusion

Starting with a general description of the necessities of optical (or optoelectronic) interconnections, its possible practical implementations into OE-MCMs is briefly described. The key challenges in the development of OE-MCMs are the ability to couple light in and out of these MCMs without having to actively align fibres (or other waveguides) to the optical components within the MCM or the waveguides that carry information to and from these optical elements, and quick insertion and de-coupling of the MCMs to boards. The methods of passive coupling, with ability to withstand repetitive operations of coupling and de-coupling, are not yet fully developed, nor is any one method universally accepted. Only continued development, and sub-systems demonstrations leading to systems use will determine what is best or acceptable. It is generally accepted, however, that optical polymers will play a lasting role in this technology arena, due to their compatibility with electronic MCM manufacturing methods and global knowledge-base for their application and processing methods. A continued dialogue between the optical application engineers and electronic MCM producers is therefore needed to accomplish this joint development.

Tables:

1. Demonstrated optical paralle links, 1995 (Courtesy O. Wada, Fujitsu Recearch Labs, Japan).

Figure captions:

1. Fundamental limits of electrical method of information transfer drawn in the x-t plane (x and t defined in 1(b)). From ref. [1].

2. Advantages of optical paths over electrical lines in transferring information. The long-haul capability of optical fibres used in communication is a combined result of low-optical losses in fibres, a technological achievement, and lack of electro-magnetic emission/absorption from/to fibres, a fundamental property of information transfer using light. In optoelectronic interconnections, however, almost all the properties listed here are harnessed. (b) Graphical explanation of two-times power compression in dB scale, the last item in (a) above. Semiconductor laser diodes exhibit linear relationship between output light power and signal electrical current, while optical detection circuits detect the square of the detected current which is linearly proportional to the optical power.

3. A schematic of the VCSEL-based POLO module [14]. POLO is a US-ARPA funded consortium-program led by HP.

4. Schematic of EU funded SPIBOC demonstrator [16].

5. A graded index (GRIN) rod lens pair forms the basis of a parallel board-to-backplane connector offering compatibility with electrical connector tolerances. (a) Planar experiment that yields < -34 dB crosstalk. (b) Final right-angle implementation [17].

6. Honeywell-GE solution to backplane-to-board optical interconnects (Courtesy J. Bristow, Honeywell, 1994).

7. Board-to-MCM optical connector in Optical Interconnection Technology (OIT) project of Honeywell and GE. Project funded by US-ARPA [17, 27]. Advantages of this approach are: it uses same equipment to install/replace optically interconnected MCMs as electrical MCMs, and, electrical and optical connections are made simultaneously to minimize cost.

8. Schematic cross-section of experimental demonstrator of LD attachment to MCM with the top-most layer being a polymeric optical waveguide. Demonstrator under development under EU-funded DONDOMCM project which ends in mid-1998.

9. Schematic of NTT's OE-MCM showing the inclusion of both ceramic and copper-polyimide electronic MCM technologies [36].

10. Coupling between optical waveguides over the electronic MCM substrate (see Figure 9) and optical fibre ribbons is accomplished using precision-machined and assembled glass capillaries and guide pins, standard methods in optical couplers. Part of OE-MCM solution of NTT [36].

Table 1   **Demonstrated Optical Parallel Links**

Group	Fujitsu		Hitachi	NEC	NTT	Siemens	AT&T	OETC	POLO	Motorola
Channels	4	20	10	12	12	12	9	32	10/10	10/10
Data rate (Mbps/ch)	1200	156	200	200	700	1000	500	625	625	150
Throughput (Gbps)	4.8	3.2	2.0	2.4	8.4	12	4.5	20	12	3
Length (m)	400	400	100	100	250	–	–	100	300	30
Power (mW/ch)	700	320	280	580	1140	100	1280	310	< 300	170
Size (cm^3)	13	20	0.76	37	149	2	38	7.3	–	18.8
Light source	LD	LD	LD	LED	LD	LD	LED	VCSEL	VCSEL	VCSEL
Wavelength (μm)	1.5	1.3	1.3	1.3	1.3	0.85/1.3	1.3	0.85	0.98	0.85
Fiber mode	SM	SM	SM	MM	MM	MM	MM	MM	MM	MM

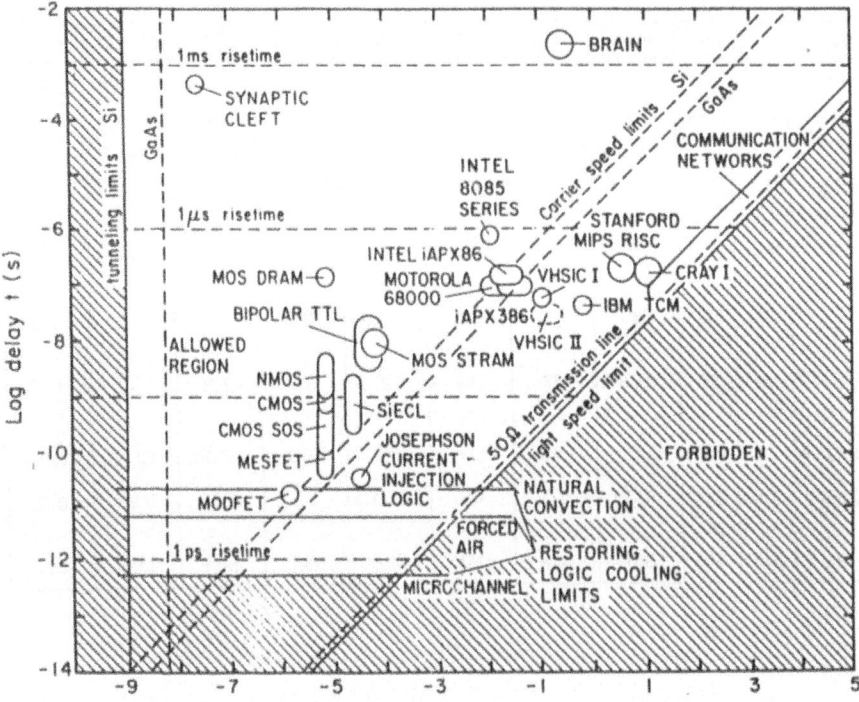

Figure 1(a) Log subelement separation, x (m)

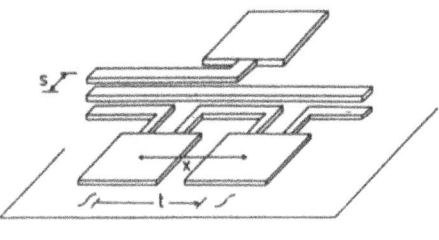

Figure 1(b)

Figure 2(a)

Advantages of Optics in Transfering Information.

Single Channel

No return (ground) path.

No ground loop.

Crossing Channels

Free space: zero crosstalk

Waveguide: <-35dB, determined by crossover imperfections.

Free-space broadcast

Extremely high bandwidth medium.

Multiple crossovers.

No EM Radiation Loss

Very closely packed parallel paths; zero crosstalk.

Loss independent of frequency

No EM Radiation Pickup

$h\nu$ EM

No interference.

Immune to EMP threat.

Independent of Source to Destination Potential

+100,000 V

0 V

No isolation required.

Provides isolation.

No Impedance Matching

1,000,000 Ω

50 Ω

Frequency independent I/O.

Future expandibility in bandwidth.

Power Compression (by Laser / Photodetector)

I^2R (10 dBm)

(5 dBm)

IV

dB(electrical) = 2 x dB(optical)

u001\tronf006.cdr

128

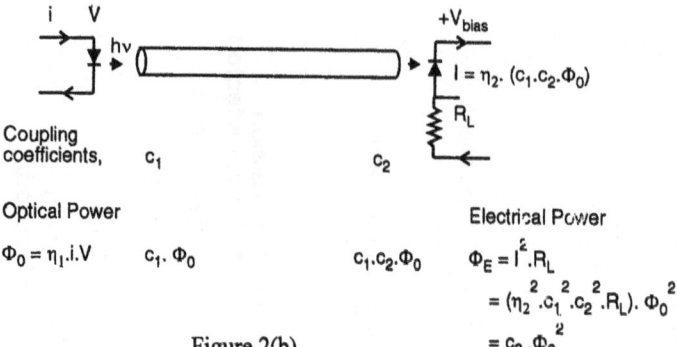

Coupling
coefficients, c_1 c_2

Optical Power   Electrical Power

$\Phi_0 = \eta_1.i.V$ $c_1.\Phi_0$ $c_1.c_2.\Phi_0$ $\Phi_E = I^2.R_L$

$= (\eta_2^2.c_1^2.c_2^2.R_L).\Phi_0^2$

Figure 2(b) $= c_3.\Phi_0^2$

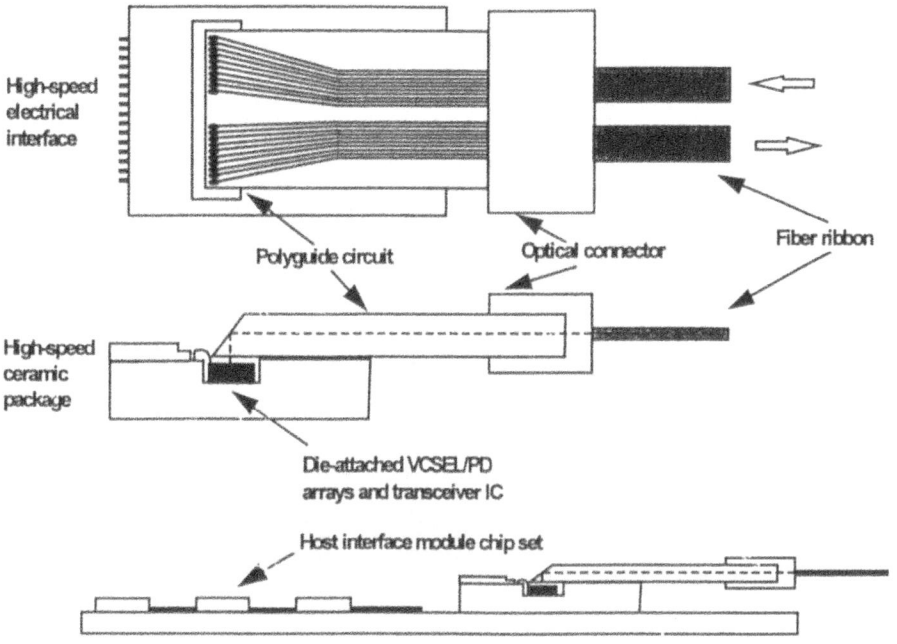

High-speed electrical interface

Polyguide circuit

Optical connector

Fiber ribbon

High-speed ceramic package

Die-attached VCSEL/PD arrays and transceiver IC

Host interface module chip set

Printed circuit board

Figure 3

130

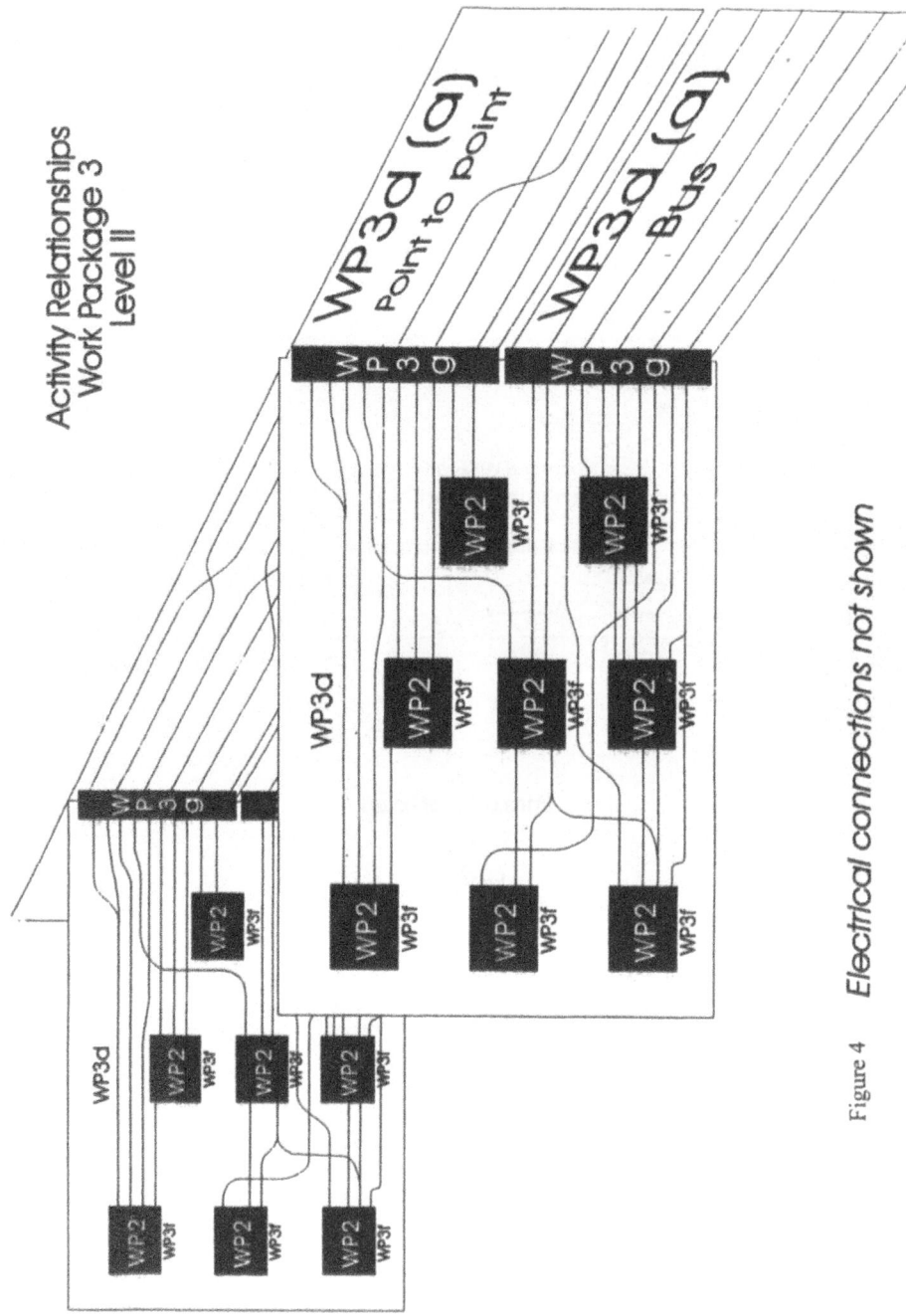

Figure 4   *Electrical connections not shown*

131

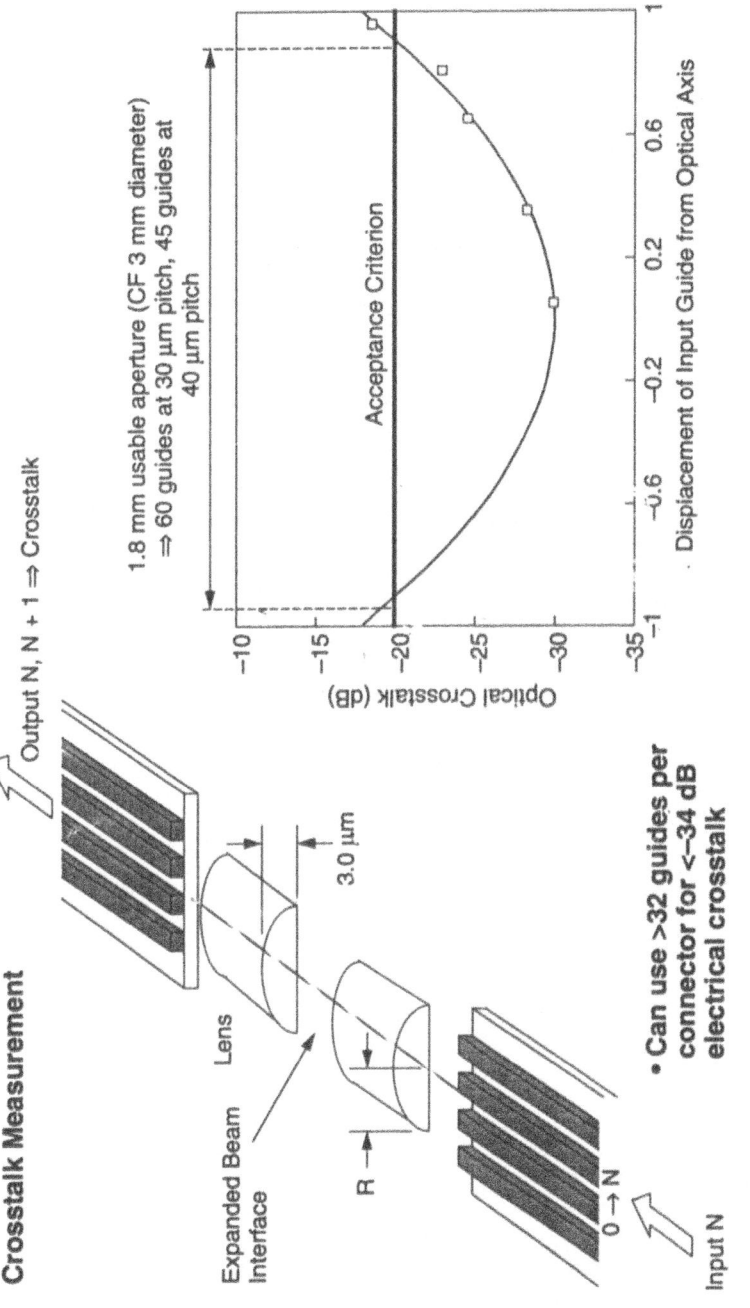

Figure 5(a)

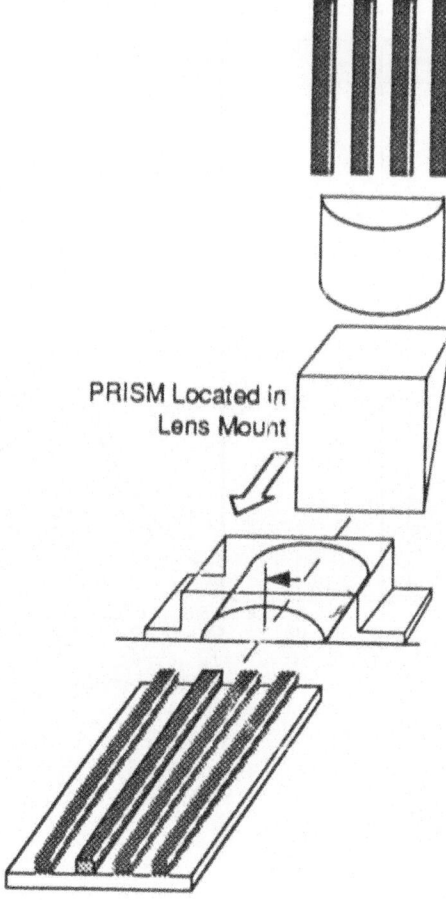

PRISM Located in
Lens Mount

Figure 5(b)

133

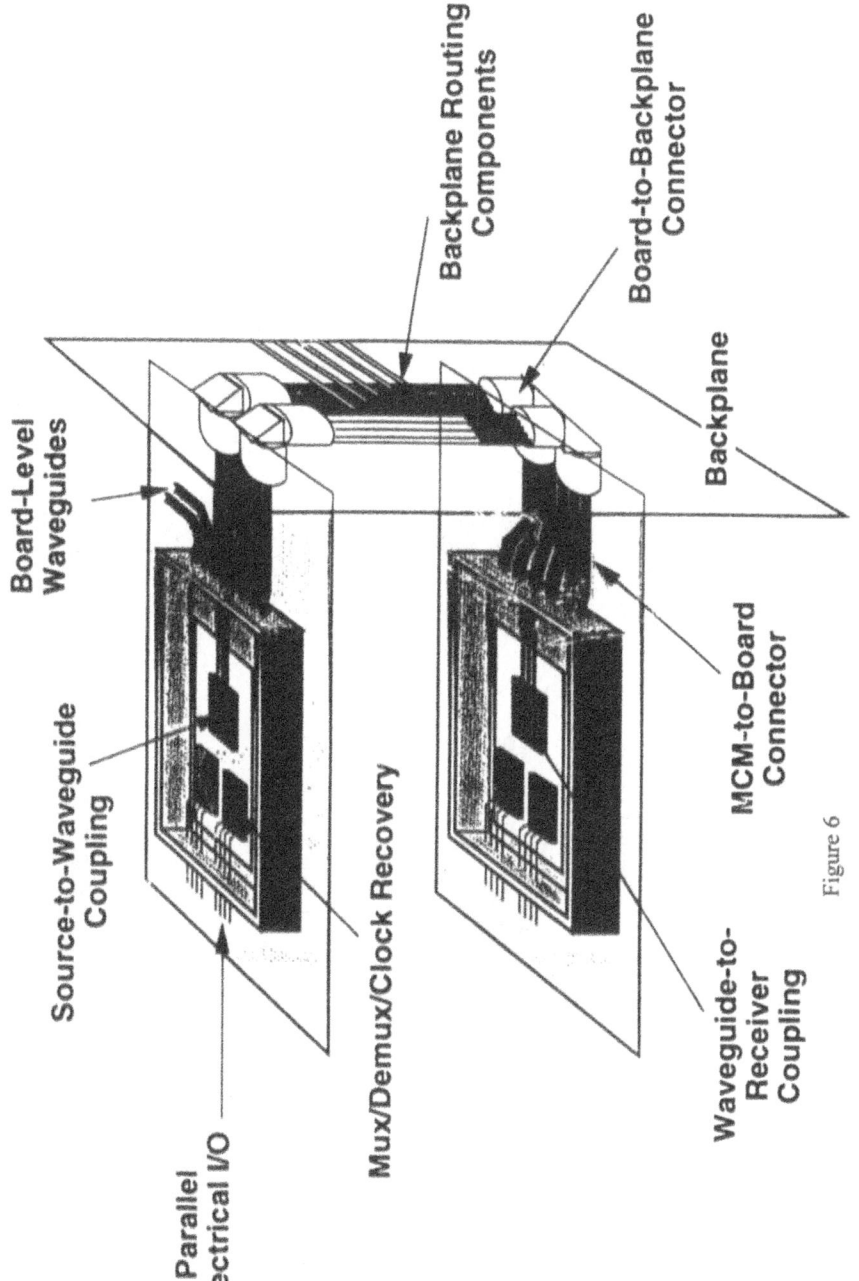

Figure 6

134

Figure 7

Board-to-MCM Connector
Honeywell Patent

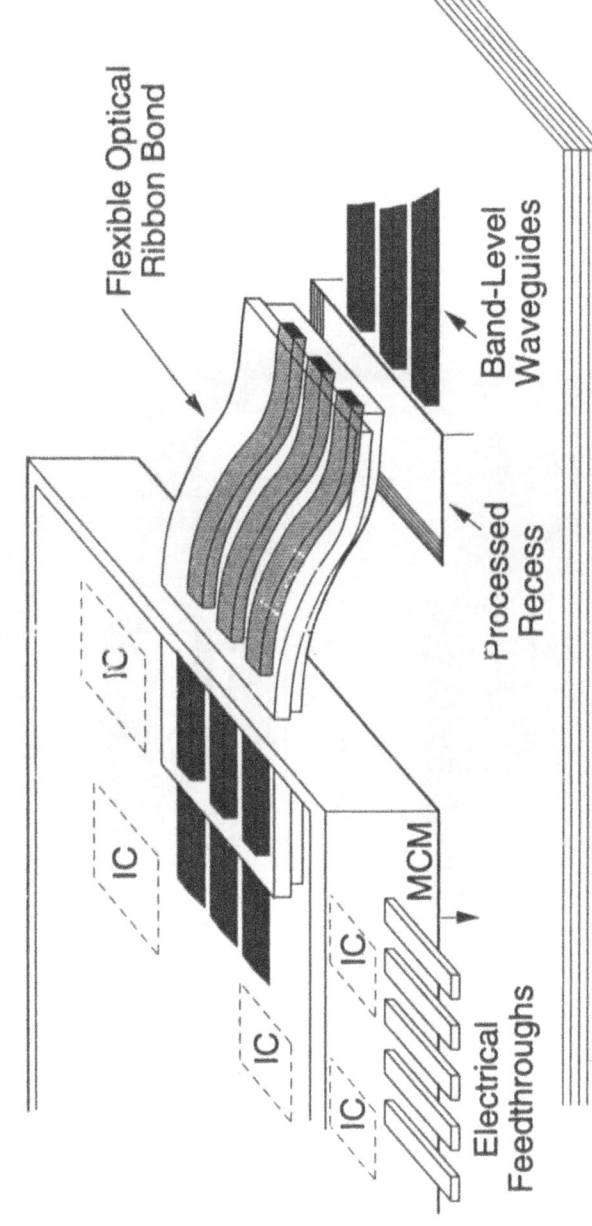

Flexible Optical Ribbon Bond

Band-Level Waveguides

Processed Recess

MCM

Electrical Feedthroughs

IC

- **Uses same equipment to install/replace optically inter-connected MCMs as electrical**

- **Electrical and optical connections made simultaneously to minimize cost**

Figure 8

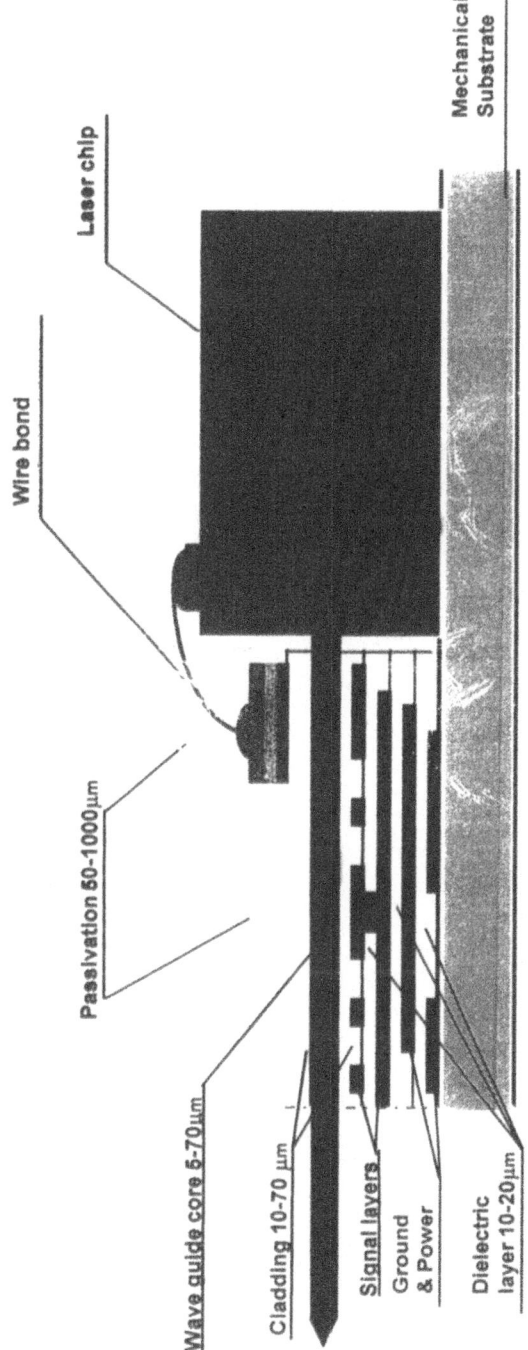

136

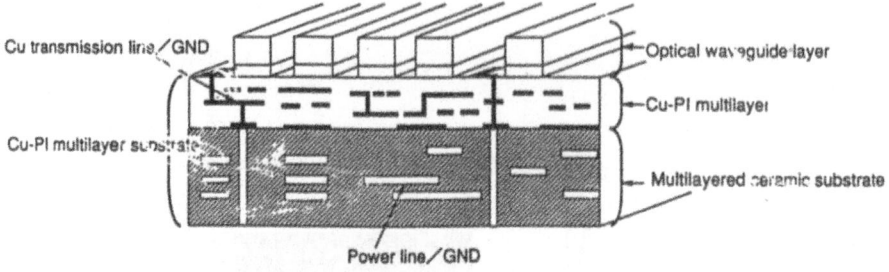

Cu transmission line/GND

Optical waveguide layer

Cu-PI multilayer

Cu-PI multilayer substrate

Multilayered ceramic substrate

Power line/GND

(a) Proposed OE-substrate

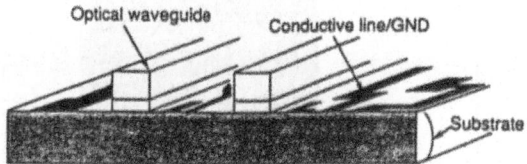

Optical waveguide

Conductive line/GND

Substrate

(b) Conventional substrate

Figure 9

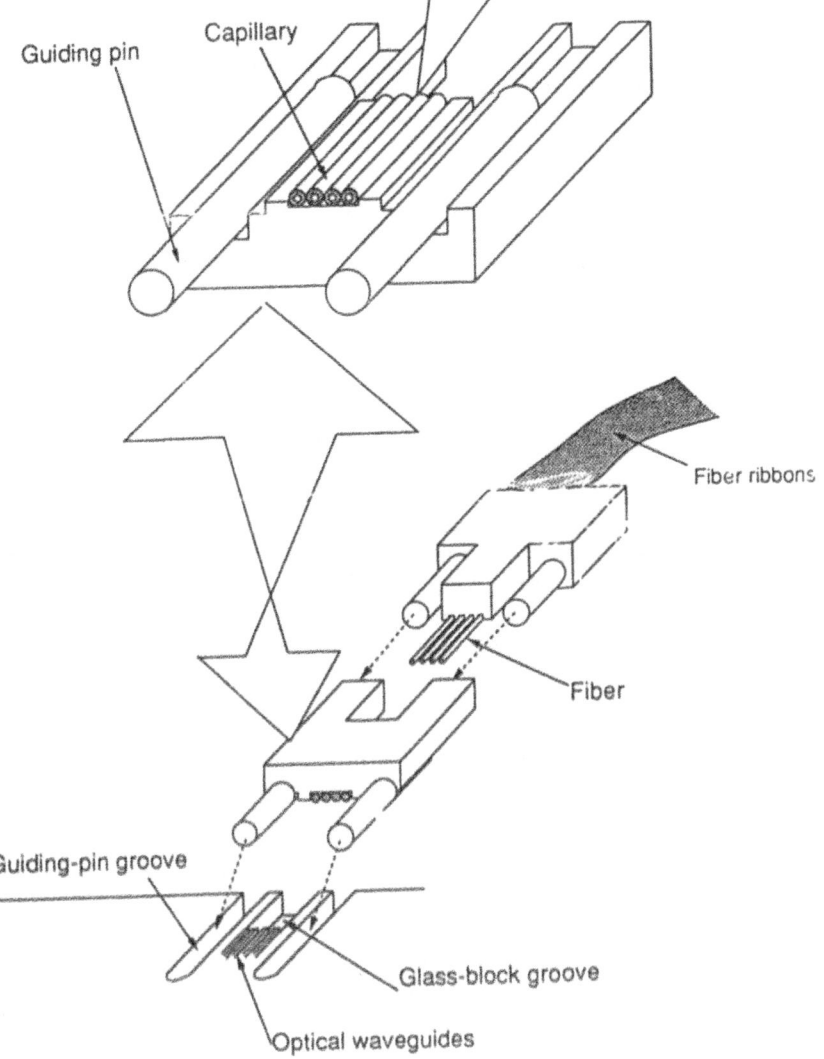

Guiding pin

Capillary

Fiber ribbons

Fiber

Guiding-pin groove

Glass-block groove

Optical waveguides

Figure 10

138

References:

1. W.E. Pence and J.P. Krusius, "The fundamental limits of electronic packaging and systems," IEEE Tr. Components, Hybrids and Manufacturing Technology, CHMT-10 (1987) 176-183. See also, Chapter 32, pp. 907-936, in "Principles of Electronic Packaging", D.P. Seraphim, R. Lasky and C-Y. Li (Editors), McGraw Hill, New York (1989).

2. (a) J.D. Crow, "Optoelectronic integrated circuits for high speed computer networks," Optical Fiber Communications Conference (OFC'89), (1989) paper WJ3.
(b) J.D. Crow, "Optical interconnects speed interprocessor nets," IEEE Spectrum Magazine (March, 1991) pp. 20-25.
(c) C.J.Anderson, et.al., "A GaAs MESFET 16x16 crosspoint switch at 1700Mbits/sec," 1988 GaAs IC Symposium Technical Digest (1988) pp.91-94.

3. T.A. Lane, M.P.Bendett, C.T.Sullivan and J.P.G.Bristow, "Digital system applications of optical interconnections," Proc. SPIE Vol. 991 (Fiber Optic Datacom and Computer Networks) (1988), pp.42-49.

4. T. Lane, B.Floren, J.Bristow, A.Wilson and J.Kessinger, "Gigabit optical interconnects for parallel processors," US Government Microcircuit Applications Conference (GOMAC), Orlando, Florida, November, 1991, Digest of Papers (1991), pp.567-568.

5. S.D. Mukherjee, "OEICs for optical interconnections," Chapter 9 in "Optoelectronic Integration: Physics, Technology and Application," edited by Osamu Wada, (Kluwer Academic Publishers, Norwell, MA, U.S.A.), (1994), pp.321-374.

6. C-S. Li, Y.Kwark, P.Lim, C.M.Olsen and H.S.Stone, "Chip set for fully differential board-to-backplane optical interconnects," Optical Fiber Communications Conference (OFC/IOOC) '93 Technical Digest (1993) pp.169-171.

7. D.L. Rogers, S. Walker, S. Gowda and H. Ainspan, "Design, fabrication and automated testing of 32 channel integrated MSM/MESFET OEIC receiver arrays," SPIE Vol 2400 (1995) 296-300.

8. Y-M. Wong, et. al., "Technology development of a high-density 32-channel 16-Gb/s optical data link for optical interconnection applications for the Opto-Electronic Technology Consortium (OETC)," IEEE J. Lightwave Technol. 13 (1995) 995-.

9. C. Tocci and H.J. Caulfield (Editors), "Optical Interconnection: Foundations and Applications," Artech House, Boston (1994).

10. S.K. Tewksbury (Editor), "Microelectronic System Interconnections - Performance and Modeling," A selected reprint volume, IEEE Press, New York (1994).

11. J. Nishikido, S. Fujita, Y. Arai, Y. Akahori, S. Hino and K. Yamasaki, "Multigigabit multichannel optical interconnection module for broadband switching system," IEEE J. Lightwave Technol. 13 (1995) 1104-.

12. L.A. Coldren and B.J. Thibeault, "Vertical-cavity surface-emitting lasers for free-space interconnects", SPIE Proc. CR62 (Critical Review: Optoelctronic Interconnects and Packaging) (1996) pp. 3-34.

13. R.A. Morgan, "Advances in vertical-cavity surface-emitting lasers," SPIE Vol. 2147 (1994) pp. 97-119.

14. W.S. Ishak, et. al., "Optical interconnects - the POLO approach," SPIE Vol. 2400 (1995) pp. 214-221.

15. B.L. Booth, "Recent developments in polymer waveguide technology and applications for data link and optical interconnect systems," SPIE Vol. 2691 (Optoelectronic Packaging) (1996), pp. 2-8.

16. P. Eriksen, K. Gustafsson, M. Niburg, G. Palmskog, M. Robertsson and K. Akermark, "The Apollo Demonstrator - new low-cost technologies for optical interconnects," Ericsson Review, Vol. 72, No. 2 (1995).

17. J. Bristow, Y. Liu, T. Marta, S. Bounnak, K. Johnson, Y, Liu and H. Cole, "Cost effective optoelectronic packaging for mulltichip modules and backplane level optical interconnects," SPIE Vol. 2400 (1995) 61-73.

18. M.M. Oprysko, "Optical waveguides in the computer environment - a packaging perspective," SPIE Vol. 1213 (Photopolymer Device Phyiscs, Chemistry and Applications) (1990) 76-84.

19. S.E. Schacham, et. al., "Waveguides as interconnects for high performance packaging," Proc. 9th International Electronic Packaging Conf., Vol II (1989) pp. 1003-1013 (copyright International Electronic Packaging Society, Inc.).

20. A. Buehler, et.al., "Optical polyimides for singel mode waveguides," SPIE Vol. 1849 (Optoelectronic Interconnects) (1993) 92-103.

21. C.F. Kane and R.R. Krchnavek, "Photodefinable benzocyclobutene as an optical waveguide material," SPIE Vol. 2153 (Optoelectronic Interconnects II) (1994) 200-207.

22. S. Kolodziej, "Plastic fiber edges closer to networking," Lightwave Magazine (September, 1992) pp.11. Summarizes highlights of International Conf. on Plastic Fibers, Paris, France (1992).

23. R.J.S. Bates, S.D.Walker and M.Yaseen, "The limits of plastic optical fiber for short distance high speed computer data links," Fiber and Integrated Optics (U.S.) Vol.12, No.2 (special issue on "Plastic Fibers"), (1993).

24. (a) P. De Dobbelaere, et.al., "Fabrication of vertical light couplers in polymeric waveguides by embossing," Paper CTuR5, 1994 Conference on Lasers and Electro-Optics Europe (CLEO-E'94), Technical Digest (IEEE Catalog no. 94TH0614-8) (1994), pp.178-179.
(b) R.T. Chen, et.al., "Compression-molded polymer based optical bus," SPIE Vol 1849 (Optoelectronic Interconnects) (1993) 59-67.

25. A. Neyer, University of Dortmund, EU-funded POPCORN project and other work on electro-plating based transfer of 3-D patterns onto plastics and polymers using injection and compression molding techniques (1995).

26. L.A. Hornak (Editor), "Polymers for Lightwave and Integrated Optics: Technology and Applications," Marcel Dekker, Inc., New York (1992). There are 24 chapters here all worth a glance, if not detailed study.

27. (a) J. Bristow, C. Sullivan, S. Mukherjee, Y. Liu and A. Husain,"Progress and status of guided

wave optical inteconnection technology," SPIE Vol. 1849 (Optoelectronic Interconnects) (1993) pp.4-10. (b) J. Bristow, C.Sullivan, S.Mukherjee, Y.Liu, Y.Liu and H.Cole, "Polymeric optical waveguides for multichip modules and board to board interconnects," the Fifth Meetings of the Du Pont Symposium on High Density Interconnect and Thin Film Polyimide Technology, book of abstracts, Wilmington, Delaware, October 4-6, 1993.

28. C.P. Wong (Editor), "Polymers for Electronic and Photonic Applications," Academic Press, Boston (1993).

29. B.L. Booth, "Polymers for integrated optical waveguides," Chapter in: C.P. Wong (editor), "Polymers for Electronic and Photonic Applications," Academic Press, Boston (1993), pp. 549-600.

30. R. Moosburger, et.al., "Digital optical switch based on 'ovesized' polymer rib waveguides," Electronics Letts. 32 (1996) 544-545.

31. R. Lytel and F. Lipscomb, "Packaging and applications of active polymer optical switching arrays," SPIE Critical Review CR62 (Optoelectronic Interconnects and Packaging) (1996) pp. 431-441.

32. J. Parker, "Optical interconnects and packaging: current trends and future perspectives", paper presented at SPIE Critical Review CR62 (Optoelectronic Interconnects and Packaging) (1996) conference, but not included in Proceedings. E-mail: j.w.parker@bnr.co.uk for copy.

33. R.F. Carson and W.J. Mayer, "Sandia's photonics program and its changing national role," SPIE Vol 2153 (Optoelectronic Interconnects II) (1994) 169-179.

34. H.F. Lockwood, "Hybrid wafer scale optoelectronic integration," Proc. SPIE Symp. Advances in Interconnects and Packaging, Nov., 1990, Boston.

35. Y.S. Liu, H.S. Cole, J. Bristow and Y. Liu, "Hybrid integration of electrical and optical interconnects," SPIE Vol 2153 (Optoelectronic Interconnects II) (1994) 337-343.

36. H. Takahara, S. Koike, S. Yamaguchi and H. Tomimuro, "Optical waveguide interconnections for opto-electronic multichip modules," SPIE Vol 1849 (Optoelectronic Interconnects) (1993) 70-78.

On Thin Film MCM-D Interconnects

J. Roggen, E. Beyne, C. Truzzi, E. Ringoot, P.Pieters
IMEC
Kapeldreef 75, 3001 Leuven, Belgium

1   Introduction

In the first part a MCM-D switching unit specifically designed for performance analyses of thin-film interconnection substrates is described. The module consists of 4 unpacked 0.7-μm CMOS ASICs (100 I/O's, 64 mm^2, Standard-Cell technology) transmitting digital signals at 200 Mbit/s on a 5-layer thin-film substrate (1-by-1 inch, 2 interconnection layers).

In the second part the integrated microwave filters in MCM-D technology will be addressed. Filters, consisting of components such as spiral inductors and capacitors, are an essential element of microwave circuits. On Monolithic Microwave Integrated Circuits (MMIC's) these structures require a large area, resulting in a high cost. The filters described here are made by integrating lumped passive components MCM-D in such a way that the desired filter characteristics are obtained. The bulky filters can now be developed very cheaply, providing the extra advantage that the filter can interconnect the different elements of GaAs devices, reducing the cost of the MMIC. Furthermore, the line widths used in MCM-D technology are broader than with MMIC, giving lower resistive lines, less parasitic effects and higher quality factors.

2   The testing methodology

MCMs are complex microsystems. Assembling "Known-Good Dice" (KGD) on a "Known-Good Substrate" is not sufficient to assess the module's level of goodness. In digital systems, an important parameter is the bit error rate, or the maximum frequency that guarantees an error-free transmission. How is this parameter influenced by propagation delays, signal integrity, power dissipation, crosstalk or switching noise, which are linked at their turn to geometrical, technological or electrical quantities?

The testing methodology presented here provides a possible way to answer this question experimentally.

This approach is based on the development of a special test module, specifically designed for such purpose (fig. 1). In the case study presented here, 4 identical unpacked 0.7-μm CMOS test ASICs (100 I/O's, 64 mm^2, Standard-Cell technology) are assembled on a 5-layer thin-film substrate (1-by-1 inch, 2 interconnection layers).

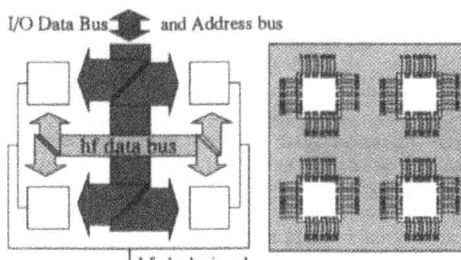

Figure 1. Schematic representation of the MCM switching units

G. Harman and P. Mach (eds.), Microelectronic Interconnections and Assembly, 141-144.
© 1998 *Kluwer Academic Publishers.*

3   IC layout

Fig. 3 shows the schematic logical layout of the chip. In fig. 4 the actual layout is shown. The structure of the chip is simple, consisting of 8 transmitting units, an equal number of receiving units, an addressing block, a clock generating unit and an 8-bit data bus. Each transmitting unit drives two high frequency transmission channels, while each receiving unit can be programmed to be connected to one out of two input channels. A 9-bit address bus, a 5-line control bus and some clocks complete the I/O signal list.

The I/O pads of the high-frequency drivers and receivers are organized in 8 similar groups, each of them containing 2 transmitters and 2 receivers, together with some decoupling GND and PWR pads. The structure of each group is shown in fig. 2.

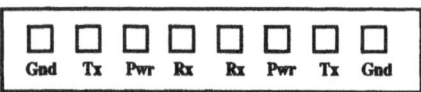

Figure 2. Structure of the high frequency I/O pad groups

Fig. 4 illustrates also the mapping of the I/O pads, showing the position of the 8 high-frequency groups (1÷8), the data bus, the address bus, the high-frequency clock, the control signals (C) and the switching noise drivers (SN). These are programmable drivers, enabled by an opportune signal, used for module ground-bounce noise analysis. With the help of these drivers, a maximum of 124 drivers can switch at the same time on the module.

Patterns to be transmitted are stored as words into each transmitting unit (and into each related receiving unit) as sequences of bytes. The word length is programmable, and can consists of any number of bytes from 1 to 8.

For a given transmission experiment, however, the length is fixed. Data to be sent are stored as words, transmitted as words and compared word by word. Before the storing phase, the transmitting and receiving units of the channel being programmed are informed about the length of the following word.

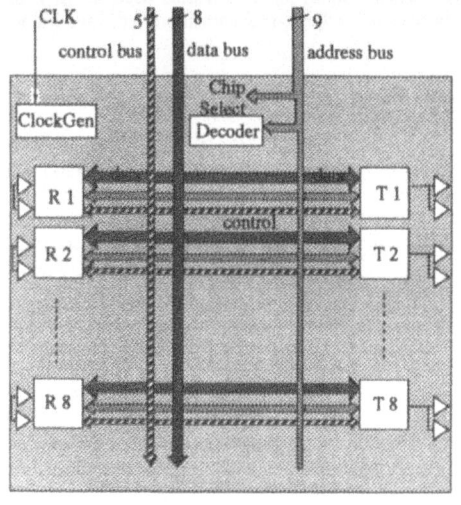

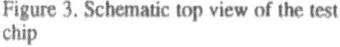

Figure 3. Schematic top view of the test chip

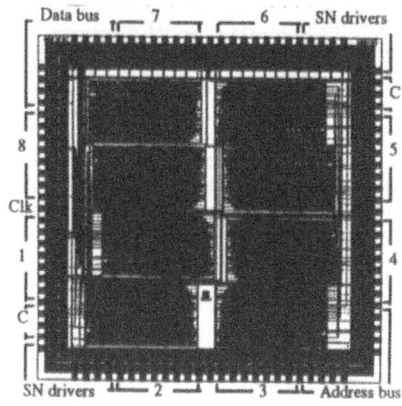

Figure 4. Layout of the test chip. Dimensions are 8x8 mm^2

Afterwards, a pointing logic stores automatically the data bytes in subsequent memory locations. During the testing phase, the word is cyclically transmitted and received and possible transmission errors are counted and stored using "byte-oriented" counters (each byte of the word is checked independently for errors). The results of each transmission experiment depend on the selected error detection mode. If the deterministic mode is selected, the first detected errors causes the data reception to stop at the end of the current word, and the faulty word is memorised.

4   Module layout

The thin-film MCM-D technology developed at IMEC uses a five layer structure, as shown in fig. 5. The dielectric layers are deposited by spin coating and consist of photo-sensitive BCB, Cyclotene^{TM} from DOW. This material has a low dielectric constant of 2.7. Layers with thickness up to 10 μm and via openings of 30 μm are used. The metal layers are deposited by sputtering and are patterned using wet etching. The metal composition is 2 μm Ti/Cu/Ti for the inner layers and 2 μm Ti/Au for the top layer. Substrates may be either 4" Si or Alumina wafers . Thermal via's are used to improve the thermal conductivity of the MCM-D multilayer.

Design rule specifications and transmission line characteristics are for X and Y layers are given in table 1 and table 2.

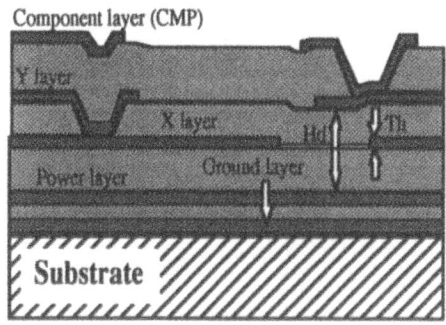

Figure 5. Cross section of the high speed MCM-D technology

Conductor:	Width (μm)	Spacing (μm)	Pitch (μm)	Thick. (μm)
X-layer microstrip	20	30	50	2
Y-layer microstrip	30	45	75	2
X- or Y- layer stripline	14	36	50	2

Table 1. Design rule specifications

Parameters	X	Y
Line Width: W	20 μm	30 μm
Line Thickness: Th	2 μm	2 μm
Diel. thickness: Hd	9 μm	13.5 μm
Rel. diel. constant	2.7	2.7
Capacitance p.u.l.	107 pF/m	73 pF/m
Inductance p.u.l.	280 nH/m	280 nH/m
Delay p.u.l.	5.5 ns/m	5.5 ns/m
Characteristic Impedance: Zc	51Ω	51Ω
Metal resistivity	2 μΩ/cm	2 μΩ/cm
Line DC resistivity	500 Ω/m	333 Ω/m
Line attenuation	$1.4 \frac{mdB}{\sqrt{m \cdot Hz}}$	$0.9 \frac{mdB}{\sqrt{m \cdot Hz}}$
Dielectric attenuation	$.75 \cdot 10^{-9} \frac{dB}{m \cdot Hz}$	$.75 \cdot 10^{-9} \frac{dB}{m \cdot Hz}$
Dielectric loss (tanδ)	0.5%	0.5%

Table 2. Transmission line characteristics for X and Y layers

5. Integrated microwave filters

Using MCM-D technology, all classic hybrid Microwave Integrated Circuit (MIC) structures may be realised. This technology is also very flexible and makes it possible to integrate microwave devices made with different technologies (e.g. microwave amplifiers and mixers in GaAs, oscillators in InP, filters in

MCM-D substrate, ...) on the same substrate with negligible transition effects using flip-chip, epitaxial lift-off or simple wire-bonding. The filter structures we develop are realised on alumina wafers and consist of two 2 mm thick conductive Ti/Cu/Ti layers, separated by 10 mm Cyclotene^{TM} dielectric with relative permittivity of 2.7. With this technology a wide range of filters can be designed.

As an example, the following figures present a 5 to 10 GHz third order Chebyshev bandpass filter. Figure 6 shows the equivalent circuit of the filter. In figure 7a we display a photograph of the filter, while figure 7b shows a comparison between S21 transmission MDS simulations and network analyser measurements. A good agreement can be seen. This shows that excellent and cheap microwave filters can be easily made by integration of lumped passive components in MCM-D.

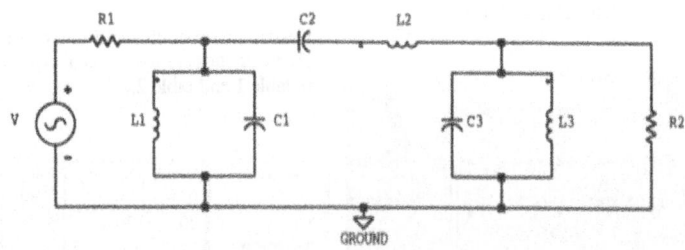

Figure 6. The equivalent network for the 3th order Chebyshev BPF.

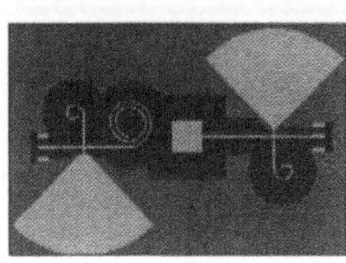
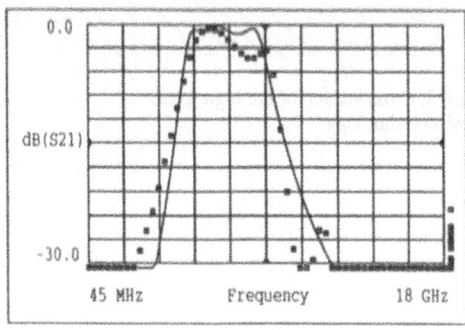

Figure 7a (left). Photograph of MCM-D 3th order Chebyshev band pass filter.
Figure 7b (right). Comparison of S21 transmission measurements (o) and MDS simulations (—) of a MCM-D 3th order Chebyshev band pass filter.

VLSI INTERCONNECTION BY BUMPLESS TAB

Gérard DEHAINE, Patrick COURANT, Karel KURZWEIL
Bull S.A.

ABSTRACT

TAB has always been associated with bumps and gang bonding. This situation has been an advantage for people requiring high throughput in production as demonstrated in the Japanese electronic industry; it has been a difficulty for other people willing to start prototyping with TAB or needing only small quantities of components.

Among the drawbacks, the non recurring costs associated with a TAB configuration are not acceptable for small runs; the same way, delays in obtaining the tape, bumping, and the associated tooling are often considered as too long.

An other difficulty comes from the increasing size of VLSI chips which are no longer easily compatible with the gang bonding operation.

A new ILB technique called single point bonding, has been applied in the industry starting five years ago; it consists in replacing the collective bonding by a point by point bonding operation performed on a modified wire bonder. This has first allowed to handle large size chips which were not mountable by gang bonding; but the most interesting of it is that it has allowed to give up the need for wafer bumping of aluminum pads.

Several companies have developed this process and some of them have applied it in production.

This paper describes the development which has been made at BULL, through European Community sponsored packaging programs and which is being now supported by the DGA/ DRET French administration in association with a large French company for military MCM applications.

Examples of memories and of high performance ASIC's are described, as well as the associated reliability evaluations.

Through these developments, bumpless ILB has shown to be a reliable alternative for bumped TAB devices; in conjunction with an effort to speed up design time and tape procurements. It appears to be an attractive solution for newcomers in TAB applications.

G. Harman and P. Mach (eds.), Microelectronic Interconnections and Assembly, 145-155.
© 1998 *Kluwer Academic Publishers.*

INTRODUCTION

The European Union - EU - brings its support to a variety of new developments in the field of microelectronics. The assembly of multichip modules is one of the areas supported by Esprit program.

The Esprit program CHIPPAC - Cost Effective High Performance Packaging - is developing new packaging approaches for applications of multichip modules in computers and telecommunications. The VLSI - ASIC's - used in the applications are typically large, complex semiconductor devices to be mounted in the multichip modules. The conventional , the most frequently used assembly by wire-bonding, perfectly convenient for single chip package assembly, is somewhat limited in complex multichip applications. There are various alternative assembly methods of bare chips mounting in multichip modules.

Tape automated bonding - TAB - is used since several decades. TAB simplifies chip handling and allows extensive electrical test prior assembly on MCM, contributing to an increase of overall assembly yield. The conventional TAB requires, however, a special finish on chip bonding pads which results in an increased device cost, which can be penalizing especially in case of mounting a short series of chips.

The bumpless TAB further described in this paper allows to use standard aluminium chip pads for chip mounting on flexible tapes for testing and handling purposes. The bumpless TAB is already applied in large volume applications while appearing as particularly attractive for small series of complex ASIC's. The European Union and the french Ministry of Defence have supported the bumpless TAB development and investigations further described in this paper.

GANG BONDING

Gang bonding is the fastest way of bonding TAB devices to a tape frame; a couple of second is needed for attaching all the connections of a chip even if the chip is complex. This was true for small devices but shows technical limitations when chip size increases. In fact for a given category of bonding equipment (quality of parallelism at bonding temperature), the credit in microns for the lack of flatness is the same and is in the range of 2 to 5 μm. When chip size increases modification of material hardness i.e. copper of leads and bump, is needed; when chip size becomes larger than 10, 12 mm, bump thickness should be increased in order to have more flatness tolerance.

For thermocompression and eutectic formation, bonding parameters are pressure at pad level and temperature; both are provided by the bonding tool. As seen on figure 1, a lack of flatness can make impossible the optimization of bonding conditions because pressure is not evenly distributed.

This is worse on complex VLSI, because usually when chips are large, they have a large number of I/O's to connect and the pitch is often in the range of 100 μm or less, yielding a shape factor height/ width difficult or impossible to achieve at bump level.

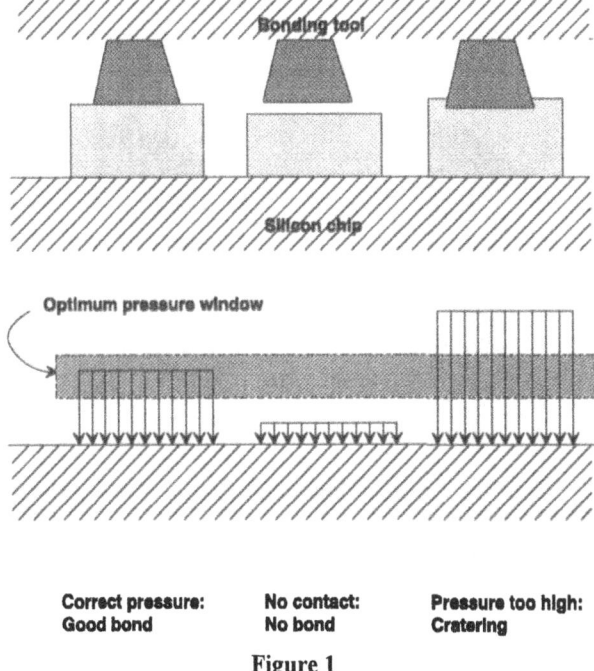

Correct pressure: No contact: Pressure too high:
Good bond   No bond   Cratering

Figure 1

Effect of lack of parallelism on gang bonding

> When bump plus lead total thickness is not homogeneous within the
> chip, optimization of bonding conditions can be impossible. When most
> of the bonds can be good, some high thickness sites will receive a too
> high pressure inducing silicon cratering while low thickness sites will
> not be bonded.

When we combine this difficulty with the cost of making bumping at wafer level, it is easy to
understand that industrials have tried to find alternate solutions to conventional bumping.

SINGLE POINT BONDING

Single point bonding has been used for years on wire bonding equipments for connecting chips
in packages or chips in multichip modules; it is in fact the most standard packaging
interconnection technique for integrated circuits. Single point TAB bonding has been
developed in some companies at the end of the eighties [1] as adaptations of wire bonding
equipments. These bonding machines have improved a lot during the last ten years especially at
the level of automation; improvements have been made also in increasing the speed of bonding
and several bonders can demonstrate bonding speed of 15 to 20 bonds per second.

Obviously this technique is slower than the equivalent TAB technique when it is performed in
gang bonding, but it offers the significant advantage of not being sensitive to normal
geometrical dispersions along the Z axis. A wire bonder detects when the tool touches chip
surface and then generate the amount of energy needed for making the bond; this energy is
applied locally and is not dispersed in the neighbor connections.

Although single point TAB bonding can be performed on " normal thickness gold bumps", its
main advantage is to allow decreasing bump thickness . This appears as obvious when we

consider the geometrical analogy between wedge bonding of a wire (or a ribbon) to an aluminum pad and the equivalent structure of a TAB copper lead on the same pad; in fact we do not need a gold pad to perform wire bonding (figure 2)

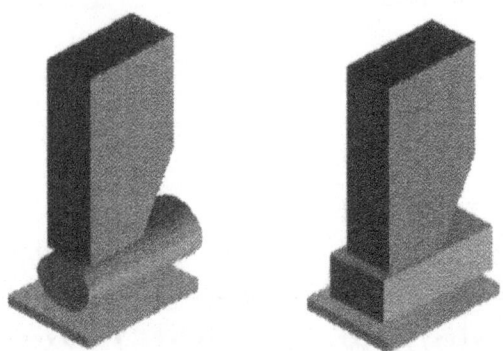

Figure 2

Analogy between wire bonding and TAB bonding of a lead. Wire
diameter is 30 μm, TAB finger thickness is 35 μm.

If we consider the metallurgical structure, a gold plated TAB finger is very similar to a massive gold wire and at the interface, gold will diffuse approximately the same way with the aluminum of the pad; lets apply, in both cases thermosonic energy, and we have realized a metallurgical bond.

Several pad configurations may anyhow be of interest since the gold bump has definitely the advantage of protecting the aluminum from being corroded by an external contaminant:

- electrolytic low bump thickness (3 to 5 μm) is easier to realize when the pitch is small
- electroless Nickel Gold deposition (as offered by Picopak in Finland) has shown good behavior during and after TAB assembly and can be performed at wafer level as well as on a single chip.

Bonding directly to the aluminum pad is possible in most cases; it has the strong advantage of allowing the use of bare chips, the way they are when going out of the wafer foundry. A protective coating will, in that case, be deposited on the chip surface and on the pad area.

Compared to the situation with normal bump thickness, when applying TAB to low bump constructions, more care should be taken for avoiding the risk of short circuit between leads and the edge of the chip since the distance is equivalent to bump thickness. This can be achieved by a forming operation of the leads before or after ILB, and /or by application of a low viscosity junction coating fixing the distance between lead and chip edge at around 5 to 10 μm.

CHIP CONFIGURATIONS

At BULL, several configurations of VLSI chips have been tested since the beginning of this development in 1987: first of them were related to ASIC VLSI's used in mainframe CPU's on which, the no bump TAB was qualified and in particular:

- 125 μm pitch on a 316 lead and 428 lead configurations (chip size 12 and 15 mm)

Most of the basic developments were performed within the EEC sponsored program APACHIP, on high complexity components [2][3]:

- 12 mm chip at 125 µm, 316 I/O's
- 15 mm chip at 100 µm, 548 I/O's
- 11 mm chip at 75 µm, 564 I/O's , see figure 3

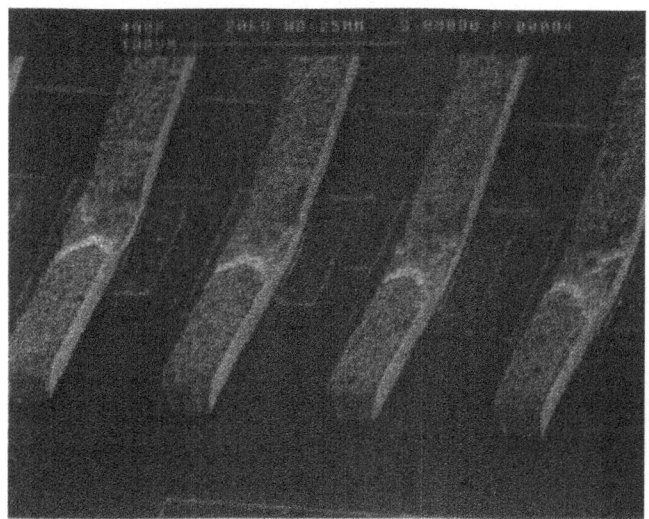

Figure 3
ILB connection at 75 µm pitch on a Siemens Nixdorf integrated circuit
(APACHIP Program)

Then, we are now in a phase of qualifying the low bump approach for military components for MCM constructions. This is undertaken in a program (PACEO) sponsored by the French Ministry of Defense, DGA / DRET; in this program BULL is associated with Dassault Electronique for validating TAB technology in military environment.[4]

Four representative chip configurations (2 memories and 2 ASIC) are being tested at that phase of the program:
>   a static RAM 256 Kb from MHS
>   a dynamic RAM 4 Mb from IBM
>   a CMOS ASIC from ES2
>   an analogic ASIC from Thomson

Both bumping configurations, low bump with electroless nickel / gold and native aluminum bump have been tested and evaluated.
Examples of ILB and OLB assemblies are shown in figures 4 and 5, for each configuration between 100 and 400 components have been assembled and electricaly tested.

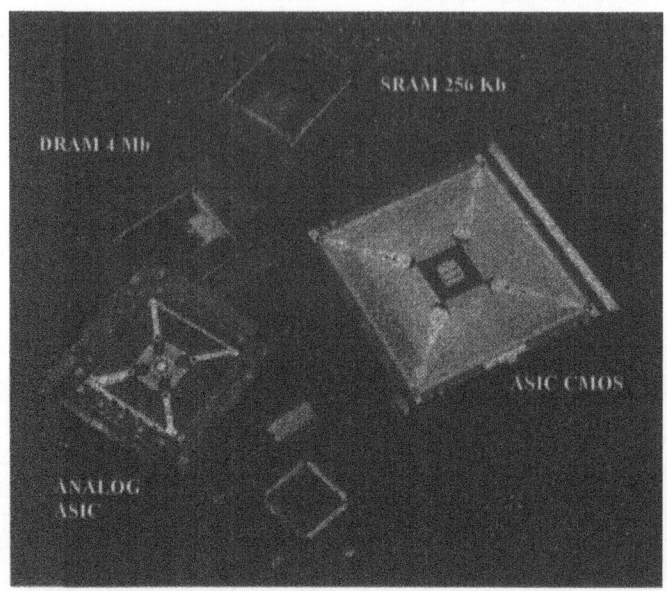

Figure 4
Samples of integrated circuits on tape; CMOS ASIC has 548 I/O, Analog ASIC works
at more than 1 Ghz.(ASIC are components designed by Dassault Electronique)

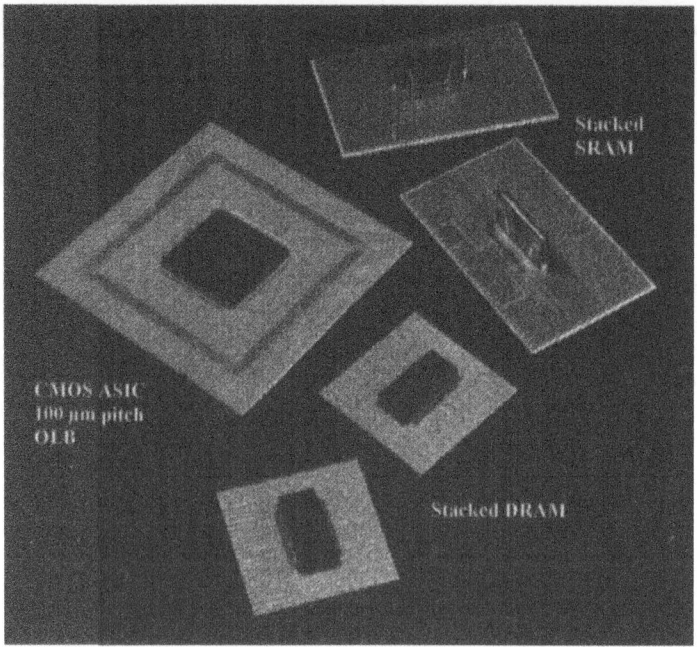

Figure 5
Components after OLB.

RESULTS, RELIABILITY

The four configurations were successfully assembled at ILB and OLB and went through qualification tests with success. In order to simplify the presentation of results, only the 100 µm configuration will be described in this paper.

On this configuration, electroless Nickel Gold bumps and aluminum bumps were tested against the military test procedures. Optimization of bonding conditions is performed on a modified wire bonder with TAB bonding software (Hughes 2460) using Taguchi procedure [5].

Testing conditions for ILB components are:
> 100 thermal shocks -55°C +125°C (MIL STD 883, method 1010/C)
> High temperature storage, 150°C, 1000 H

Test procedures includes functional electrical test at -50°C, ambient and 125°C.
Pull test is performed on non coated devices after high temperature storage.

Chips are then OLB mounted on test substrates (figure 5) and receive the following constraints:
Testing conditions for TAB chips on substrate:
> 200 thermal shocks -55°C +125°C (MIL STD 883, method 1010/C)
> High temperature storage, 125°C, 1000 H
> Burn-in 125°C, 200 H
> 85°C/85 RH 2000 H

Figures 6 and 7 shows detail of lead/pad area for electroless bump and aluminum pad.

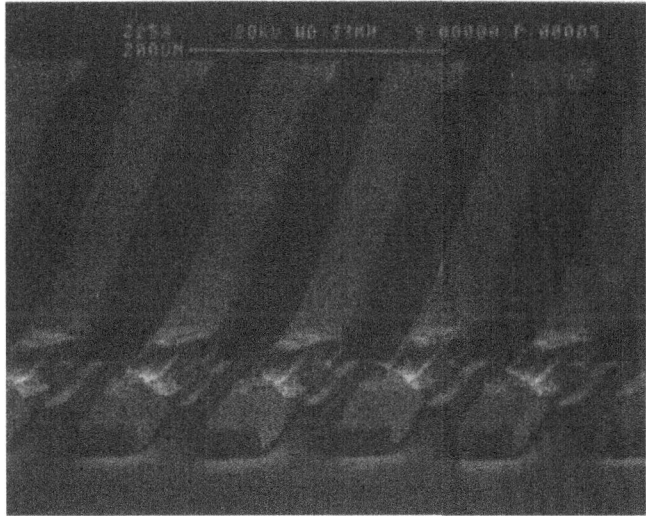

Figure 6
ILB at 100µm pitch on electroless bumping
Bump is made of 5 µm of Nickel and 0,5 µm of gold.
(ASIC 548 I/O)

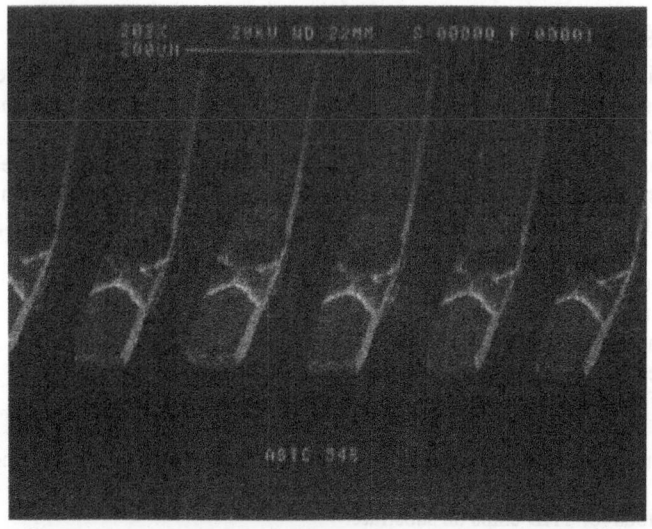

Figure 7
ILB at 100μm on aluminum pad

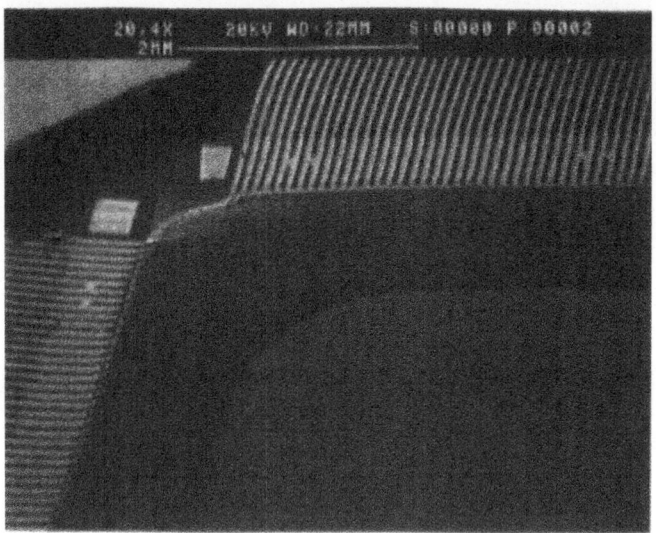

Figure 8
ILB at 100μm with coating ring on ILB area

Joint integrity

Pull test is the most convenient way for assessing the mechanical stability of a joint;
conventional pull testing has been applied on 100 μm configurations after aging tests. Results
are summarized on figure 9 and 10. The values obtained show a slight decrease in the average
pull strength values that corresponds to a modification of the copper structure of the leads with
temperature.

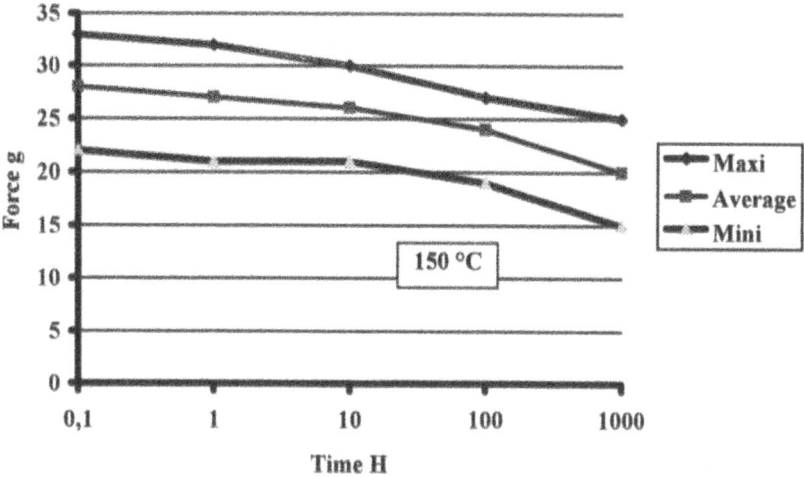

Figure 9
Evolution of pull test on 100 μm device
aluminum pad with no bump

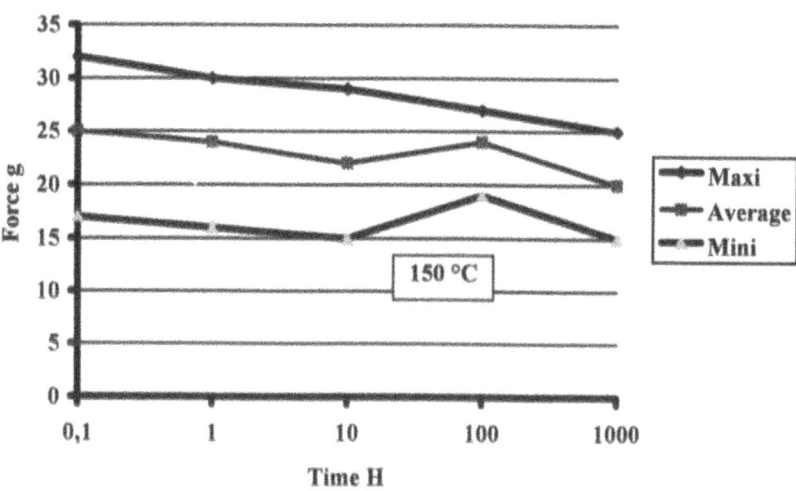

Figure 10
Evolution of pull test on 100 μm pitch device with electroless bump.
On these curves evolution of the strength is moderate and comes mostly from
copper strength evolution (more than 95% of break mode is copper lead).

No evidence of fragile Al-Au intermetallics has been detected even after aging at high
temperature; the metallurgical analysis is still in progress [6] but it seems that the amount of
gold present on the TAB lead is low compared to gold ball bonding, and that the temperature

of aging is low compared to what is needed for initiating Kirkendal effect. Some parts have been tested at 300 °C, 200 H without a significant change in pull strength values compared to the 150 °C 1000 H test.

ILB joint resistance has been monitored on Kelvin structures provided around the chip, the results are shown in figure 11; the average value of 12,5- 13 milliohm includes the resistance of a small length of the lead. We estimate ILB interface resistance itself around 3 milliohm.

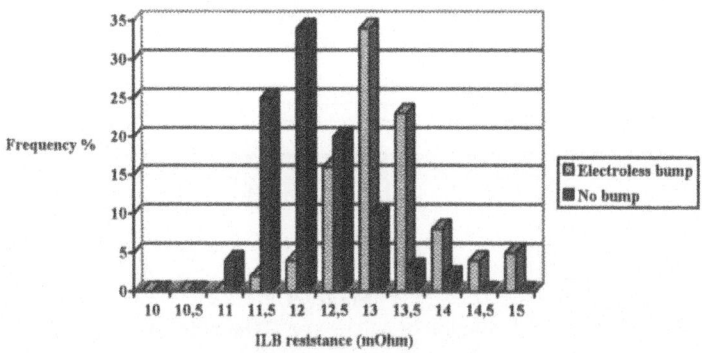

Figure 11

Distribution of ILB resistance measured at 25 °C.
No significant variation is obtained after 1000 H at 125°C and thermal shocks; electroless bump is slightly more resistive than the no bump configuration.

FUTURE DEVELOPMENTS

Most of the chips available on the market, as far as they are available in wafer form, can receive conventional or low thickness gold bumping. Bumping is a normal situation for high volume manufacturing and provides the lowest cost of assembly.

Anyhow when entering in TAB business or for potential users like defense manufacturers requiring fast prototyping and small quantities of parts, the initial setup cost is high and cannot be spread over a large production quantity. For these manufacturers, skipping of the traditional bumping process is an advantage.

In the PACEO program, we are focusing on other aspects of the TAB assembly steps that are still representing a difficulty; these points are:

- TAB tape manufacturing: artwork and hard tooling are bringing a cost per frame which is too important for small quantities; prototyping with soft tooling or by the sheet is under investigation.
- Lead punching and forming: still requires specific tooling; we are investigating flexible point by point operation which appears to be promising for prototype quantities. [7]
- OLB: single point OLB is also a good way for assembling TAB devices to a substrate; this operation can usually be made with the same kind of equipment as ILB.

All these aspects will be taken care of, during the PACEO program; and may end to the setting up of a structure offering TAB services to companies interested in entering in TAB assembly in 1997 on an industrial basis, and in the mean time for prototyping of TAB configurations.

CONCLUSION

TAB seems to start again for industrials requiring high testability and high reliability; its draw-backs may be overcome by simpler procedures allowing to reduce the associated costs to an acceptable level. When everybody speaks of KGD (known good die)., we need to remember that TAB is still the most obvious way of obtaining KGD.

The possibility of giving up some of the difficulties associated with bumping as described in this paper, is a step towards simpler TAB which may convince potential users to start working with it.

ACKNOWLEDGMENTS

These developments have been supported by the European Community in ESPRIT APACHIP program and in the on going CHIPPAC program. It is supported also by DGA/ DRET in the PACEO program in cooperation with DASSAULT ELECTRONIQUE for the military applications of TAB.

REFERENCES

1. G. Silverberg, " Single point TAB (SPT): A versatile tool for TAB bonding" presented at ISHM Symp, Sept 1987.

2. G.Dehaine, P.Courant, " TAB developments in European community, APACHIP," ITAP symposium San Jose CA, February 1992.

3. G.Dehaine and P.Courant, " Single point ILB at narrow pitch , " IEEE Transactions on Components, Packaging, and Manufacturing Technology, November 1994, Volume 17.

4. G.Dehaine, P.Courant, A.Dravet, R.Even, " Technologie TAB pour applications militaires et industrielles", ISHM Forum Microelectronique, PARIS February 1995.

5. Brian Lynch, " Integration of an advanced Silicon process with thermocompression TAB bonding, using robust design methods," ," ITAP and Flip chip symposium San Jose CA, February 1994.

6. Private discussions with E.Zakel from TU Berlin, 1994

7. French patent n° 94 15885, P.Courant, Single point TAB lead forming feature

Critical Issues in Wire-Bonded Chip Interconnections to the Year 2001

George G. Harman
National Institute of Standards and Technology
Gaithersburg, MD 20899, USA
Ph = 301-975-2097, FAX = 301-948-4081

Abstract - The current and future issues of materials, reliability, and yield of wire bonded interconnections used in microelectronics are described. These issues will affect the future production and use of such interconnections over the next 5 years.

I. INTRODUCTION

Currently, there are about 4×10^{12} wires bonded per year on the planet. The plot of wire bond production (as gold ball bonds) is given in Fig. 1. The infrastructure required to produce such numbers of bonds is so extensive that no other chip interconnection method can displace wire bonds for the foreseeable future, although other technologies, particularly flip-chip, will experience increasing utilization. In 1996, there are ~45 billion ICs produced, of which about 0.5 billion are flip-chip interconnected. The IC industry is driving wire bonding technology towards: (1) increased yields (<25 ppm defects), (2) decreased pitch ($\leq 70 \mu$m for ball bonds), and (3) lowest possible cost(!). However, many specific technical and material issues are involved in achieving these goals. Some examples are new bond pad metals, higher frequency ultrasonic (US) energy, and lack of quantitative understanding of the US bonding mechanism for both 60 kHz and higher frequency energy.

II. FINER PITCH AND HIGHER YIELD BALL AND WEDGE BONDS

Currently, most bonding pad pitch for ball bonds is still in the 100 to 115 μm range. However, 90 μm pitch is in current production. Eighty micrometer pitch processes are becoming robust and will be in production within a year. Seventy micrometer pitch should be in production in a year or two. The latter await chips designed with pads having such fine pitches. These will be made with ~25 μm diameter Au wire. Fig. 2 gives an example of current 70 μm pitch ball bonds made with 25 μm diameter wire. These ball bonds appear different from coarser pitch ones of the past. Finer pitch and wire diameters will be more subject to wire-sweep failures during plastic molding. Unless a solution to that problem is achieved, it may be the limiting pitch for high-volume production of plastic molded packages, although insulated wire [1]-[2] and lower viscosity molding compounds may help. Finer ball bond pitch (which can be achieved), therefore, may not be used except for special purposes. Some typical bond parameters for fine pitch ball bonding are given in Table 1. Note that the bonded ball size for the 70 μm process is only 47 μm, less than twice the wire diameter. If the pitch became much smaller (e.g., ~60 μm), the bond pull test would adequately evaluate the weld strength of ball bonds and the shear test would be used only for setup purposes.

Wedge bonding has been experimentally demonstrated at 40 μm pitch using 10 μm diameter Au wire [3], but can be readily made at 60 μm pitch using 25 μm diameter Al or Au wire, although chips having such fine pitch pads have not been made, and test probes are not currently available either. Wedge bonding at such fine pitch is performed using high frequency (>60 kHz) US energy, (see III).

G. Harman and P. Mach (eds.), Microelectronic Interconnections and Assembly, 157-165.
© 1998 *Kluwer Academic Publishers.*

TABLE 1

AUTOBONDER MACHINE PARAMETERS FOR FOUR FINE-PITCH BALL
BONDING PROCESSES USING 25 μm DIAMETER GOLD WIRE[a]

Machine or Test Parameter ↓	100 μm Process	90 μm Process	80 μm Process	70 μm Process
Free-Air Ball Dia. (μm)	50	_	43.2	40.6
Bonded Ball Dia. (μm)	74	61.3	55.8	47
Bonded Ball Height (μm)	16.1	13.5	12.5	5.9
Shear Force (gf)[b]	35.4	32.4	25.7	19.2
Shear Strength (SS) (gf/mil^2)[b]	5.36	7.06	6.9	7.04
Intermetallics Under Ball (% of interface area)	65.6	_	79.5	>80

[a] Data are typical, and were collected from the literature and from private communications.
[b] English units are used in the industry for SS. SI units (μm or mm) result in values so high
or low that they are not intuitively comprehensible (1 gf= 9.8 mN. Also, 1 mil = 25.4 μm).

III. BONDING WITH HIGH(ER) FREQUENCY ULTRASONIC ENERGY

The original reasons for choosing 60 kHz for US bonding are obscure, but fine-wire
bonding machines have used that frequency from the 1960s to the present. This frequency
resulted in transducers and bonding tools that were appropriate to microelectronic dimensions
and were stable during the bonding load. However, other frequencies (e.g., 25 kHz) have been
used for bonding large diameter Al wires in power devices. The possibility of US welding
over a wide frequency range has been known for some time. The Welding Handbook [4] cites
US welding frequencies as ranging from 0.1 up to 300 kHz. The basis for our understanding
of the welding mechanism of an US (metal) softening mechanism is given by Langenecker
[5]. His Al single crystal softening curve is given in Fig. 3. He stated that the ultrasonic
softening process was verified up to 1 MHz, so it is not surprising that higher frequencies
would find use in microelectronics bonding if an advantage could be demonstrated.

The present interest in using high frequency (Hf), >60 kHz, for microelectronics wire
bonding was started by Ramsey and Alfaro [6]. They studied thermosonic Au ball bonding on
IC pads using US energy in the range of 90 to 120 kHz, and reported that such frequencies
produced better welding, at lower temperatures, and in shorter bonding times than 60 kHz.
This frequency also resulted in more complete Au-Al intermetallic formation and thus, more
complete weld formation.[1] Also, others have given additional advantages for using Hf for

---

[1] The amount of Au-Al intermetallic formation is often used to demonstrate the extent of welding [6].

wedge bonding [7]. The first large-scale implementation is occurring with auto-wedge bonders, and most auto-wedge bonders sold in the last year use US frequencies of 100 to 120 kHz. The lower mass (inertia) of Hf transducers should allow wedge bonding machines to run faster. The other advantage is that strong Hf wedge bonds can be made with lower deformation and with shorter welding time. All of these are desirable for high-speed and fine-pitch bonding. However, a very recent work [8] compared Al wedge bonding at 60 and 120 kHz in a DOE study, and found no low-deformation advantage. The authors did find that less metal splash occurred around the bond perimeter at 120 kHz, achieved higher yields, and in shorter bonding times. In other studies [9], a range of higher frequencies, 100, 140, 250 kHz, was reported to produce good Au ball bonds at low temperature (50 °C) compared to 60 kHz. The best shear strength and the shortest bonding time were obtained at 250 kHz. However, the bonding window was very narrow (at such low temperatures), requiring closer control over machine parameters and materials in order to assure reproducability. Other workers reported that 90 to 120 kHz US energy resulted in better ball bonds to pads on polyimide that were placed over active areas of IC chips [10]. Since the use of such Hf is relatively new, one can expect that limitations as well as advantages will appear in the future. This is a very dynamic area.

One explanation of the differences in Hf *wedge bonding* was given by Shirai [7] (and incorporated in [11]). He suggests that the Hf tool-to-wire vibration produces a higher strain-rate (than 60 kHz), and therefore a much higher stress in the Al wire. The wire thus becomes strain-rate-hardened, deforms less, and more US energy is transmitted to the weld interface. This results in a strong Al to Al wedge bond with lower deformation. From [6], the higher frequency increases the rate of metallic interdiffusion in Au-Al ball bonds, and thus makes a better metal weld. This approach appears to be reasonable; however, there are many unanswered questions, and much more must be learned about the Hf bonding mechanism. The explanation used by Shirai is based on a strain-rate model for *single-crystal* LiF, and Ge [12]. These are brittle ionic and covalent materials. Wire bonding uses soft, polycrystalline, face-centered cubic metals (Au and Al) that respond to stress easily by deforming, although they do have some (unknown) high-strain-rate-modified response. An example of strain-rate hardening in various Al samples is given in Fig. 3. On this scale the frequency change from 60 to 120 kHz shows little difference in strain rate effects. Such data should be measured specifically on bonding wires in order to verify or reject this theory.

Our understanding of US mechanism (which is included as part of most modern US bonding explanations) is based on a 1968 paper by Langenecker [5]. He showed that the strain (deformation) of Al single crystals under ultrasonic excitation was comparable to that at elevated temperatures (see Fig. 4). (His work has never been verified[2].) However, if we accept this work as valid, then we must also accept the fact that the US softening mechanism is valid up to 1 MHz, implying that the strain-rate hardening is not significant below that frequency. However, that paper omits many measurement details and itself leaves questions. Shirai's proposed mechanism, which may be qualitatively correct, should be further studied, and in particular, its underlying theory rederived around the known properties of soft, polycrystalline, face-centered cubic metals. The strain rate explanation has not been applied to Au-Al ball

---

[2] It is ironical that so much of our understanding of US bonding is based on this work, but such effects have never been verified on Au and this material is used for ~95% of all wire bonds.

bonding, and some empirical evidence suggests that the improved intermetallics obtained with Hf may in fact be related to increased temperature generated by Hf during bonding. Experiments are underway to determine the validity of this proposal.

Most of our understanding of the US welding mechanism is based on fundamental studies carried out in the early to mid 1960s. These experiments need to be repeated with current measurement methods and high-speed computers for mathematical computation (verification) of models. Full understanding is the necessary basis for continued advancement of US bonding, as we push towards the limits of speed, fine pitch, and high yield. Our present understanding as well as the design of US bonding equipment is mostly empirical, and that knowledge has been pushed to its limits.

The resonant vibration modes, during bonding, of a 60 degree (thin) wedge bonding tool, with 60 kHz excitation, are given in Fig. 5. This author has not measured such vibration modes for higher frequency excitation (e.g., 120 kHz). However, since the same tools are often used, one would assume that there would be twice as many maxima and nodes if the tools are operated in a resonant mode. If different tools are used, then these characteristics would be dependent on the resonant frequency (design and material) of the specific tool. Loading of the tip, during bonding, have a large effect on movement of the node at any frequency. As the drive frequency increases, say to 300 kHz, the tip motion should be strongly affected by this nodal shift. Even though welding (bonding) would take place, slight loading differences from bond-to-bond (with resulting nodal shifts) could affect reproducibility from one bond to the next, resulting in lower yield. It is not possible to predict the frequency that this would become significant since it would be dependent on the characteristics of the particular US system and especially the tool material and its design. The bonding tool's amplitude loading effects are equally unpredictable. Bottleneck capillaries, used for fine pitch ball bonding, have a mechanical discontinuity near the tip, and the vibration modes under various bonding loads could also be unpredictable. Such tools might be designed to operate in a non-resonant mode; however, no information is available on such. Holographic interferometry would appear to be the best method of studying such phenomena.

IV. NEW METALLIZATIONS FOR LEAD FRAMES AND MCMs

The vast majority of all wire bonding today is made to Cu lead frames that have spot-plated Ag bond pads. This is reliable, although for the highest bonding yield the temperature is usually $\geq 200^\circ$C. Recently, Pd-plated Cu leadframes have been introduced into the IC industry as a replacement for spot-plated Ag bond pads, to promote adhesion of the plastic molding compounds, and to serve as a noble metal surface to enhance surface mount soldering to the external leads [13]. The Pd film is plated over a 1.5 μm (60 μin) Ni film on the Cu lead frame. The Pd is so thin, 0.076 μm (3 μin) that it dissolves in solder without forming brittle Pd-Sn intermetallic compounds. Thermosonic Au wire bonding to these Pd films is similar to bonding to the usual Ag plating, except that reoptimization of the bonding parameters is required (higher power, force and/or time). The reliability of this metallurgical bonding system for commercial grade product has been established by volume production. Palladium and Au are completely miscible and no intermetallic compounds exist. Both Au and Pd have strongly positive electrochemical potentials, so bond interface corrosion is unlikely. Palladium will slowly form a green oxide in air at about 400 °C, which presumably would reduce bondability. Therefore, it would be safest to bond Pd at interface temperatures <300 °C or in a neutral atmosphere. Palladium is mildly subject to corrosion by halogens and sulphur, so lead frames should be protected from them.

Palladium diffuses rapidly into Au by grain-boundary-diffusion and might have reliability problems similar to Au bonds on thick-film Ag. Therefore, a potential problem *could* result if the device was in a long-term, high-temperature environment. The thin Pd layer could be absorbed (diffused) into the Au of the wire bond, resulting in de-wetting under the crescent bond at the Ni interface. Such has not been studied and probably will not occur in the thermal environment of typical commercial plastic-encapsulated devices. Also, Pd absorbs ≈ 900 times its volume of hydrogen, and in the process it expands and becomes brittle and could separate from the Ni base. Any hydrogen will diffuse out with heat and thus would be liberated, if used in a sealed hermetic package, and possibly combine with oxygen and to form water vapor.

When a **gold bonding surface** is required, as on MCM metallization or on ceramic packages, electroplating is usually used. This has many disadvantages, the most important of which is the requirement for a continuous electrical path. Nickel underplating has long been applied by one of several electroless (autocatlytic) plating processes. Recently, processes for producing high-quality autocatalytic Au have been reported, one from a high pH solution [14], and the other from an approximately neutral one [15]. These are capable of depositing ~1 μm of high-purity Au, and should replace electroplating for many microelectronic applications. Work is currently underway to verify the reliability of these metallizations.

V. SMALL SAMPLE STATISTICS METHODOLOGY FOR MCMs

High-performance MCMs, using wire bonded interconnections, are often made in low volumes. The main production problem is assuring yield and reliability in the parts-per-million (ppm) defect range. One effort at high-reliability bond-production control with small sample sizes was described by Heleine et al. [16]. They used the deformed width of wedge bonds (which fitted a *normal distribution)* in conjunction with reliability test data to predict bond reliability (lifetime) of chip-on-board (COB) devices on PC boards. While not a generic solution, this concept can be used as a starting point to develop more universal small sample bond-process controls. An example of fine-pitch bond-yield prediction was used by Shu [17], who measured the gaps between adjacent ball bonds on fine-pitch bonding test die. *He* **assumed** *that they were normally distributed*, and used that assumption to calculate the probability of shorting between adjacent balls in a variety of fine-pitch wire bonding situations. Based on these examples, one possible approach to small sample yield or reliability control would be to study the failure modes, establish and then combine the *normal* distributions of the most likely failures, and use the result to predict yield or reliability. However, to prove that such distribution assumptions are valid in the ppm range requires very large numbers of test bonds, which contradicts the low volume production needs of MCMs.

The above methods assumed, or established normality of, the control parameter using a limited number of bonds. However, this cannot be a general assumption for all wire bond parameters. For instance, it has been demonstrated by Owens [18] that the bond pull force may not be normal, see Fig. 6. This parameter, as with all others, will depend on other events, such as wire break modes, pad surface conditions, contamination, etc. At the very least, the normality of the control parameter must be established first. The future of high-yield and high-reliability bonding for limited production runs needs further work. A statistical methodology must be developed that can be applied to a wide range of *low-volume* wire bonding situations.

VI. CONCLUSIONS

Many of the critical issues affecting wire bonding over the next 5 years have been discussed. These included the thrust towards fine pitch (towards 70 μm for ball bonding), lack of a quantitative understanding of the US bonding mechanism, the use of high frequency for US bonding, new metallizations for bond pads, and the need for high yield and reliability for MCM production. Other issues (e.g., means of obtaining high yield into or below the 25 ppm range, continued problems of Au-Al intermetallic failures in plastic devices, limitations on the usefulness of in-process bond monitors) could not be addressed because of time and space.

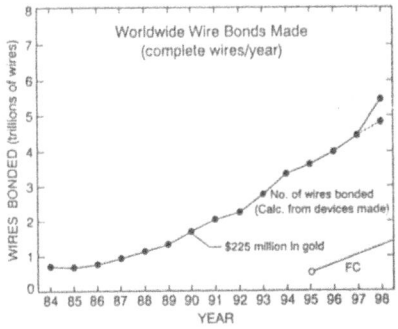

Fig. 1. The number of gold wires bonded per
year (projections past 1995). Data from K&S, Inc.

Fig. 2. An example of gold ball bonds with 70
μm pitch, courtesy of K&S, Inc.

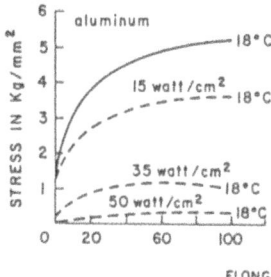
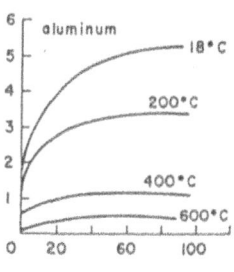

ELONGATION, %

Fig. 3 Stress vs. elongation for Al single crystals. The left *dashed* curve indicates straining
during irradiation at 20 kHz ultrasonic energy. The right curves show comparable
stress-strain behavior resulting from heating alone. (Langenecker)

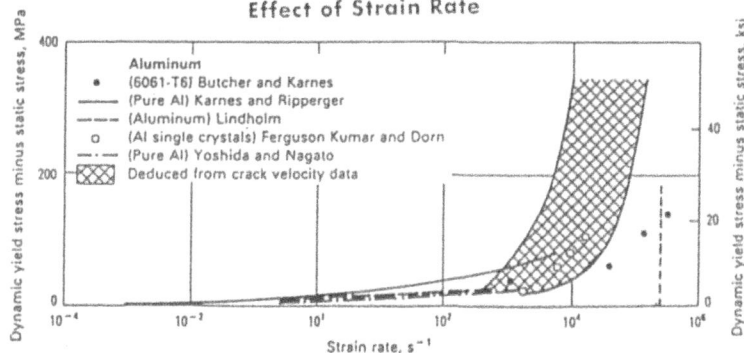

Fig. 4 Effect of strain rate on Al flow properties (Metals Handbook, 9th Edition, V-8,
Mechanical Testing). The vertical *dashed* line on the right is approximately at the
position of 120 kHz induced strain rate.

164

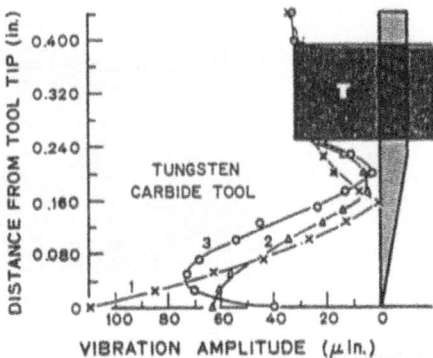

Fig. 5   The resonant vibration modes, during bonding, of a 60 degree (thin) Al wedge bonding
tool, with 60 kHz excitation. Curve 1 (x) is for free vibration. Curve 2 (∆) is taken 7
ms after application of US energy. Curve 3 (o) is 5 ms before the end of the bonding
cycle. Note the amplitude loading effect as the bond matures (Harman, NBS Spec.
Pub. 400-2 (1974)).

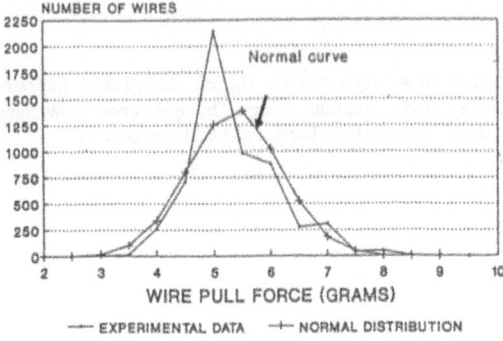

Fig. 6   Aluminum wedge-bond pull-test data compared to a normal curve based on the
empirical mean and standard deviation (Owens).

REFERENCES

[1]   Otto, A. J., "Insulated aluminum bonding wire for high lead count packages," *ISHM Journal*, vol. 9, no. 1, 1986, pp 1-8.

[2]   Okikawa, S., et al., "Development of a coated-wire-bonding technology," *Proc. 39th IEEE ECC*, May 1989, pp. 736-742.

[3]   Ohno, Y., Ohzeki, Y., Yamashita, T., Iguchi, Y., Kanamori, T., and Arao, Y., "Development of ultra fine wire for fine pitch bonding," *Proc. 41st IEEE ECTC*, May 1991, pp. 519-523

[4]   "Ultrasonic welding" in *The Welding Handbook*, Eighth (and earlier also) Edition, V-2, Chapter 25, 1991, pp. 784-812.

[5]   Langenecker, B., "Effects of ultrasound on deformation characteristics of metals," *IEEE Trans. Sonics and Ultrasonics*, SU-13, pp. 1-8, 1966.

[6]   Ramsey, T. H., and Alfaro, C., "The effect of ultrasonic frequency on intermetallic reactivity of Au-Al bonds," *Solid State Tech.*, V-34, Dec. 1991, pp. 37-38.

[7]   Shirai, Y., Otsuka, K., Araki, T., Seki, I., Kikuchi, K., Fujita, N., and Miwa, T., "High reliability wire bonding technology by the 120 KHz frequency of ultrasonic," *Proc. 1993 Intl. Conf. on Multichip Modules*, April, 1993, pp. 366-375.

[8]   Gonzalez, B., Knecht, S., and Handy, H., "The effect of ultrasonic frequency on fine pitch Al wedge wirebond," *Proc. 46th IEEE ECTC*, May 1996, pp. 1078-1087.

[9]   Jaecklin, V. P., "Room temperature ball bonding using high ultrasonic frequencies," *Proc. Semicon/Test, Assembly & Packaging*, Singapore, May 1995 pp. 208-214.

[10]   Heinen, G., Stierman, R. J., Edwards, D., and Nye, L., "Wire bonds over active circuits," *Proc. 44th IEEE ECTC*, May 1994, pp. 922-928.

[11]   Levine, L., "The ultrasonic wedge bonding mechanism: two theories converge," *Proc. 1995 ISHM Symp.* Oct. 1995, pp. 242-246.

[12]   Johnson, W. G., "Yield points and delay times in single crystals," *J. Appl. Phys.* V-33, Sept. 1962, pp. 2716-2730.

[13]   Abbott, D. C., Brook, R. M., McLelland, N., Wiley, J. S., "Palladium as a lead finish for surface mount integrated circuits," *IEEE Trans. on CHMT*, vol. 14, Sept. 1991, pp. 567-572.

[14]   Gaudiello, J. G., "Autocatalytic gold plating process for electronic packaging applications," *Proc. 45th IEEE ECTC*, May 1995, pp. 534-537.

[15]   Inoue, T., and Ando, S., "Stable non-cyanide electroless gold plating which is applicable to manufacturing of fine pattern printed wiring boards," *Proc. 45th IEEE ECTC*, May 1995, pp. 1059-1067.

[16]   Heleine, T. L., Murcko, R. M., and Wang, S-C., "A wire bond reliability model," *Proc. 41st IEEE ECTC*, May 1991, pp. 378-381.

[17]   Shu, B., "Fine pitch wire bonding development using a new multipurpose, multi-pad pitch test die," *Proc. 41st IEEE ECTC*, May 1991, pp. 511-518.

[18]   Owens, N. L., "Wire pull and normality assumptions," 9th Ann Proc. IEPS, Sept. 1989, pp. 595-601.

Bonding of Al Wires to Copper Contacts on PCBs and Alumina Substrates - a Comparative Study

Michał Cież

Research and Development Centre for Hybrid Microelectronics and Resistors

30-701 Kraków, Poland

The use of copper conductive paste and copper laminates in manufacture of electronic circuits is getting wider and wider. It is stimulated both by economical and technical factors. The electric connections between semiconductor chips and substrates are usually realized by wire bonding. One of the most important problem, which must be solved properly is the suitable quality of bonds. .

This paper presents the results of investigations corried out in R. and D. Centre for Hybrid Microelectronics for the optimization process of ultrasonic bonding aluminium wires ($25\mu m$) to copper contacts on ceramic substrates and PCBs. In our experiments different layer structures were investigated on alumina and PCB to find the right parameters of wire bonding process.

The investigations started in two regions:

- selection of optimal bonding parameters (frequency, sonotrode pressure, time) to achieve the highest initial pull - strength of bonds,
- performing of electrical and environmental tests for evaluation of long term quality of bonds and eneabling the verification of wire bonding process parameters.

The measure of bonds quality were figures of pull strength and contact resistance.

The tests showed distinct difference between the choosed bonding parameters for alumina substrates and PCBs, when this some level of initial pull-strength was required.

Evaluation of quality and reliability of ultrasonic bonds and comparision of both technologies were carried out on a base of following trials:

- thermal treatment (48h, $200^{\circ}C$)
- endurance test (1000h, $125^{\circ}C$, 0.1A)
- thermal shocks (-40, $+125^{\circ}C$, 300 cycles)
- optical examination.

G. *Harman and P. Mach (eds.), Microelectronic Interconnections and Assembly,* 167-177.
© 1998 *Kluwer Academic Publishers.*

The paper lists the results achieved and presents them in many diagrams.

On the base of this investigations it was possible to optimize the setting of bonding process parameters in such a manner, that very good bond between copper contact and aluminum wire is possible without additional metal - platting of copper contact.

INTRODUCTION

In many applications where cost is an important factor, copper thick film and copper claded laminates replace gold and Pd/Ag thick film. It arises from relatively low price of copper on the one hand and its improved electrical and thermal conductivity on the other hand. To ensure the high quality of electronic circuits manufactured on the base of copper layers the reliable connection between semiconductor dies and substrate contacts must be performed. From long time it is in use well recognized and established wire bonding process using gold wire in automatic ball - wedge bonding. However in certain situations the use of gold wires causes some disadventages:

- rapid increase of price, when a great number of connections must be bonded,
- the danger of purple plague in the gold-aluminium joints on the semiconductor surface, especially when circuit works at high temperatures or large temperature variations.

When high bond quality is required aluminium wire will normally be used. To perform the bonds the ultrasonic bonding process can be applied. As defined in [4] in ultrasonic bonding, an electrical sine wave of fixed frequency is converted into a harmonic mechanical motion which in turn generates a standing wave in an acoustically coupled steel horn. A bonding tip mounted transversely to the horn transmits the wave trough a bonding wire to the film contact, finally dissipating in the underlying substrate and steel base of bond tool. The elastic wave absorbed at the wire/contact interface induces interfacial fusion, difussion, intermolecular attraction and mechanical deformation. These phenomena results in a temperature increase which can approach the melting points of the materials involved, as well as actual transfer of electrons. Generally, the welding process is divided into three distinct phases:

- firstly the surface cleaning phase,
- secondly the breakdown of the oxide layers,
- thirdly the joining together of the pure metals.

The metals are pressed together until there is less than one atomic lattice distance between them and the resulting weld is characterized by high quality and extreme stability.

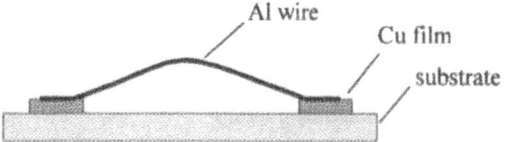

Fig.1. Wedge - wedge bonded wire.

The ideal bond contact on a PCB consists of a layer of copper covered with a 4 to 5 μm layer of nickel on which a 0.1 to 0.2 μm layer of gold has been deposited. This gold layer does not affect the bonding process, only protects the contacts against impurities and chemical reactions that could occur during the manufacturing process.

One of the purpose of this investigation was to finde the answer on the question does the metalization of copper contact on PCB is absolutely necessary for proper bonding process of aluminium wire or not.

EXPERIMENTAL PROCEDURE

Materials.

In this investigation work were used the mixed bonded copper conductive pastes from three manufacturers. The pastes were screenprinted trough 260 mesh stainless screen on a 96% alumina substrate 30x50mm forming a test pattern of rectangular copper contacts (Fig.2a).

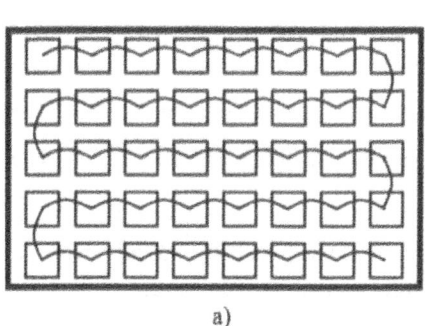

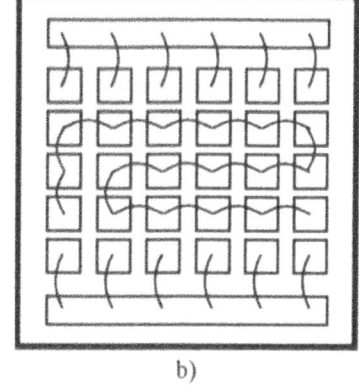

a)

b)

Fig.2. Test substrates a) alumina, b) FR-4 laminate

The firing of thick film layer was performed in multizones (with inert atmosphere) BTU furnace QA41-654.

For PCB test pattern the FR-4 copper laminate 40x40x1.5mm (Fig.2b) was used. The copper pads on the laminate were not metallized. There were mechanically cleaned and washed with deionised water direct before ultrasonic bonding.

The aluminium bonding wire 25μm diameter was obtained from Heraeus as type AlSi1-USB. The essential electrical and mechanical properties of wire are shown in Table 1.

Table 1

Physical Properties of AlSi1 Aluminium Alloy

		Unit	
1	Melting point	$^{\circ}$C	600
2	Thermal Conductivity at 20°C	$\frac{w}{mk}$	195
3	Coefficient of Thermal Exp.	10^{-6}K^{-1}	25
4	Resistance per 1m at 25μm diam	$\frac{\Omega}{m}$	61.15
5	Thermal voltage vs Cu	mV	-0.36
6	Burn - out current at 25μm diam	A	0.39
7	Elongation	%	1-4
8	Breaking load at 25mm diam	mN	110-150

Technological equipment and testing apparature.

Taking into consideration that many parameters affect the bonds quality, we assummed that thick film test substrate with copper pattern should be fired in the inert gas atmosphere with controlled composition. For this reason the automatic gas analyser Simac DS 2500 was applied to BTU furnace. The analyser enabled very precise (0.02ppm accuracy) measurement and control of oxygen content in the furnace.

The bonding operations were carried out on a semiautomatic ultrasonic wire bonder Unitek 8-166. Figure 3 illustrates the bonds of AlSi1 wires to copper contacts.

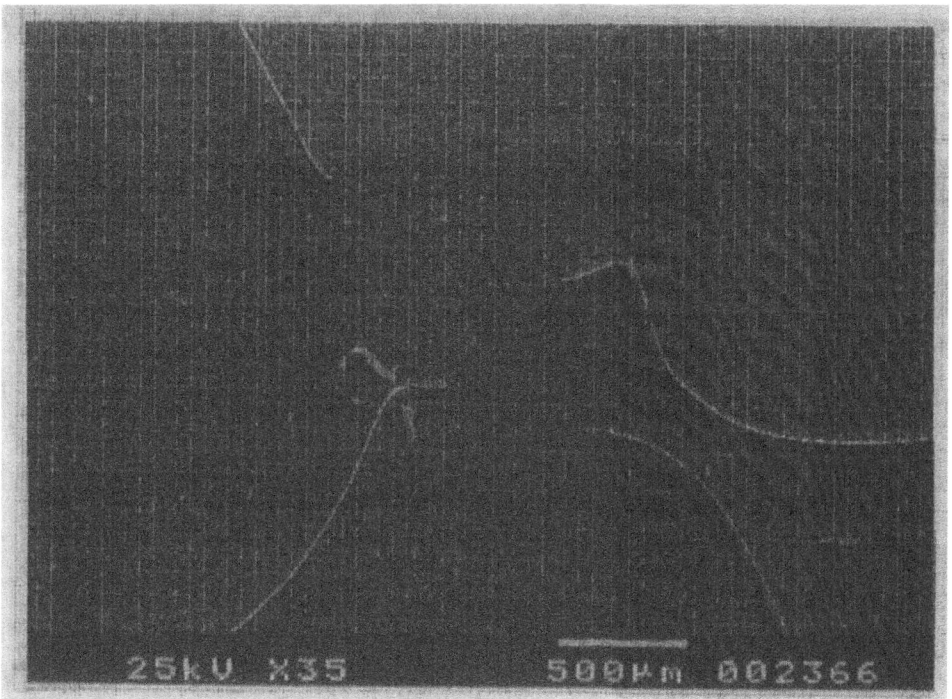

Fig.3. SEM of Al/Si bond to copper contact.

The wedge bonding conditions were examined by applying the loop pull test method with the automatic pull strength tester HT-3. Also the contact resistance of bonds was measured with a four - point method ensured by miliohmmeter Hioki 3226.

RESULTS

Many variables affect the quality and reliability of wire bonds. The most important are operating equipment parameters as:

- ultrasonic power,
- duration of time the ultrasonic energy is applied,
- bond force on the wirebonded junction.

Also significant are the parameters such as:

- surface cleanliness,
- contact layer material,
- substrate material,
- wire uniformity,

- metallurgical properties of wires such as hardness and tendency for intermetallic formation,
- climatic conditions in bonding area.

In order to measure the effect these factors and influence of elevated temperatures have on wire bondability and wirebond reliability the destructive pull test was used and as well as nondestructive test of contact resistance measurements. The experiment have started with screen print and firing of three various copper pastes on ceramic substrates in order to choose the paste which enables to achieve the highest initial pull - strength of bonds. The content of oxygen in the inert gas significantly influences on wire bond pull strength as shown in figure 4.

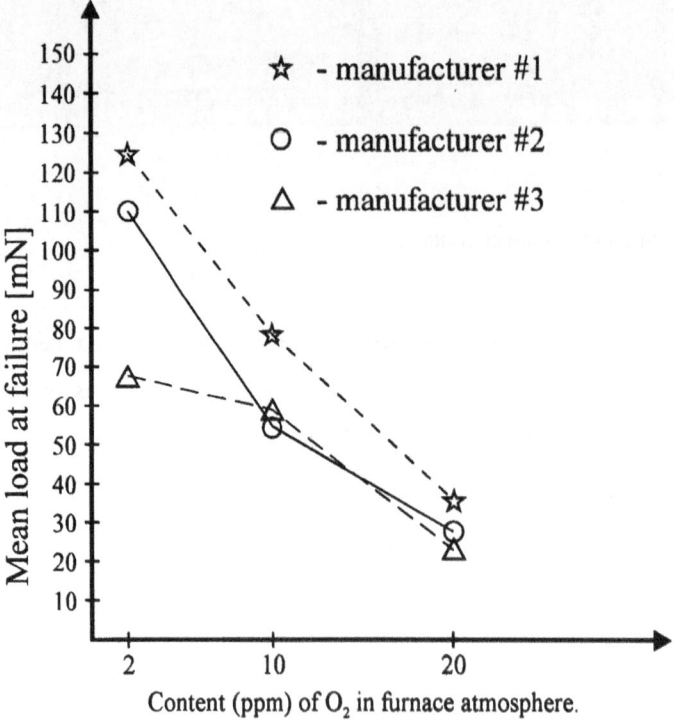

Fig.4. Influence of O_2 content in an inert atmosphere on bond strength.

To next experiment steps the paste from manufacturer No2 was chosen. The effect of bond time, power, ageing at elevated temperature resulting pull strengths are shown in figures 5, 6 and 7 respectively for both ceramic and PCB substrates.

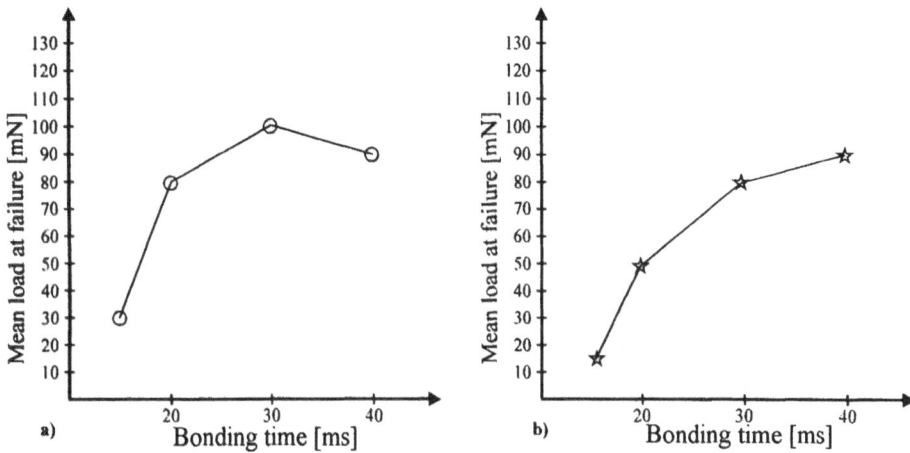

Fig.5. Influence of bonding time on bond strength
   a) ceramic, b) FR-4 laminate

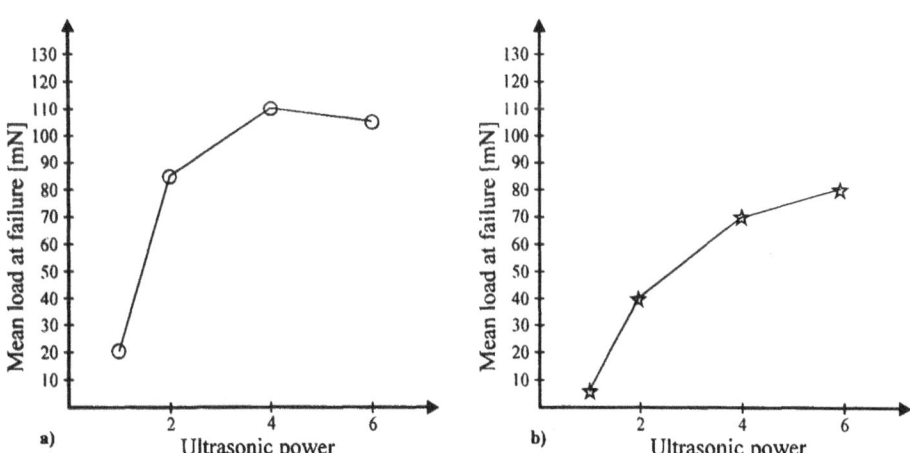

Fig.6. Influence of ultrasonic power on bond strength
   a) ceramic, b) FR-4 laminate

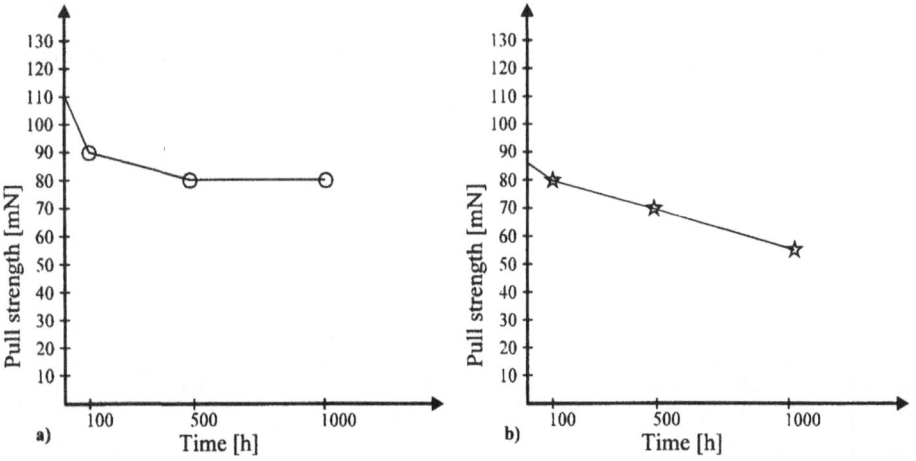

Fig.7. Pull strength of bonds aged at 125°C

a) ceramic, b) FR-4 laminate

Each data point on the diagram represents the average of 50 to 70 bonds. Base - line bonder parameters used for the parameter effect study are shown in Table 2.

Table 2

Base-line Bonder Parameters Used for the Parameter Effect Study

Parameter	Ceramic	PCB
Bonding Force	200mN	200mN
Time	20ms	30ms
Power	4.5 units	3 units

In every experiment wedge bond parameter of interest was varied sequentially and bonding performed at each new value. After completing a range of setting the initial conditions were reset and another bond parameter was chosen and varied in a similar manner.

Diagram on figure 8 showes the influence of thermal treatment on pull strength and contact resistance for bonds on ceramic substrate.

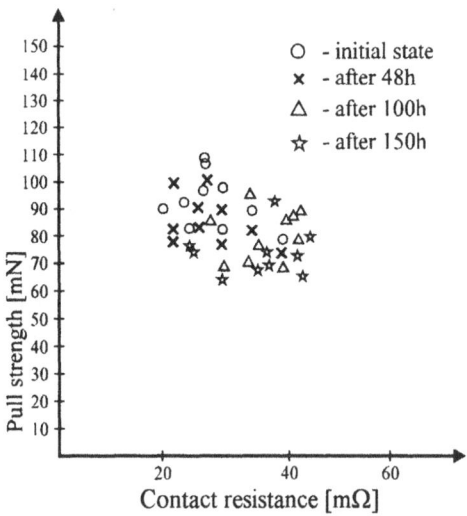

Fig.8. Influence of thermal treatment (150h, 200°C) on pull strength and contact resistance.

The slowly in time increas of contact resistance value and pull strength value decreas is observed. In comparision to aluminium wire wedge joints to gold layer [7] the pull strength decreasing is significantly slower. The lack of Kirkendall voids can be an essential explanation of this behaviour.

On the figure 9 the influence of long time electrical load and elevated temperature on pull strength is shown.

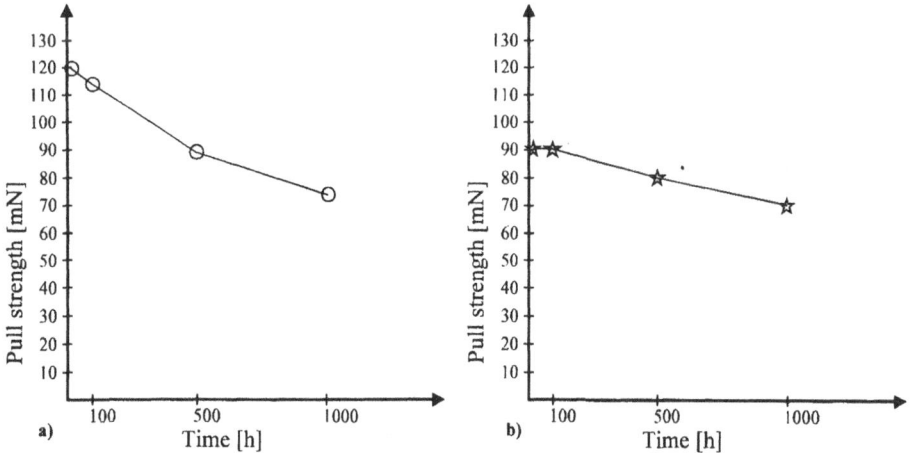

Fig.9. Influence of electrical load and temperature on pull strength (I = 0.1A)
   a) ceramic, b) FR-4 laminate

The tendency of pull strength changes are similar for ceramic substrates and PCBs, only the levels of values are different.

Figure 10 shows the morphology of pull tested wire.

Fig. 10. Morphology of pull tested wire bonds on alumina substrate

The presented datas were completed by examine the influence of 300 thermal chocks (-40, +125°C) on pull strength of bonds on ceramic substrates and FR-4 laminate. The results were

Pull strength average	Initial	After thermal shosks
Alumina substrate	110mN	70mN
FR-4 laminate	95mN	70mN

CONCLUSION

To obtain good wire bonds on both ceramic substrate and FR-4 laminate the optimising procedure in establishing of bonder parameters should be carried out, separately for both materials. It was possible to establish the bonding windows with strictly chosen power, force and time.

The carried out bonding experiments showed that wire bonding of aluminium wire to non metallised copper layers is possible, but on condition that wirebonding process is performed directly after exact cleaning of copper surface.

It seems that contact resistance measurement could be a perspective method for nondestructive testing of bonding quality to contacts on substrate.

ACKNOWLEDGMENTS

The author would like to express the deepest thanks Mr.Gandurska and her team for valuable research on copper thick films. Special thanks to Mr.Boba for his major contribution in bonding operationes and bonds testing, much appreciation is extended to Mr.Polak for printing and graphic arrangement of this paper.

REFERENCES

1. "Badanie współpracy past i warstw przewodzących na bazie Cu z pastami i warstwami dielektrycznymi w układach wielowarstwowych". Report from research work nr 773669203. OBR Kraków 1994
   J.Gandurska, S.Nowak, I.Śnieżyńska and al.
2. T.J.Green, R.G.Lausby, "Using DOE to Reduce Costs and Improve the Quality of Microelectronic Manufacturing Processes". The ISHM International Journal of Microcircuits & Electronic Packaging No 3/1995
3. M.Sheaffer, L.R.Levine, B.Schlain, "Optimizing the Wire-Bonding Process for Copper Ball Bonding, Using Classic Experimental Designs". IEEE Transaction on Components, Hybrids and Manufacturing Technology, No 3, September 1987
4. T.A.Wassick, L.Economikos, "Open Repair Technologies for MCM-D", IEEE Transactions for Components, Packaging and Manufacturing Technology, Part B: Advanced Packaging, No 1, February 1995
5. Ch.A.Harper, Handbook of Electronic Packaging, Mc Graw-Hill 1969
6. R.L.Shook, Thermosonic Wire Bonding on Single-layer Polymer Hybrid Integrated Circuits (POLYHICs), Hybrid Circuits No 13, May 1987
7. J.F.Haag, On the Resistance Increase of Gold-aluminium Wire Bonds, Hybrid Circuits No 26, September 1991.

MICROWELDING OF LEADS TO THE FILM STRUCTURES WORKING AT ELEVATED TEMPERATURES

T. Pisarkiewicz, T. Habdank-Wojewodzki and M. Ciez[*]

Dept. of Electronics, University of Mining and Metallurgy

[*] Res. and Dev. Center for Hybrid Microelectronics and Resistors

Krakow, Poland

Abstract

The authors investigated microwelding of Pt wires and ribbons to PdAg, PtAg and Au thick film contact pads deposited in a form of single, double or triple layer structures. The weldings were carried out using modern Hughes MCW 550 microwelder. The welding current pulse power and duration as well as welding head electrodes distance and force were varied to obtain the maximum adhesion of the lead. In order to evaluate the contacts performance mechanical strength tests and aging tests were carried out. Scanning electron microscope investigations of the connection area allowed for optimization of the contact pad thickness and lead wire diameter.

Obtained results for Pt wire of 0.08 mm in thickness welded to the double PdAg/Au thick film contact pad seem to fulfill the requirements of durable and stable electrical connection to the microstructures working at elevated temperatures.

INTRODUCTION

The common problem for thin and thick film microstructures, mainly sensors, working at elevated temperatures (of about 350°C or higher) is the connection of lead wires [1,2]. Increased temperature very often requires the structure to be suspended on the leads to avoid power dissipation. In effect, the heated structure working in aggressive atmosphere must be connected to the pins using ambient resistant and mechanically durable leads and contact pads.

Connection of the sensor microstructure to the measurement electronics can be achieved in a number of ways, including soldering, wirebonding, snap-in connection and welding [3]. Elevated temperature requirements promote microwelding as a promising technology since a selection of wire diameters or ribbons can be employed and the increased temperatures do not pose undue problems.

G. Harman and P. Mach (eds.), Microelectronic Interconnections and Assembly, 179-184.

TECHNOLOGY OF THE TEST SAMPLES

Electrical contacts for testing experiments were fabricated by depositing thick films onto alumina substrates 0.6 mm in thickness. The thick films were screen printed and fired forming single, double or triple layer structures. The structures thus obtained are specified in Table I.

Table I. Thick film structures used as contacts in microwelding experiments.

Single layer	double layer	triple layer
PdAg	PdAg/PdAg	PdAg/PdAg/PdAg
PtAg	PtAg/PtAg	PtAg/PtAg/PtAg
	PdAg/PtAg	
	PdAg/Au	

The successive layer was printed after firing of the previous layer. Typical noble pastes compositions were used for the printing. The topology of the contact pads is shown in Fig.1. The lead wires were microwelded to the contact pads in a manner enabling pull strengths tests.

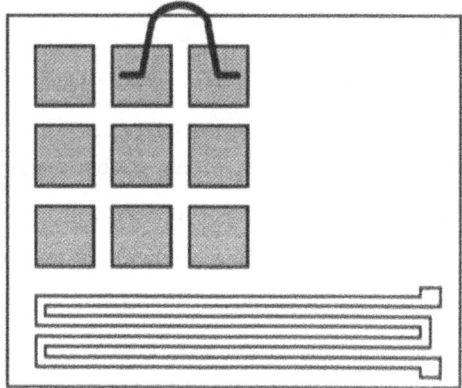

Fig.1. Layout of the test substrate with contact pads and welded Pt wire (0.08 mm in thickness)

LEAD CONNECTION PROCEDURE

The platinum wire leads were microwelded to the contact pads using Hughes MCW 550 welder. The welder was equipped with microweld head using parallel gap welding electrodes as shown in Fig.2.

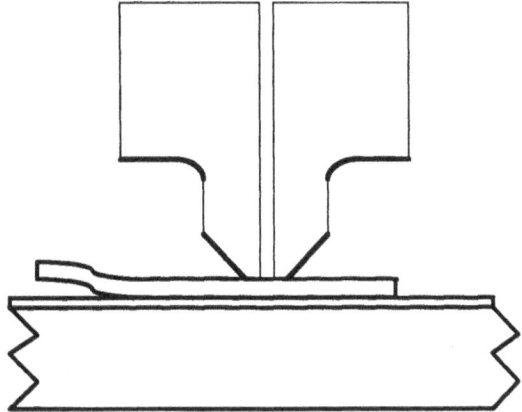

Fig.2. Parallel gap welding of a wire to the thick film contact pad. The microweld head with vertical electrode clamping was used.

In order to optimize the quality of a joint, the welding conditions were established experimentally for a given welded wire and contact pad used. For the 0.08 mm Pt wire and PdAg contact pad the average welding conditions were as follows:

- welding voltage 0.8 V
- welding duration 500 msec
- electrode gap 0.3 mm
- welding force 700 gm.

The conditions had to be adjusted individually when the contact pad changed its resistance.

PULL STRENGTH TESTS

The pull strengths tests were carried out automatically by measuring the pulling force as shown in Fig.3.

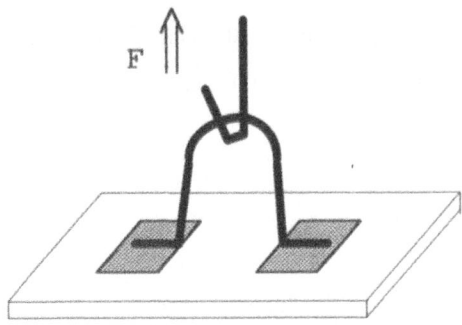

Fig.3. Pull strength test evaluating the adhesion of a wire to the selected contact pad.

The tests have been done for all contact pad structures with microwelded Pt wire. In some cases the adhesion was so good that the wire broke before pulling force damaged the junction.

The influence of elevated temperature on the junction quality was also investigated. In this case the samples were aged at 350°C for 72 hrs. The results of pull strength tests for the investigated connections before and after aging are given in Table II.

Table II. Pull strength tests for the connections: contact pad /Pt wire carried out before and after aging of the structure at 350°C for 72 hrs.

Type of contact		Pull strength (G)	Pull strength after aging (G)
single layer	PdAg	40	40
contact	PtAg	40	40
double layer	PdAg x 2	50	100
contact	PtAg x 2	50	80
	PdAg/PtAg	150	150
	PdAg/Au	wire break	wire break
triple layer	PdAg x 3	150	70
contact	PtAg x 3	150	70

As can be seen from the table the best connections were obtained for the double layer contact pad PdAg/Au and microwelded Pt wire. The aging process during 72 hrs had no influence on the junction durability. Mechanically strong connections enable the suspension of the whole sensor structure to avoid power losses during operation of the sensor at elevated temperatures, as shown in Fig.4.

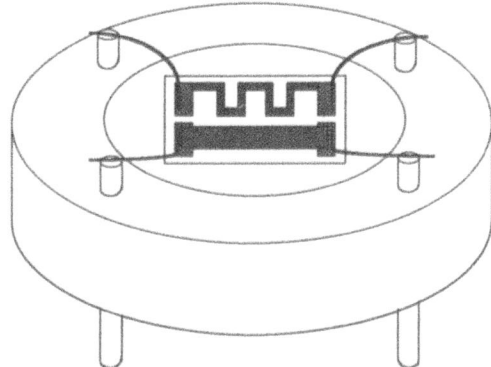

Fig.4. Gas sensor working at elevated temperature [4].

The quality of the microwelded connections was also investigated by using Scanning Electron Microscopy (SEM), Fig.5.

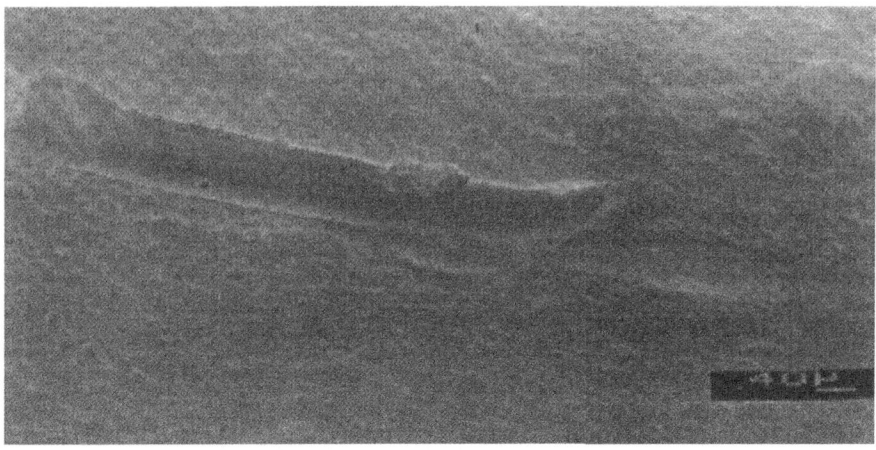

Fig.5. SEM picture of the contact area: PdAg/Au contact pad and broken Pt wire after the pull strength test.

As can be seen from Fig.5 the adhesion of the double layer PdAg/Au to the alumina substrate is so good that during the pull strength test the wire breaks without destroying the contact area.

CONCLUSIONS

Elevated temperature working conditions of the sensor structure, of order 400^0 C or higher, promote microwelding as a technique for connection of lead wires.

Thick film contact pad on a ceramic substrate and a Pt wire can be joined firmly after selection of appropriate welding parameters.

The most durable connection obtained by the authors was:

double layer PdAg/Au as a contact pad and thin (0.08 mm in dia.) Pt wire as a lead.

REFERENCES

1. S. Nowak and T. Pisarkiewicz, "Thick and thin film heaters for gas sensors", Proc. SPIE , Int. Conf. of Microelectronics, 21-23 Sept. 1992, Warsaw, Poland, vol. 1783, p.315.

2. T. Pisarkiewicz, S. Nowak, B. Kic, D. Lusniak-Wojcicka and J. Gandurska "Thick film technology in manufacturing of heaters and contact pads for oxide semiconductor gas sensors", Proc. 16th Conf. of ISHM Poland Chapter, Krakow, Sept. 28-29, 1992

3. S.J. Stein, C. Huang, T. Grunstein and S. Sykora "Thick film sensors for methane detection", ISHM 93 Conf. Proc., p.1.

4. T. Pisarkiewicz, M. Jachimowski, S. Nowak, T. Stapiński, K. Zakrzewska and T. Habdank-Wojewódzki, "Thin film semiconductor ozone sensor", Int. Conf. on Metallurgical Coatings and Thin Films ICMCTF96, April 22-26, 1966, San Diego, CA.

NEW TRENDS IN THE INTEGRATION AND EDUCATION IN MICROSYSTEMS TECHNOLOGY *

Prof. Dr. Juraj Banský
Technical University Košice
Slovak Republic

During the last 10 years, the intellectual foundation for the field we call materials science and engineering began to take shape and to achieve recognition. In the field of materials science and engineering as well as in the microelectronics industry the preparation of advanced materials is in nearly all cases a step in the process from raw material to electronic end products. This step is very important and may be the bottleneck or the essential discovery in an innovation. Generally can an advanced material be decoupled from its past and future, i.e. the processes of synthesizing the material and of shaping it into its final state in the device that forms the final end products. This field has achieved its special self "push" and "pull" mechanism, in which "push" is coming from increasing ability to make anything one can think of, and "pull" comes from electronic devices in which the specification are becoming more and more stringed.

The fine structures in microelectronics devices have decreased in size less than 0.25 um. Technology development over the last couple of decades boils down to the more efficient implementation of the central hardware concepts devices and the connection between them. The integration level of chips continues. The technologies and facilities are in the development stage, and in the year 1996 prototypes of 16 to 64 Mb DRAM chips, CMOS circuits with 600K logic gates and bipolar circuits with 30K are expected. This fact leads to the continuous increasing the input/output pincount for each new generation, that means to solve the problems with packaging and interconnections.

One of the main trend for the future will be "integration": how to solve the problems which are connected with shrinking dimensions in microelectronic circuits, microstructure engineering, development in material science and problems with the production of many so-called high-tech products. This means the preparation and the application of the advanced materials will have to be an integrated process of several at present different technologies. Such approach requires coordinate interdisciplinary efforts spanning basic science to real devices based on the synergism between advanced instrumentation, material synthesis and subsequent application.

To realize such new trends in materials science and engineering, in microelectronics and in microsystems technology our society has to be much more active in the education at both the undergraduate and graduate levels, mainly in the area of synthesis and processing. Industry, universities and research institutions should take a more active role to stimulate interaction on each level. We must significantly increase the number of well-educated, well trained persons who are able to cooperate in multidisciplinary teams and to identify the perspective needs and opportunities in material science and engineering with respect to the emerging technologies. The goals of this activity should be the protection of our nature, the safety of human beings and a continuous improvement of our standard of living.

---

* Paper was presented. Written paper never received.

G. Harman and P. Mach (eds.), Microelectronic Interconnections and Assembly, 185-186.
© 1998 *Kluwer Academic Publishers.*

MCM Implementation of a 2.5 Gb/s ATM switch

Pierre Plaza, Jose Luis Conesa
I+D Telefonica
Madrid, Spain
Phone: 34 1 337 4359; Fax: 34 1 337 4212
Email: pierre@tid.es.

Abstract

The ATM switch is a 4 input per 4 output device with a total througput of 10 Gb/s. It features 24 bare dies (ASICs), 8 SOJ commercial memories and capacitors. The module dissipates a power of 150 to 200 W. It is being implemented in 2 different MCM technologies: MCM-L and MCM-C with identical dimensions and pining.

The paper will focus on the electrical design, the MCM substrate technologies, the KGD solution using bumpless TAB, thermal management, board interconnect solution using an interposer, module test methodology and a cost analysis.

Due to its large size, high speed, high power, this switch solution represents a challenge. This development is a part of CHIPPAC : Cost Effective High Performance Packaging for Electronic Systems, an Esprit (European Union sponsored) Programme.

Introduction

The development of process technology has enabled a new generation of printed circuit boards to emerge to meet the demands of electronic systems. Evolution of photolithography plating and mechanical drilling has reduced feature sizes down to those that match the I/O pitch of silicon devices. New techniques show a new route forward laminated board limitations and new constructions leading to lower costs for very high density interconnect competing in the MCM market for information technology applications.

High speed, high power, high performance multichip modules traditionally used MCM-C or MCM-D technologies. The need of a low cost effective solution for the development of the ATM switch presented in the following section, has lead to 2 MCM solutions where area is reduced drastically on the application equipment, increasing as well the performance. The 2 different MCM technologies considered for the switch are: MCM-L and MCM-C LTCC.

The following issues will be described: the electrical designs which represent a challenge in terms of assembly yield, the substrate technologies (a comparison), the use of bumpless TAB as a KGD solution, the thermal management and board interconnect solutions.

ATM Switch Architecture

The switching function is performed using self routing spatial techniques [1,2]. An input cell is routed to an specific output depending on the contents of the cell header (VPI/VCI). Therefore several cells may be routed at the same time to a given output. To solve the contention problem, queues of cells are implemented at the outputs of the switching fabric, on the ICM circuit [3].

In general ATM is so attractive because of its flexibility and bandwidth on demand philosophy. The switch that is here described obtains statistical gain by incrementing substantially the link bit rate with respect to the typical source rates. Fig. 1 shows the architecture of the 4x4 high speed 2.5 Gb/s per I/O switch.

The switch I/O processing is implemented by the CMC circuit which can be configured in two modes : the CM or MC mode as input and output processor respectively (see the corresponding blocks in Fig.1). In the CM mode, ATM cells are paralleled and sliced, producing information packets called Microcells (Fig.2) at the CM outputs. In the MC mode the opposite function is performed.

The switching core is scaleable and constructed by interconnecting several ICM ICs. Cell parallelism is used to increase the throughput : 8 bits wide ATM cells are converted to a 32 or 64 bits in parallel format that is processed by the switch core. These formats

G. Harman and P. Mach (eds.), Microelectronic Interconnections and Assembly, 187-191.
© 1998 *Kluwer Academic Publishers.*

188

are programmable on the CMC and lead to two possible throughput values: 1.6 Gb/s and 2.5 Gb/s respectively. The CM and the MC blocks parallel I/O (Microcell) interface to the switch fabric can thus have 8 or 16 four bit active buses.

The switch incorporates interesting operation and maintenance (OAM) functions: statistical measures, traffic control, insertion and extraction of ATM cells via the external controlling microprocessor .

L (bits)	W (active I/O's)	Micro-cell length (bits)	Frame length micro cells	Equivalent throughput per I/O (Gb/s)
32	16	20	96	1,55
64	8	12	160	2,60

Table 1

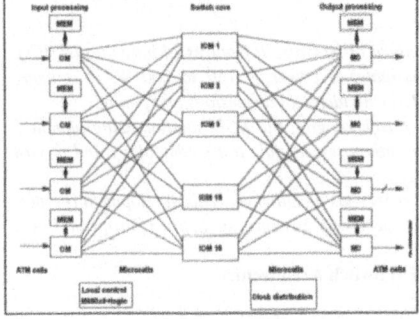

Figure 1: Functional architecture of the high speed switch demonstrator

W		4
(L/K) = 4	Cell slice	Routing tag

Microcell = Routing tag + Cell slice

16 bits of routing tag:
1 bit: microcell assigned/unassigned
1 bit: microcell
12 bits: destination
2 bits: destination coding

Figure 2: Microcell format.

Switch Core

The core of the high speed switch, is based on the ICM, an ASIC developed within a former ESPRIT project. The ICM main function is to switch microcells (output buffer architecture). It was initially designed to provide 8 inputs and 8 outputs, but for implementation reasons the final architecture is 8 inputs and 4 outputs, being possible to construct a 8x8 microcell switch with two ICMs.

Every incoming microcell flow is provided together with its own clock, but after the input FIFO blocks, a single common processing clock is used. Input clocks can have different phases and even their frequencies can vary up to 0.1%. At the output of the circuit, the microcells are generated with their clocks obtained from the common processing one.

Each logical output FIFO is implemented with 8 physical ones (1 Kbit each). Eight microcells can be written simultaneously (each into its corresponding FIFO), but only one FIFO is enabled for reading at a given microcell time. Every output FIFO is able to store up to 168 microcells (12 bits long, 4 bits wide) for L = 64 or 96 microcells (20 bits long, 4 bits wide) for L = 32. One block that manages the FIFO filling level, generates the sub-FIFO address, discards ordinary microcells when filling level exceeds a programmable threshold and generates the synchronisation signals for the output microcell flow control (frame and microcell indications) is included.

The ICM layout and main statistics of the realised IC are reported in figure 3.

Complexity: 650 kTransistors
Speed: 74 Mhz
Die Size: 12.8x12.1 mm^2
Technology: 0.7 Micron CMOS, 2 level metal
Test Fault Coverage: 93,76%
Operating temperature: 0 to 70 degree Celsius
VDD values: 4.75V to 5.25V
Power dissipation: less than 3 watt

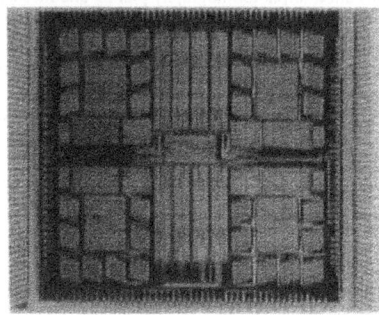

Figure 3. ICM die picture and characteristics

Input and Output Processing

The switch function performed in the switch core is based on the routing tag that contains all the required information to select the output or outputs to which the microcell (switch core data unit) must be delivered. Taking into account that in an ATM switch this function will operate on the content of the cell header (VPI/VCI) and the switch data units must be ATM cells instead of microcells, some input and output processing is necessary.

At the inputs, ATM cells (53 bytes CCITT format or 56 bytes -53 bytes CCITT format + 3 bytes header extension-) are received (8 bit parallel, 311.04 MHz). These cells are internally expanded to 64 bytes/cell, parallelized (64 bits) and sliced (each cell slice is 8 bits long, 4 bits wide). Additionally the cell is identified (VPI/VCI) and a routing tag (used in the switch core) is obtained from a external memory as a function of the cell identifier (VPI/VCI). This routing tag is replicated and appended to each cell slice to form a microcell (cell slice + routing tag). All these functions (and some additional ones such as cell insertion from the microprocessor, cell counting, ..) are incorporated in the input processing basic component: the Cell to Microcell converter (CM).

Each output processing building block (Microcell to Cell converter -MC-) receives the 16 microcells generated from every incoming cell and directed towards the output the MC is associated to. Microcells are reassembled into cells (monitoring the process in order to detect mistakes), cell extension (8 bytes) is discarded and serialised (8 bits), obtaining at the output ATM cells with the same incoming format (53/56 bytes, 8 bit parallel, 311.04 MHz). Additionally each cell is identified (VPI/VCI) and a new CCITT header (header translation) is obtained from a external memory as a function of the cell identifier (VPI/VCI). All these functions (and some additional ones such as cell extraction to the microprocessor, cell counting, ..) are incorporated in the MC.

Both CM and MC functions (but not simultaneously) are implemented in a single ASIC: the CMC. According to the operating mode, it generates microcells starting from ATM cells (CM mode) or assembles microcells into ATM cells (MC mode), supporting in both cases 2.5 Gb/s of throughput.

The CMC layout and main statistics of the realised IC are reported in figure 4.

ECL part complexity: 8.3 kTransistors
ECL part speed: 311.04 MHz
CMOS part complexity: 300 kTransistors

CMOS part speed: 74 MHz
Die size: 219.48 mm^2 (14.28mm*15.37mm)
Technology: 0.7 Micro BiCMOS, 2 level metal
Design for testability: full scan + custom BIST
Power consumption (max.): 8.3 W

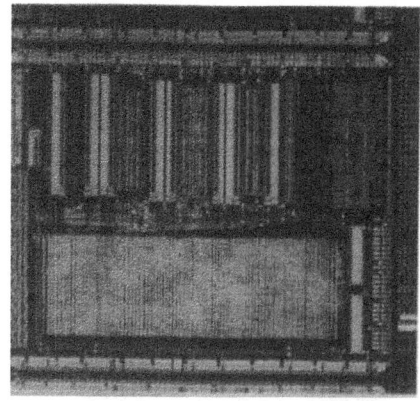

Figure 4: CMC Layout and main characteristics

MCM Technological Solutions

The laminated MCM is built with a High Density Printed Circuit Board (HDPCB) BULL Substrate Technology. Whereas the Ceramic substrate is implemented with a LTCC technology from Elektrokeramic GmbH. There were 1500 nets to route on the module. The Substrate size is 140x140 mm^2 in both cases. Signal lines are adapted to 50 Ohms.

Eighteen layers were needed to complete the layout for the MCM-L implementation and 21 layers for the LTCC technology. Table 3 illustrates the main features of both processes. These processes are able to use IC's interconnected by wire-bonding or fine pitch TAB. The laminated module has been built with a BT resin to overcome the need for high temperature resistant material. PCB finish include tin/lead for TAB and SMT devices and a nickel/gold selective plating for connector pads.

A flying probe tester was used in order to select known good substrates before components assembly.

The modules are mounted on a PCB (Fig5) which is part of a system build with a rack (Fig.6) that is used to implement the ATM switch demonstrator. In the rack there is also a control board where the controlling microprocessor is implemented. The latter takes care of the operation control, test and maintenance functions on the switch.

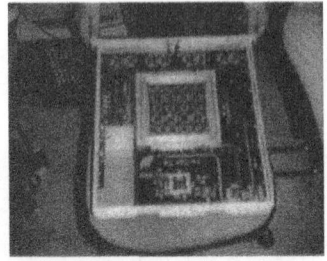

Figure5: PCB used to mount the ATM switch

Figure 6: Rack of the switch demonstrator

Thermal Management

The modules will dissipate between 150 W and 200 W. Heat dissipation is done by a thermal contact between the back of the dies and a metal plate which support a heat sink. Measured thermal resistance between dies junction and thermal plate is under 1,5°C/W. A Metal plate fixture on the substrate allows to press the substrate against the connector. Figure 7 shows the principles of thermal assembly.

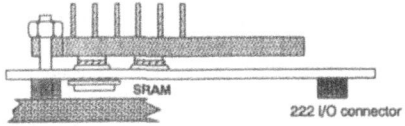

Figure 7: HAM Thermal Assembly

KGD Solution

In order to ensure a high assembly yield, it was decided to mount the ASIC bare dies on TAB and to test them before assembly. The latter allows to increase assembly yield and thus reduce the assembly cost. A bumpless TAB process was used on specially designed TAB lead frames [2]. Figure 8 shows an

estimate of assembly cost of HAM if using wire-bonded dies compared to TAB mounted dies.

Bumpless TAB process uses a single point bonder which bonds directly TAB leads on the aluminium pads of the silicon chips. After TAB bonding (Inner Lead Bonding - ILB), the tape lead frame is mounted in a standard JEDEC carrier which allows a fully functional test of dies before the assembly.

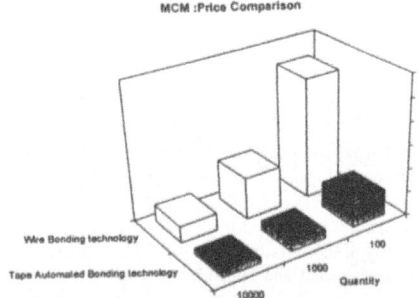

Figure 8: Cost Comparison between Wire-Bonding and TAB for HAM Assembly

Board Assembly process

The substrates were designed with SMT components mounted on one side and TAB components on the other side. Fig. 9 shows a picture of a partially assembled substrate.

Figure 9: HAM Substrate

Interconnect solution between the Substrate and the system board

HAM uses an especially designed surface mounted connector (SURFCHIP) developed by FCI (FRAMATOME Connectors International). This 220 I/O's connectors (Fig. 10) exhibit improved electrical and mechanical performances. The electrical connection is obtained by pressing the MCM substrate on the top of the connector.

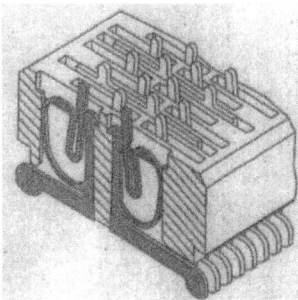

Figure. 10: Surfchip Connector

Comparison between the two substate technologies

The following two tables show the properties of the substrates and design features. LTCC represents the best solution in term of thermomechanical properties. The BT resin substrate represents the best substrate for electrical performances. The organic substrate for this application is the cheapest solution.

	BT D glass	LTCC
Mechanical Data		
Young modulus (GPa)	33 (X), 17 (Y)	91,8
Poisson ratio	0,37	0,276
Physical data		
densitiy (g/cm3)	1,677	3,095
CTE (ppm/°C)	15 (X), 18 (Y)	4,7
Electrical data		
Dielec. Const.	3,7	8
Loss tang;	0,01	0,001
Charac. Imp. (Ohms)	50 +/- 5	50 +/- 5
Line Resist. (Ohms)	0,06 -0,09	0,11
Line Capa. (pF/cm)	1,1	2,9

Table 2: Substrate Characteristics Comparison

	HDPCB	LTCC
Line Width	75 μm	145 μm
Line Space	125 μm	100 μm
Buried vias Diam.	100 μm	117 μm
Blind vias	No	Yes
PTH Diam.	200 μm	117 μm
Via grid	1 mm	0,5 mm
Routing density	23 500	20 000

Dielec.Mat.	BT - FR4	Ceramic
Dielec.Const.	3,7 - 4,8	7,8
Relative cost	*	**
Electrical perf.	**	*

Table 3: Substrate Design and Cost Comparison

Conclusion

The ATM switch MCM design described on this paper represents one of the most challenging solutions implemented up to now for this kind of applications.

The MCM-L has shown to be the cheapest solution in comparison with the MCM-C technology. Large substrates are more suited for organic technology and offer higher track conductivity with a lower dielectric constant. Bumpless TAB can be considered as a true KGD and can be considered as a reliable die attach process.

MCM technology is definitively attractive not only for automotive and process control applications but also for the Telecommunications market and should be considered as an attractive cost effective solution.

Acknowledgments

The authors wish to thank their CHIPPAC colleagues for their support in submitting this paper.

References

[1] F.A. Tobagi, "Fast packet switch architectures for broadband integrated services digital networks ", *Proceedings of the IEEE*, vol. 78, no. 1, January 1990.

[2] R. Wulleman and T. Van Landegem, "Comparison of ATM switching architectures", *International Journal of Digital and Analog Cabled Systems*, vol. 2, October - December 1989.

[3] L. Merayo, P. Plaza, G. Piccinini., "Technology for ATM multigigabits/s switches", *Globecom'94*, vol.1, pp. 117-122, Nov. 1984.

[4] P.Courant, G.Dehaine, " TAB goes bumpless ", Proceedings of the 10th *European Microelectronic Conference*, May 95, Copenhagen Denmark

[5] BPA report, " Worldwide multichip module technologies & markets through to the end of this decade ", June 93, Nice France.

Processing and Performance of High Dielectric Permeability Thin Anodised Films in Multilayer Structures for MCM(D) High Speed Digital Applications

Dr. Radosvet G. Arnaudov -- Technical University -- Sofia, Bulgaria
Nikola St. Yordanov -- Technical University -- Sofia, Bulgaria
Dr. Sc. Phillip Iv. Phillipov -- Technical University -- Sofia, Bulgaria

ABSTRACT: A comparatively untraditional manufacturing MCM(D) technology has been investigated for the purpose of achieving simplified processing, high reliability and new performance. It covers deposition of separate dielectric insulating layers of the multilayer MCM structure, using the methods of electrochemical anodisation of Al or Ta. This process has been developed for producing multilayered MCM in order to reduce the number of processing step, compared with other conventional technology routes. Also application of dielectric materials with small thickness and high dielectric permeability suggests still greater reduction of element dimensions, but strongly affect high frequency signal propagation, transmission line properties and parasitic crosstalk of interconnection. That is why a precise CAD simulation, control and analysis are carried out, different test structures are developed and some recommendations for engineering design tools are given. Practical test measurement results are also discussed.

1. INTRODUCTION

Thin film multichip module (MCM-D) technology offers significant weight, volume and potential cost savings coupled with higher performance for high-speed digital and analogue electronic systems. A critical issue in MCM manufacturing and research is achieving high reliability with cost-effective processing. Reliability is a critical point since many applications for MCM require great heat dissipation, operation from dc. to gigahertz frequencies, and high performance functionality over wide extremes of temperature, shock, vibration, humidity, and atmospheric pressure. MCM are also needed for high-speed conduction network in order to complement both CMOS-base digital computing and processing systems with advanced ECL and GaAs-based microwave integrated circuits [1].Any detailed discussion of high-performance packaging for the next generation electronic systems must include both hybrid wafer scale integration (H-WSI) and monolithic wafer scale integration (M-WSI), but our emphasis will be exposed onto the H-WSI, because of its versatility, combined technologies, performance of unique or custom specific abilities. The definition of a H-WSI is an electronic system, composed of multiple discrete unpacked integrated circuits and passive devices, connected by means of an organic or inorganic thin film deposited interconnect structures with at least 10 times the complexity of available device technologies [2]. In approach to MCM-D it's clear that thin film dielectrics are deposited on a base substrate and thin film metal connection traces are formed on the top of each insulating layer.

193

G. Harman and P. Mach (eds.), Microelectronic Interconnections and Assembly, 193-212.
© 1998 *Kluwer Academic Publishers.*

Table 1. Typical characteristic of MCM-D technologies.

Feature	Typical range
Line width	20-40 mm
Line pitch	50-100 mm
Dielectric thickness	
Polymer	10-30 mm
Inorganic	2-15 mm
Metal thickness	2-5 mm
Via diameter	15-40 mm
Number of metal layers	3-7
Number of active chips	10-50
Number of I/O	100-1500

The dielectric layers are predominantly some polymers - polyimide, benzocylobutene, polyolephene, PTFE, epoxy, etc.; in some specific applications - some nonorganic, such as polycrystalline SiO_2, Si_3N_4, SiON, Al_2O_3. Metallization is generally sputtered aluminium, plated copper or gold with proper adhesion barrier to prevent chemical reactions on the metal-dielectric interface. The base interconnect carrier substrates are usually a planar silicon wafer or some ceramic substrates - Al_2O_3, AIN, BeO, SiC, diamond, sapphire. When carefully studying the chemical and physical parameters of these materials it's obvious that there exist some mismatches and drawbacks:
- A great difficulty is to match the thermal expansion of the base substrate and the thin insulating films, which can cause stress and cracking at higher power ratios.
- Sometimes the polymers possess significant moisture uptake (2-6%) which can result in severe electromigration process through the vias or between adjacently situated parallel conduction lines leading to a short circuit and failure.
- Application of adhesion deposited layers for the metals (Au, Ag, Cu) on the interface metal-insulator, is not always an advantage - it may cause degradation of the Q-factor of microstripe lines and great increase of the insertion losses, due to the skin-effect for microwave frequencies.
- When the MCM is exposed to severe thermal shock and vibration influences, different types of ceramics or silicon are not always the proper solution, because of their fragility and possibility of nonrepairable damages.
As high speed GaAs and SiGe chips continue to advance, the MCM's operation frequency range increases, there will be a need of new improvements in thin film passive components' performance and manufacturing (signal and termination resistors, signal and decoupling capacitor, matching networks, couplers, bias filters, etc.) The emerging development trends in MCM-D show the following tendencies:
- Resistors, capacitors, inductors will be processed within the multilayer interconnect structure.
- MCM processes of assembly and chemistry must be comparable with the passive elements.
- Implementation of high dielectric permeability thin films for larger capacitor's values.
- A Tao_xN_{1-x} thin film resistor process is being pursued because of its stability, linear

TAR and inherent physical and chemical inertia to MCM processing chemistry.

- Implementation of microstripe planar components with distributed parameters, especially for matching networks. To reduce their geometric dimensions it's desirable to use insulating multilayer materials with high dielectric constants (10 - 50).

-The increasing densities and clock rates of the leading edge MCM technologies will push the thermal cooling requirements of the base carrier substrates, or looking to some metals or alloy components as base material plates.

In order to satisfy all these on-going and future trends in MCM-D development, we have proposed a new approach to the metal layer structure organisation (Fig.1), and technology production process, which are not fully depicted in the literature yet (Fig. 2). The main advantages may be described as follows:

- Application of vacuum sputtered Al (or Ta) for the conducting transmission lines, contact pads, ground planes, etc. (no need of adhesion barrier layer on the dielectric-metal interface).

- Formation of planar thin film AL_2O_3 (At_2O_5) capacitors and insulating higher dielectric layers by means of electrochemical selective anodisation of the vacuum sputtered AL, Ta.

- Application of Al plate as a base metal substrate which ensures excellent mechanical stability, thermal conductivity, matched thermal expansion with the multilayer structure materials, proper surface finish, etc.

- A good planarization of the whole thin film structure, because the metal conducting lines are fully embedded into the dielectric anodised layers (Fig. 2a).

- No need of through hole metallization and vias drilling of etching in some specific applications (Fig. 2b).

2. FABRICATION PROCESS
2.1 ANODISATION

Untraditional technology for manufacturing of MCM (D) on the base of Al-AL_2O_3 is presented in this chapter. The method includes stages of surface treatment of AL and its alloys, on one hand, and basic part of the stages of hybrid technology, namely the structuring of films. Al is a material that suits the requirements for application of effective materials in microelectronics. Optimisation of the technology and parameters of the processes of electrochemical total and local formation of AL_2O_3 dielectric layers is discussed.

Two versions of the stages of realisation of test structure of MCM based on Al-Al_2O_3 are presented in figures 2a and 2b. The substrates of a definite Al-alloy with dimensions 50 x 40 x 1.5 mm are degreased in organic solvent (trichlorethylene), at 50°C for 3 min. Then substrates are subjected to alkaline chemical degreasing in "Corostat" or "Voalan" or similar solution at $60-70^{\circ}$C for 2 min. After rinsing in water substrates are subjected to alkaline etching in NaOH at $60-70^{\circ}$C for 1 min. Then follows rinsing in water and processing in acid mixture (HNO_3, HF and water) at $18-20^{\circ}$C for 1 min. and again rinsing in water. A subsequent treatment of the substrates with HNO3 for 1 min has a positive influence on the Al surface. Follows rinsing in water. This system of preliminary preparation of the Al surface ensures its excellent cleanness and the necessary smoothness (and lustre if required).

Optimisation for total anodisation of the substrate surface includes system of sulphate electrolyte that contains 180-200 g/l H_2SO_4. This system is chosen for the purposes of the investigation because of its simplicity, manufacturability, possibility for obtaining rather thick layers even at soft conditions, with satisfactory dielectric properties. Electrochemical formation of Al_2O_3 dielectric layer was carried out at -5°C (deep anodisation) and 15°C (common anodisation).

The results of the investigation of electrochemical anodisation of Al substrates in sulphate electrolyte at anode current density 2.5 A/dm^2 and temperature - 5°C are presented in Fig. 3 and 4.

Fig. 3. characterises the output of the system at different duration of the process. It shows that the increase of the duration results in almost linear increase of the Al_2O_3 film thickness. Layer thickness is in the range between 60 and 90 mm, which means that the rate of film formation at so chosen parameters is much higher than the rate of layer's dissolving in the electrolyte. This result is in conformity with the data given in [ ], concerning the same problem.

Dielectric permeability (e_r) and breakdown voltage of the so formed layers are measured. Results are presented in Fig. 4.

The figure shows that with the increase of Al_2O_3 layer thickness from 60 to 90 mm the dielectric permeability smoothly increases from 10 to 12 (curve 1). These values are due to the great layer thickness and probably to a lower degree to additives from the ions from the electrolyte to the layer pores. With the increase of layer thickness increases the breakdown voltage (U_{BR}), which varies between 1200 and 1600 V. These values satisfy the requirements for this parameter and could be accepted as indirect index for the probable lower influence of the additives in thicker layers.

Optimisation of total electrochemical anodisation of the substrate surface includes also formation of dielectric layer at 15°C and anode current density 1, 1.5 and 2 A/dm^2 in sulphate electrolyte. Results of the investigations are presented in fig. 5, 6 and 7.

The changes of the dielectric layer thickness with the duration of the process for the given three anode current densities are presented in fig. 5.

This figure expresses the manufacturability of the system for different modes. From fig. 5 it could be seen that layer thickness increases in linear dependence from the duration of the process. The increase of the anode current density results in increase of the layer thickness, thus in increased manufacturability of the system. Al_2O_3 layer thickness varies in the range 5 - 30 mm. The higher temperature of the system increases the dissolving rate of the dielectric layer during the anodisation, which is in compliance with the references .

The change of the dielectric permeability of Al_2O_3 layers with the change of their thickness is depicted in fig. 6.

The dielectric permeability is in linear dependence from layer thickness and it varies in the range 3.5 to 8. These values are lower compared with the values obtained at deep anodisation. This is due to the thinner layers and probably to the greater influence of the additives from the electrolyte. The change of the current density from 1 to 1.5 A/dm^2 doesn't have a significant influence on the dielectric permeability because of the close values of the layer thicknesses. At 2 A/dm^2 the layer thickness is close to the values of deep anodisation so are the values of dielectric permeability.

The change of the breakdown voltage of the layers is depicted in fig. 7.

The increase of layer thickness results in comparatively sharp and almost linear increase of breakdown voltage. The values of this parameter vary between 200 to 1000 V. This wider range is due to the greater variety of thicknesses. Comparatively lower values of the breakdown voltage in this case are due to smaller thickness and probable influence of the ions in the layer pores.

Electrochemical AL_2O_3 dielectric layers are consolidated (compacted) in water at $90^{\circ}C$ for 10 min. A thermal treatment in air at $200^{\circ}C$ for 2 hours wasn't effective for the electrical parameters of the layer.

Technological process described in fig. 2 includes consolidation of the layer with vacuum deposited 2 mm polyimid layer after thermal treatment at $240^{\circ}C$ for 1 hour.

The next step for both technologies (fig. 2a and fig. 2b) is vacuum deposition of 5 mm Al layer. For this purpose the dielectric substrates of pure Al_2O_3 are heated for 30 min at $200^{\circ}C$ and Al is evaporated in vacuum of 5.10-6 Torr for 1 min. Polyimide containing substrates are heated at $50^{\circ}C$ for 30 min. Vacuum deposition of Al is carried out under the same conditions.

Photolithography is used for the formation of the test structure of resonator system. After optimisation of the process positive resist AZ1350 was chosen as most suitable for the local anodisation. When using spirit in the electrolyte a negative resist should be used. On the test structure beside the resonator system there is a combination of marks by the means of which the window for local anodisation could be controlled.

The next stage for both technological cycles is local anodisation of vacuum deposited Al layer. This is the stage that largely determines the possibility of the technological cycles so much attention was given to the optimisation of this process. Different compositions of the electrolyte were investigated besides the sulphate system: sulphate-oxalate, phosphate-citrate-oxalate. phosphate-oxalate. These types of composition are recommended in the literature for the formation of thin Al_2O_3 layers (1-5 mm) with very good dielectric properties. This motivates the implementation of these investigations. The results of the investigation with combined citrate-phosphate-oxalate system are presented in this paper. They are given in fig. 8, 9, 10. The dependence of the layer thickness from the duration of the process for anode current densities 1.5, 2 and 2.5 A/dm^2 for 10, 20 and 30 min at $15^{\circ}C$ temperature of the electrolyte for the same system is given.

It could be noted from fig. 8 that layer thickness is between 2 and 7 mm. The increased duration of the process for all anode densities leads to proportional increase of layer thickness. The tendency is to increase the layer thickness with the increase of the anode density is rather clear. Thus data is in compliance with other results.

Fig. 9 presents the variation of the dielectric permeability of the layers with the change of their thickness.

It could be noted that for 1A/dm^2 current density with change of layer thickness from 2 to 6 mm the dielectric permeability varies between 3 and 6. With the increase of the anode current density the increase of ε becomes more even and less clearly expressed and the values of this parameter vary in a narrower range: between 4 and 5. This tendency could be due to the larger porosity of the layers at higher current densities.

The values of ε for the layers obtained using this system are comparable with those obtained with sulphate system at $15^{\circ}C$ although layer thickness is lower. This is in compliance with data in [3, 4] where is underlined that the above mentioned system produces thin layers with high density and excellent dielectric properties.

The breakdown voltage of thin Al_2O_3 layers, obtained using this system vary between 130 and 500 V and evenly increases with the increase of layer thickness (Fig. 10). The low values of this parameter could be explained with the small layer thickness. A weak but clear tendency for decrease of breakdown voltage is noted with the increase of the anode current density.

The probable explanation is maybe increased porosity of layers.

The results of the investigations of anode processes of Al realised in different systems show that the most suitable system for total anodisation is the sulphate system at low and normal temperature, and for local anodisation a combined composition is recommended, or sulphate electrolyte at 14-18°C.

2.2 FLATTENING DURING FABRICATION PROCESS

It is hard to fabricate a multilayer structure on uneven Al-picture made by etching. In our case we use anodisation for the nonconductive region. So one has any time possibility to use flatten surface [6].

We have used a polyimid film either, but only to change the ε-value and not to flatten Al-configuration .

From [6] it is well known how to use the very fine pores in the anodic oxidation of the Al-film. It is possible to embed the conductive part of the Al-layer by using short anodisation process at the beginning. If the evaporated film is 5µm thick we can start the process excluding the via-places and stop by 1 µm. After applying and developing of the photo-resist, defining the conductive parts towards X or Y direction, we start the temperature process, which causes filling up the pores by resist on each place that should be remaining conductive. The vias-stoplayer must be off resistive nature during that time and removable afterwards. The rest of photo-resist in the pores does not influence the ε-value significantly.

2.3 CONSIDERATION ABOUT METALLIZATION AND INTEGRATED PASSIVES

Integrated passives development considers the realisation of capacitors and resistors. We use the metallization step by changing the metal system from Al to Al/Ta. If the connections are not very loud, it is possible to use the layer for an interconnect on this flat. A better solution is to design every passive component by the means of vias. So it is available to combine compatible inductors using multilayer structure embedded as usually by this technology.

We use an old technologies to fabricate R and C from a composite Al/Ta layer. The compromise has to found for the sheet-resistance Rs because one uses the same layer for interconnections for C and to design R.

Fig. 11 and Fig. 12 show the possibility to fabricate the layer under conditions which are available in a common installation. We involve the process by controlling of the partial pressure of oxygen and nitrogen. The thickness of the layer can be chosen to be approximately 1.5µm. Special attention is necessary in order to achieve a suitable value for -TKR and +TKC by temperature process after deposition. We involve for the close future to use Al/Ta mixed layers Fig. 13 and to combine the resistive system NiCr/CrSiO2 with two values for Rs.

3. COMPUTER SIMULATIONS AND ELECTRICAL IMPEDANCE MEASUREMENTS

Application of dielectric materials with high dielectric constant and small thickness (Chapter 1 and 2) suggests great reduction of connecting element dimensions and good simulation of dielectric properties of bulk microwave substrate materials, but requires precise computer analysis and synthesis as these dimensions become more critical to design and manufacturing failures at higher operating frequencies. These are the reasons to examine the RLGC transmission-line method (3, 4, 5) and the method of quasi-TEM field of coupled lines (4, 5), using SPICE and S-compact CAD tools, to investigate pulse-transmission line characteristics and parasitic coupling and cross-talk in close parallel situated lines (Fig. 14 and 15).

Several test structures of microstrip lines with different structures and dimensions were prepared. Electrical measurements of transmission line impedance and parasitic coupling were carried out on "Rhode-Schwarz" vector analyser ZPV, S-parameter test adapter ZPV-Z5 and sweep-oscillator SMS-2 in the frequency range 100-1000 MHz . Computer simulations cover the frequency range 0.3 - 3 GHz.

4. CONCLUSION

When discussing the simulation and measurement results, the following conclusions and recommendations could be drawn [4]:

- The increase of line length results in increase of the insertion losses and de-phasing of the output signal at unchanged wave impedance and environmental conditions.

- It could be stated that the dielectric layer thickness (h) is negligible if a proper ratio between line width and dielectric thickness (w/h) is chosen, related to the dielectric permeability (e_r) and wave impedance (Z_0).

- The attenuation of transmission line signal decreases with the decrease of Z_0 at unchanged line length and dielectric constant.

- With the increase of the line distributed bulk resistance (R_i), the insertion losses increase without changing significantly the pulse edge and plateau.

- Higher wave impedance lines are more sensitive towards variations of the load resistance (R_T), while lower wave impedance lines are more sensitive to variations in pulse form and plateau.

- It is recommended to chose high-resistance load when operating with high wave impedance lines and low resistance load when operating with low wave impedance lines.

- For a proper matching to 50 ohm generator and load, line dimensions are recommended to be as follows line width (w) in the range 10-100 mm, thickness of insulating dielectric layers in the range 5-30 mm.

- The distance between grounded reference planes in MCM-D structure should be in the range 0.3-5 mm, depending on the chosen line dimensions and materials.

- When the distance (s) between close parallel lines exceeds (3-5) times their width a sharp decoupling improvement is observed (better than K = -15dB).

The chosen fabrication process of MCM (D), transmission line simulations and measurements are suitable for further investigations in metal base substrate MCM (D) development.

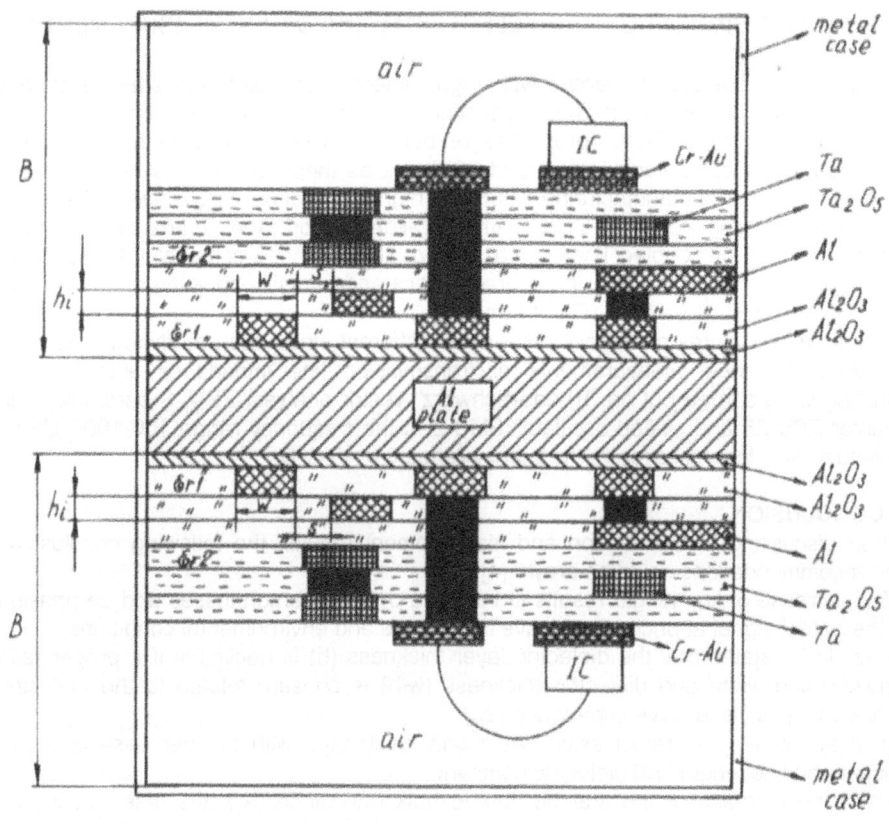

Fig. 1. Cross section of an examplary MCM-D structure, developed on ametal substrate and shielded in a metal case (combined Al/Ta technology)

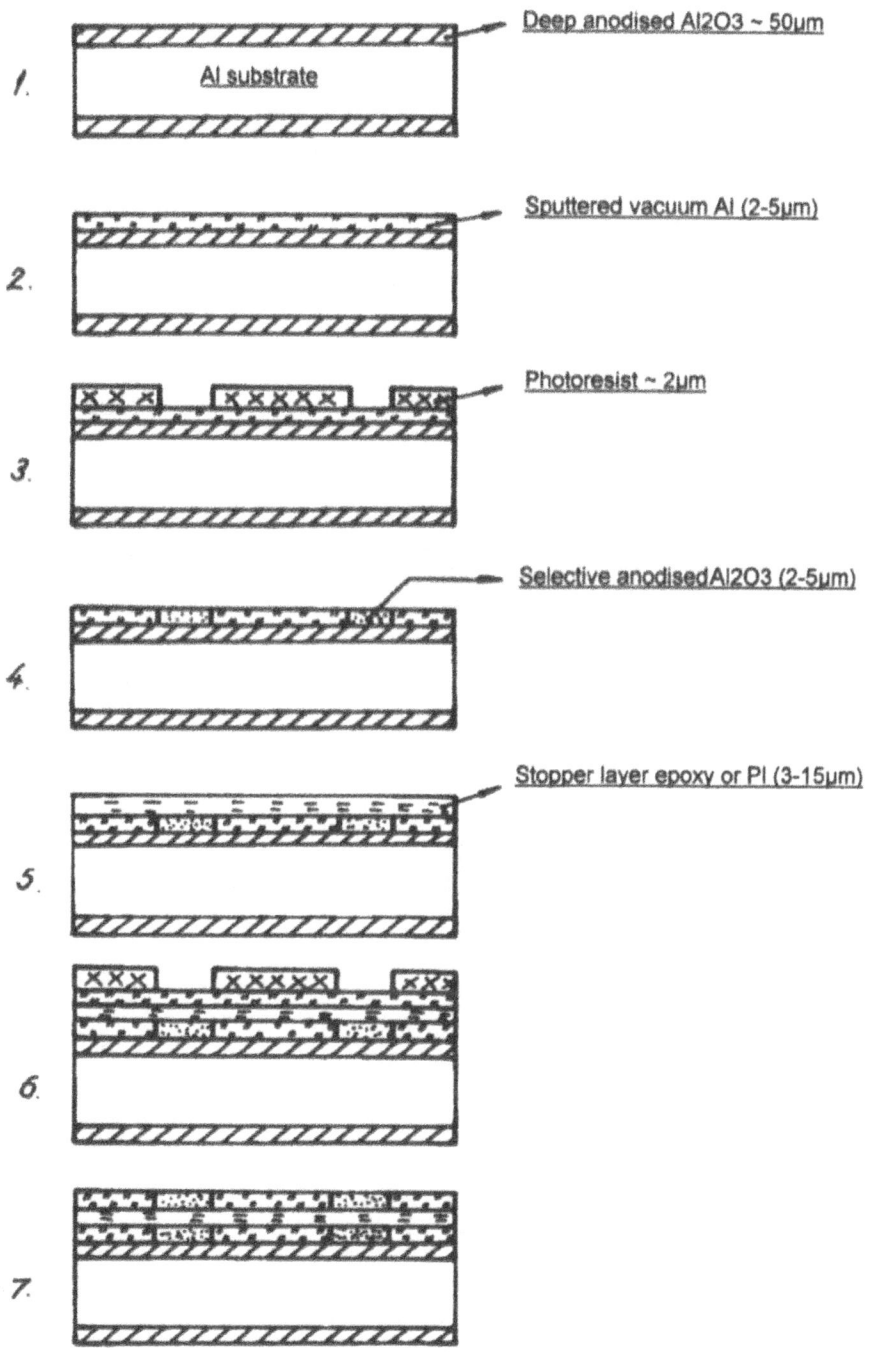

Fig. 2a. Technology steps of metal anodised MCM (D) fabrication process with stopper layer

202

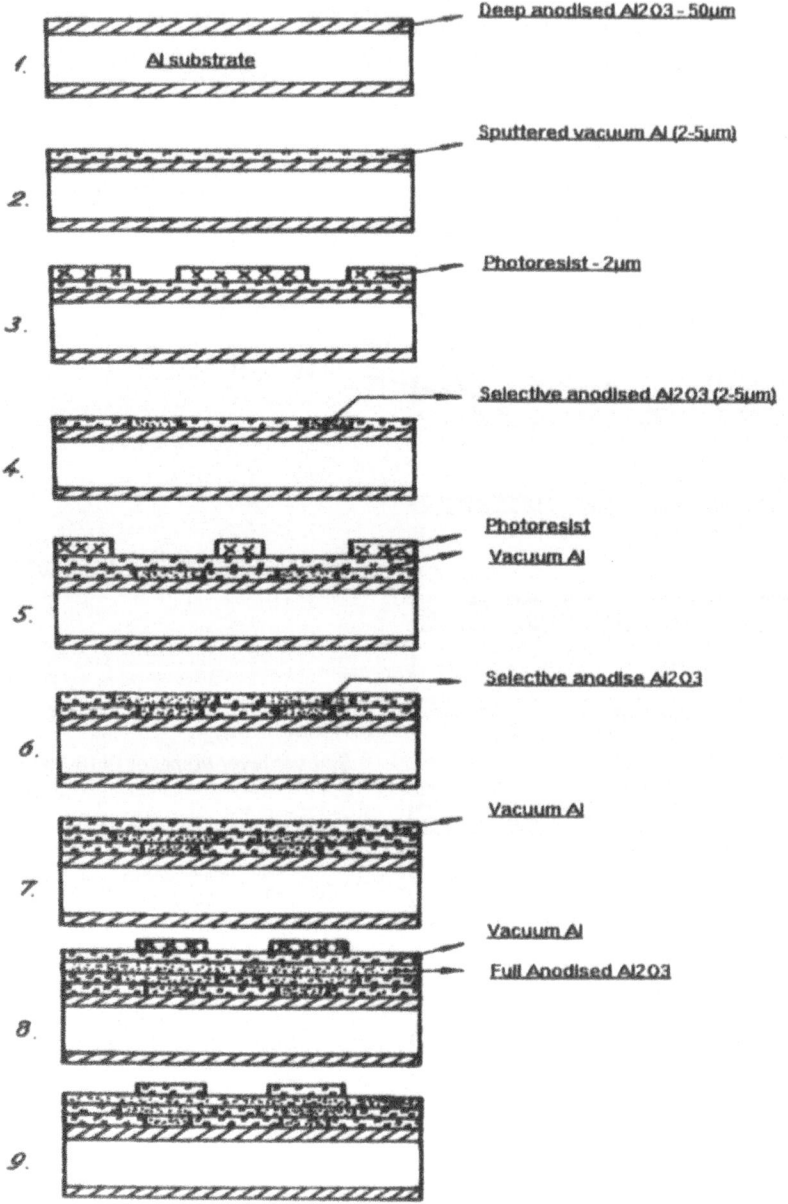

Deep anodised Al2O3 - 50µm

1. Al substrate

Sputtered vacuum Al (2-5µm)

2.

Photoresist - 2µm

3.

Selective anodised Al2O3 (2-5µm)

4.

Photoresist
Vacuum Al

5.

Selective anodise Al2O3

6.

Vacuum Al

7.

Vacuum Al
Full Anodised Al2O3

8.

9.

Fig. 2b. Technology steps of metal anodised MCM (D) fabrication process without stopper epoxyor polyimide layer

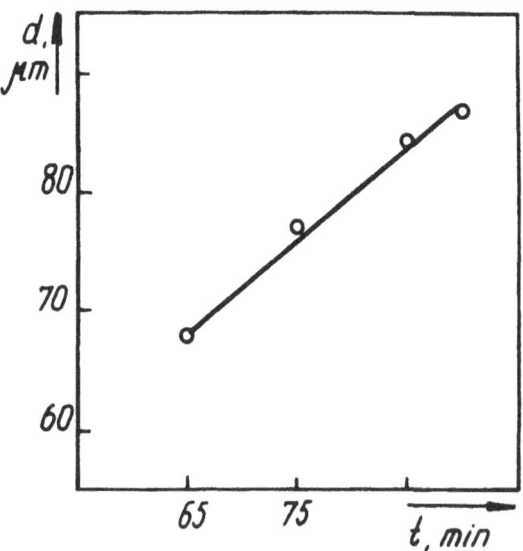

Fig. 3. Relation Al2O3 layer thickness d [μm] from the duration of the process [t,min] for deep anodisation of Al at T=-5°C and anode current density 2,5 A/dm²

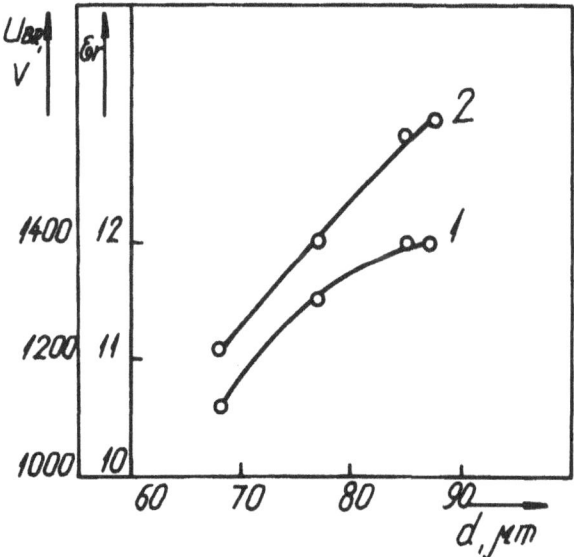

Fig. 4. Relation dielectric permeability ε (curve 1) and breakdown voltage (Ubr,V) (curve 2) from layer thickness d [μm] of Al2O3 layer for deep anodisation of Al at T=-5°C and anode current density 2,5 A/dm²

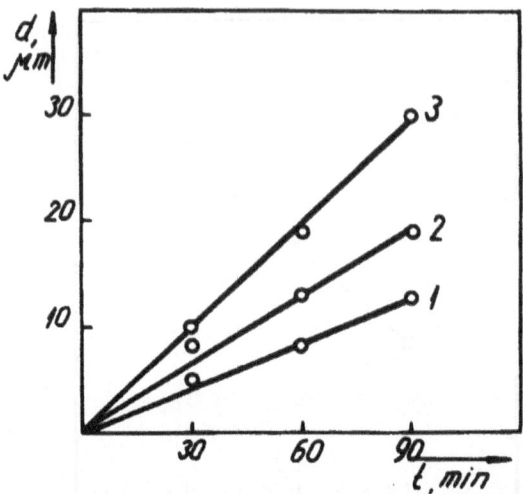

Fig. 5. Relation Al2O3 layer thickness d [μm] from the duration of the process [t,min] for Al anodisation in sulphate electrolyte at T=15°C and anode current density 1 A/dm² (curve 1), 1,5 A/dm² (curve 2) and 2 A/dm² (curve 3).

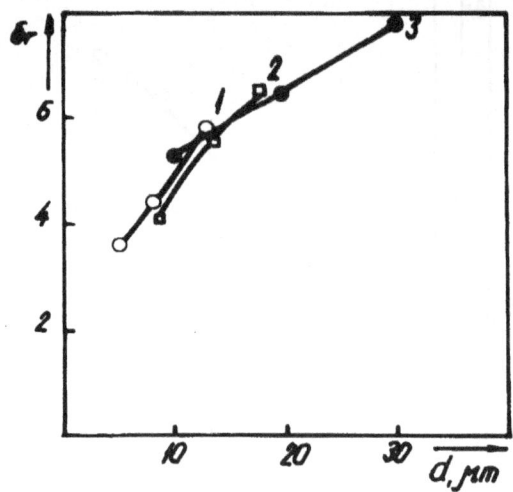

Fig. 6. Relation dielectric permeability (εr) from Al2O3 layer thickness d [μm] for Al anodisation in sulphate electrolyte at T=15°C and anode current density 1 A/dm² (curve 1), 1,5 A/dm² (curve 2) and 2 A/dm² (curve 3)

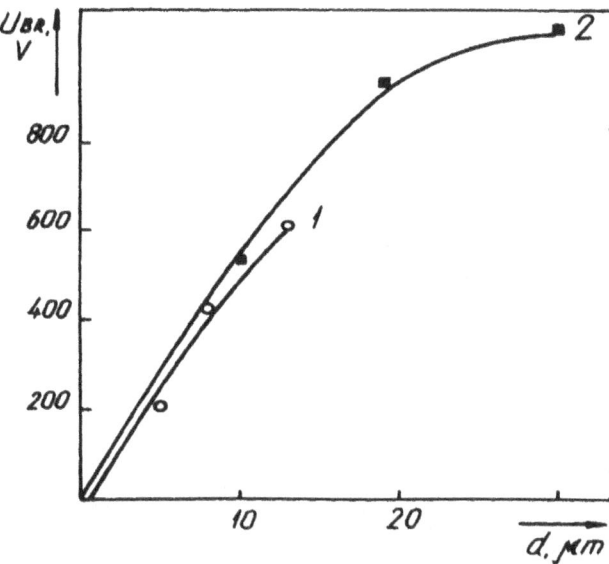

Fig. 7. Relation breakdown voltage [Ubr,V] from process duration [t,min] for Al anodisation in sulphate electrolyte at T=15°C and anode current density 1 A/dm² (curve 1), 1,5 A/dm² (curve 2) and 2 A/dm² (curve 3)

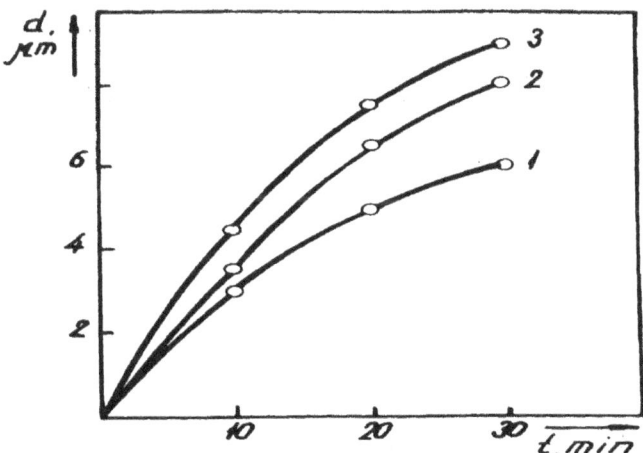

Fig. 8. Relation Al2O3 layer thickness d [μm] from the duration of the process [t,min] for Al anodisation in citrate-phosphate-oxalate electrolyte at T=15°C and anode current density 1,5 A/dm² (curve 1), 2 A/dm² (curve 2) and 2,5 A/dm² (curve 3)

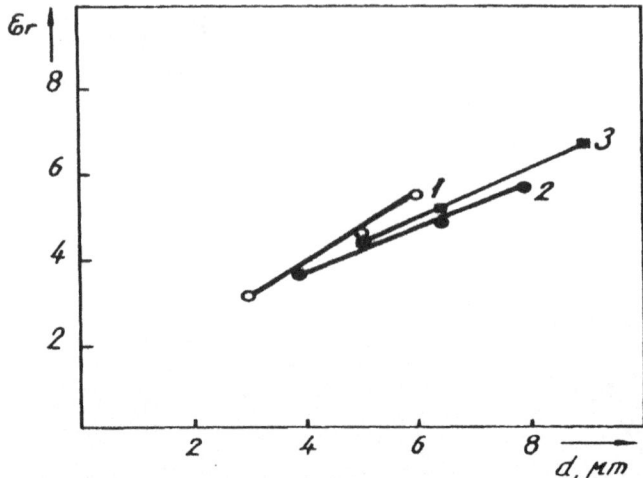

Fig. 9. Relation dielectric permeability (εr) from Al2O3 layer thickness d [μm] for Al anodisation in citrate-phosphate-oxalate electrolite at T=15°C and anode current density 1,5 A/dm² (curve 1), 2 A/dm² (curve2) and 2,5 A/dm² (curve 3)

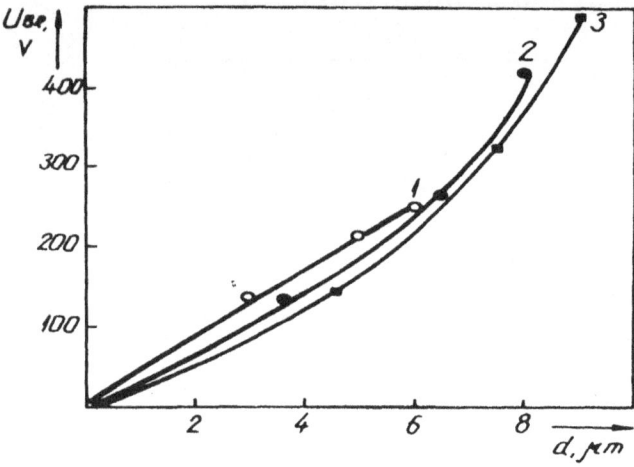

Fig. 10. Relation breakdown voltage [Ubr,V] from Al2O3 layer thickness d[μm]for Al anodisation in citrate-phosphate-oxalate electrolyte at T=15°C and anode current density 1,5 A/dm² (curve1), 2 A/dm² (curve2) and 2,5 A/dm² (curve3)

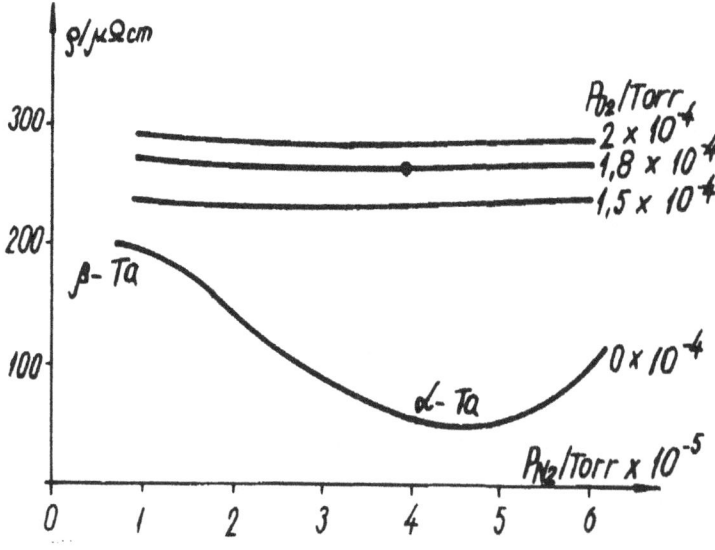

Fig. 11. Relation Rs Ta-layer for different partial pressure

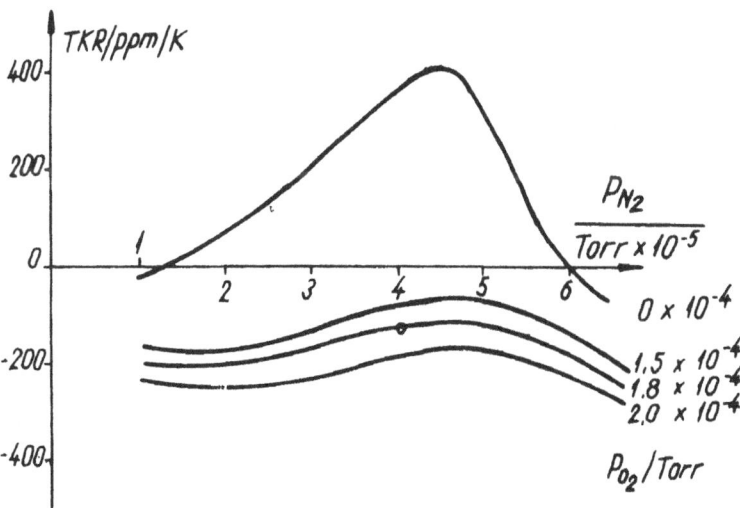

Fig. 12. Relation TKR Ta-layer for different partial pressure

208

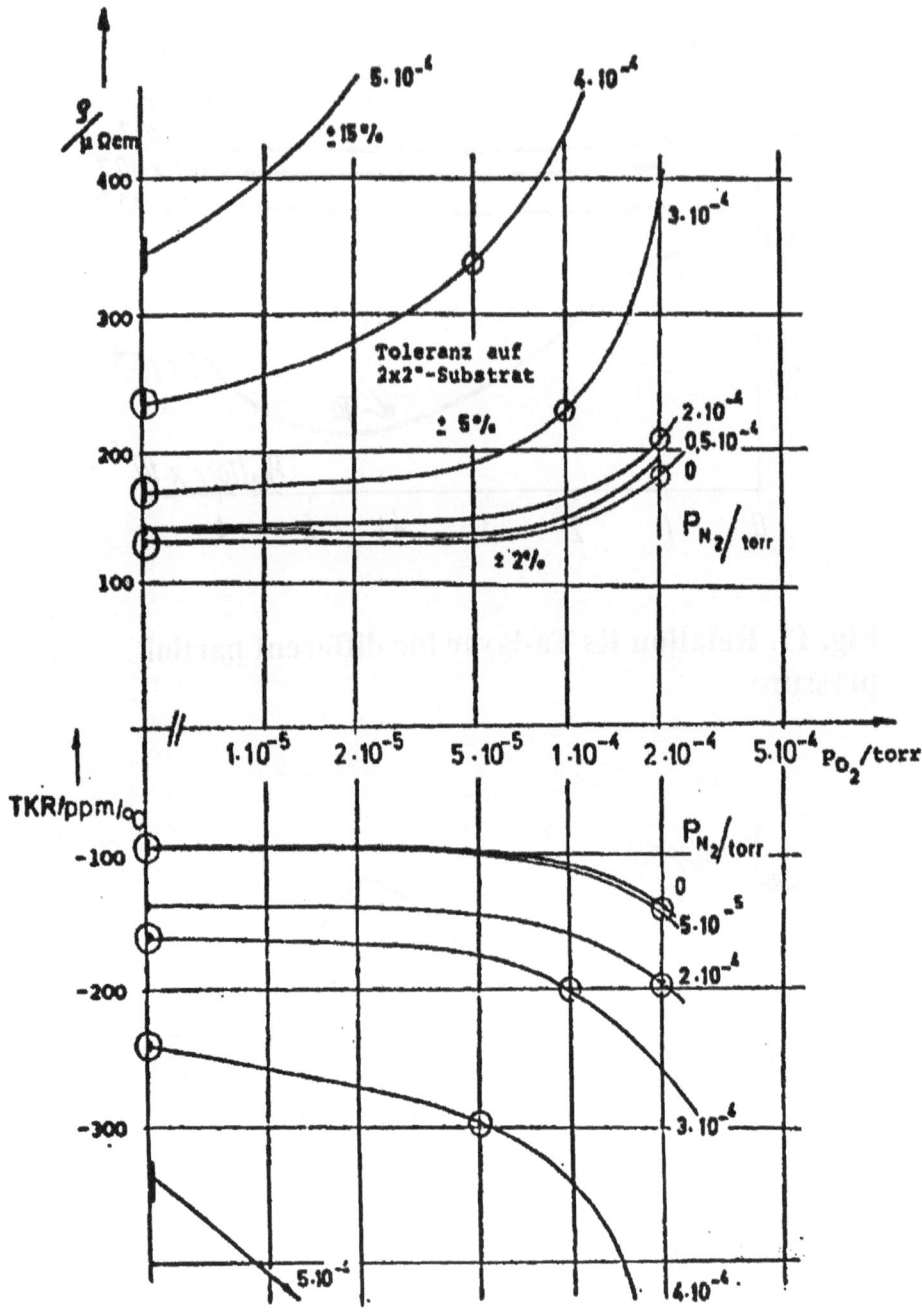

Fig. 13. Specific resistivity and TKR of TaAl-layers

Simulation results of Si-parameters and pulse transmission line characteristics of a single microstripe line (SL)

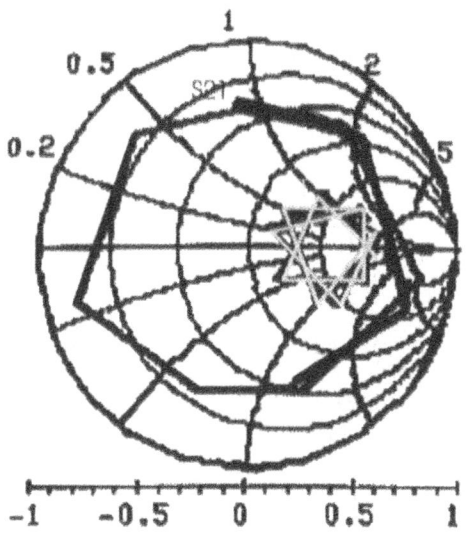

SL
Al, Al2O3
$\varepsilon_r = 7,5$
W=50 μm
B=1 mm
t=2 μm
l=8 cm

Fig. 14a.

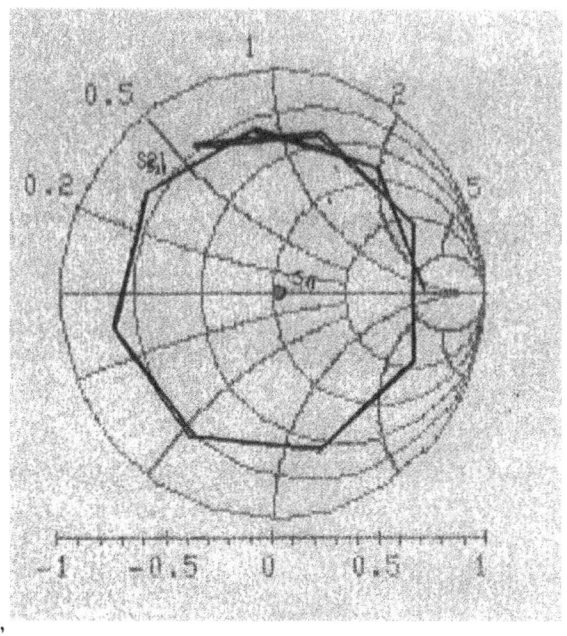

SL

Ta, Ta2O5

$\varepsilon_r = 25$

W=50 μm

B=2 mm

l=4 cm

t=2 μm

Fig. 14b.

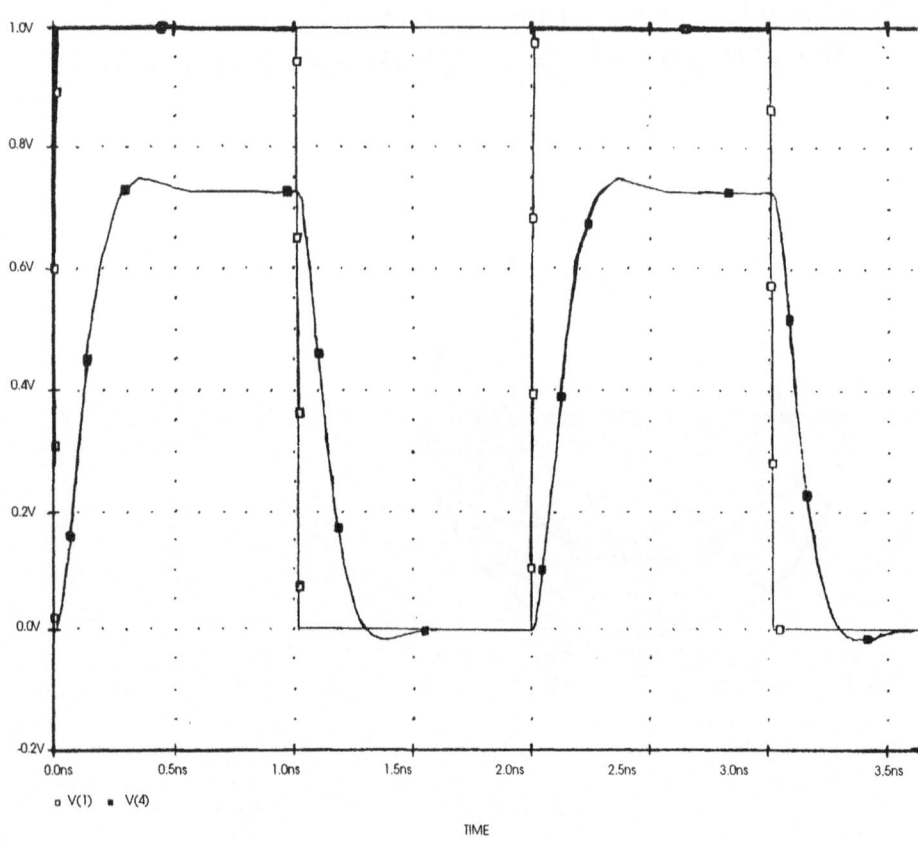

W=20 μm
h=2 μm
t=2 μm
Zo=11,5 Ω
Ri=7,3 Ω/cm
Ci=7,42 pF/cm
Li=0,98 nH/cm
εeff=6,55
Rt=50 Ω
l=1 cm
Al,Al2O3

Fig. 14c. Influence of Zo and Ri (Al)

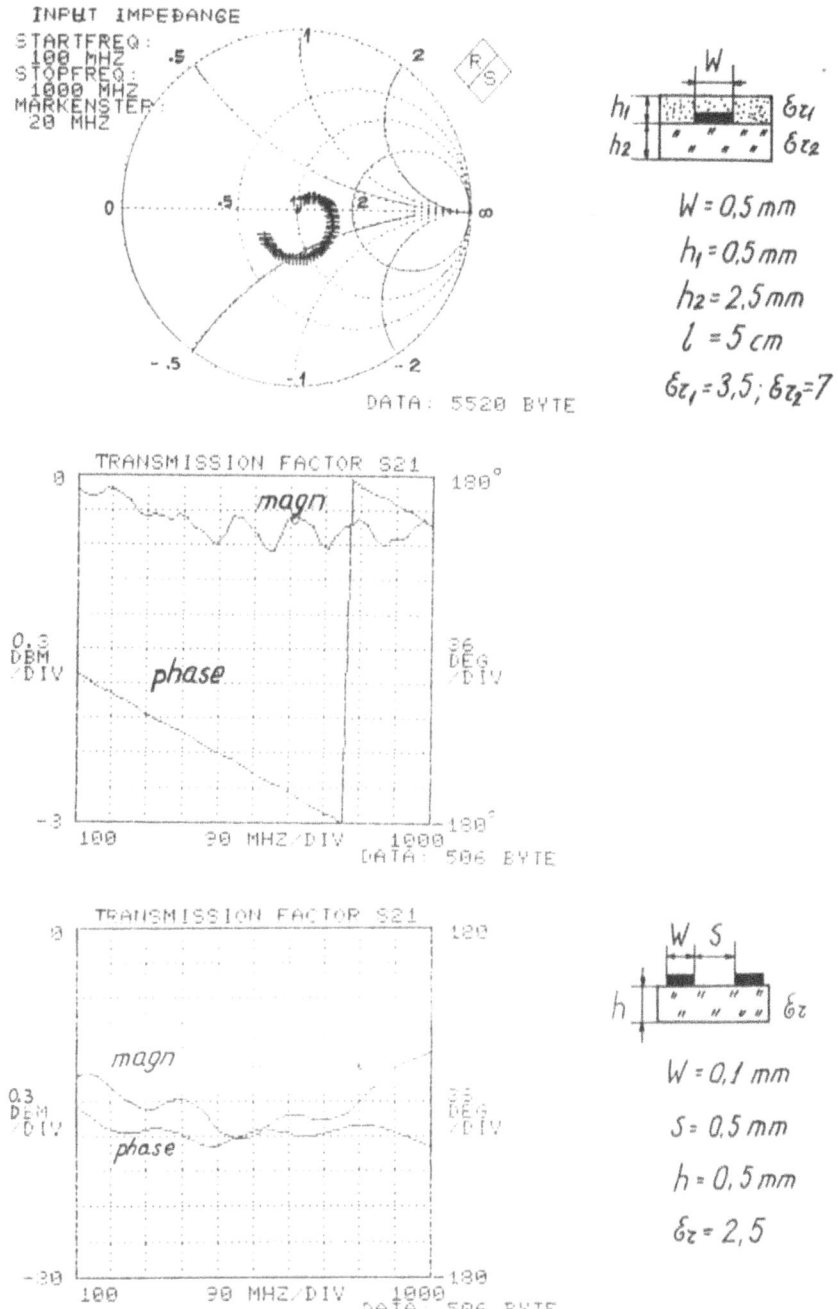

Fig. 15. Vector measurement results of input impedance S11 of a transmission line and a coupling crosstalk S21 of two close parallel lines (W=100μm, S=0,5μm)

REFERENCES

[1] Chinoy, P. B., J. Tajadod, (1993) IEEE Transactions on components, hybrids and manufacturing technology, November, vol.16, N 7, pp 714-720.

[2] Fillion, R. A., (1993) IEEE Transactions on components, hybrids and manufacturing technology, November, vol.16, N 7, pp 615-626.

[3] Yungseon, E., William R. Eisenstadt (1993) IEEE Transactions on components, hybrids and manufacturing technology, August, vol. 16, N 5, pp 555-562.

[4] Phillipov, Ph., R. G. Arnaudov, S. Tzanova, 1994-1995, 17-th ISSS HT, Bulgaria, Sozopol, pp 62-84.

[5] Ramesh Senthinadan, Srinivas Nimmagada, (1993) IEEE Transactions on components, hybrids and manufacturing technology, September, vol.16, N 6, pp 705-713.

[6] Private publication

OPTICAL INTERCONNECTION ELEMENTS INVESTIGATION IN CHALCOGENIDE GLASS LAYERS FOR INTEGRATED OPTICS

A. Kikinesshy, M. Marjan, V. Vlasov , I. Mojzes* , G. Ripka*
Uzhorod .State University, Uzhgorod, Ukrain,
*Technical University of Budapest, Hungary

ABSTRACT
The main optical parameters of photosensitive chalcogenide glass layers, photoinduced transformations in such unorganic high - resolution photoresists and methods of fabricating of waveguide elements on the bases of this materials are revieved. Our investigations and technology developments are related to As(Ge)-S(Se) - based chalcogenide glasses.

I. Introduction
Thin films of chalcogenide glasses (CG) have some properties that explain our interest to the said glasses as to the materials for integrated optics. These properties are: (I) rather low level of losses in near and middle IR region; (ii) high refractive index and its variations in broad limits depending on composition; (iii) interection possibility of surface acoustic wave with waveguide mode [1].

Fundumental property of CG thin films is light sensitivity [2,3], which is realised through the change in refractive index and solubility, and initiates such waveguide elements as three-dimentional waveguide elements and waveguide holograms [1,4,5].

Because of their waveguide properties CG are themselves interconnection elements of integrated circuits. In this paper we shall describe phenomena and technology which are in the bases of elements, working directly in CG layers, thus providing with new interconnections and interactions.

II. CG films properties and passiv holographic elements on their basis
Our investigations and technology developments are related to thin films of As(Ge)-S(Se)-based CG. Commonly, constant optical memory on CG thin films has several levels which differ in ability for erasing and in erasing temperature [2]. Dissolving rate in alkaline etchers changes in the exposed parts of CG films.

For evaluation of photoresist parameters the samples were not annealed beforehand, i.e. non-reversive optical memory was not excluded from the process; together with the reversive memory it made contribution (the same in the sign) into the change of etching rate. Some results for photoresist propertes are given in the table. Here $\gamma = (V_d/V_e)$ is the ratio of the dissolution rates of the unexposed (V_d) and exposed (V_e) areas of the films.

Table. Parameters of photoresist CG films

Glass	Solvent	V_e, µm/min	V_d, µm/min	γ	Type
As_2S_3	$K_2Cr_2O^-$	0.012	0.03	2.5	neg.
As_2S_3	$C_7H_{17}N$	0.005	0.041	8	neg.
AsSe	KOH(10%)	1	2	2	neg.
AsSe	$C_3H_7NH_2$	4	0.8	5	pos.
AsSe	Na_2S(10%)	0.03	0.006	5	pos.

Transition from the initial amplitude-phase recording in CG-films to relief-phase one is accompanied by the considerable gain in parameters, including maximum magnitude of

213

G. Harman and P. Mach (eds.), Microelectronic Interconnections and Assembly, 213-216.

diffraction efficiency and energy necessary to reach this efficiency. For example, if to reach maximum diffraction efficiency (η), which is 4%, in amplitude-phase grating, we needed 100 J/cm², then having reinforced such type of grating chemically from the initial value of deffraction efficiency 0.02%, we shall obtain relief-phase grating with diffraction efficiency 18%. This example is shown in fig.1.

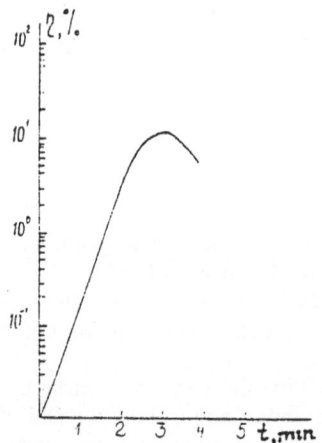

Fig.1. Diffraction efficiency versus etching time in CG thin films

We used reflecting holographic diffraction grating as a test-object that evaluates photoresist properties for possible use in integrated and fiber optics. Grating with spatial frequency 2400 groves/mm was registered by symmetric scheme by two flat waves with the subsequent chemical treatment and metallizing. Spectral dependence of diffraction efficiency in the first order (when the grating works in autocollimational scheme) is shown in the fig.2.

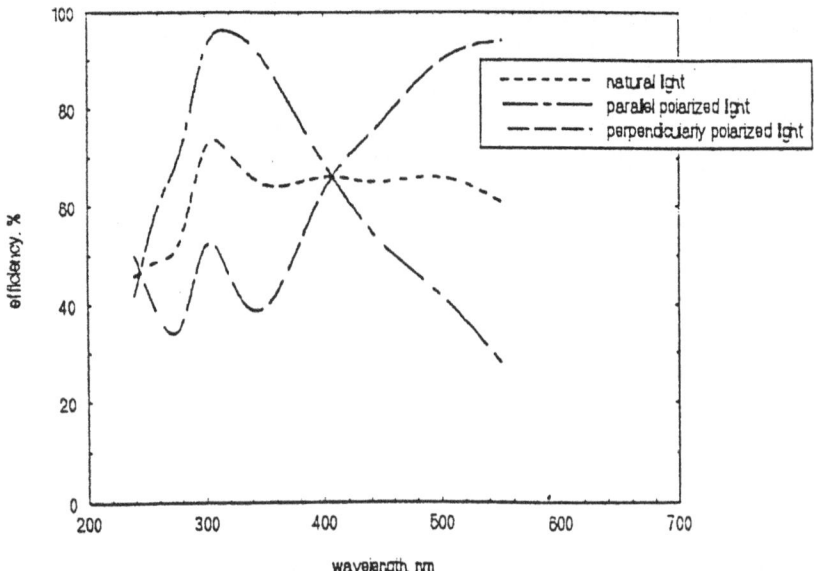

Fig.2. Diffraction efficiency in the firstorder for refractive holographic grating (2400 g/mm) in CG thin films versus wavelength.

Diffraction efficiency maximum in the samples under natural light reached 70-80%. Depending on time interval between film fabrication and exposure, the total density of exposure doze was 0.01-0.03 J/cm², i.e. it increased with relaxation processes taking place in light-sensitive layer.

The technology of hologram recording and chemical etching was applied for couples' fabrication on As-S waveguides which were intended for IR-region, but modelled also for visible λ=0.63μm.

Coupling grating was fabricated by Ar-ion laser irradiation with the wavelength λ=0.51 μm directly in the waveguide layer As_2S_3 with refractive index 2.47; grating's period was λ, grating's working order was q=-1. Input efficiency was η=9-12 % in the experimental samples.

When waveguide grating works in Bragg regime of diffraction and the surphace wave does not leave the waveguide then phase grating becomes efficient for the refractive index variations. As in the previous case, recording was carried out by Ar-ion laser with the wavelength 0.48 μm. Because of the strong absorption in waveguide material, the grating is localized in the surface layer.

Grating with the period 0.8 μm in As-S film worked as mode selector in reflection regime non-critical to the upper limit of the coupling length. Experimentally we observed deviation of zero TE-mode in the grating with efficiency up to 50%. Angle distance between zero and the first TE-modes was 2°38'. The first mode did not interact with the grating It corresponds to 70Å if calculated into spectral selectivity of the grating.

III. Optically controlled intensity modulation of waveguide mode

Possibility to control irradiation transmission in the waveguide with the help of irradiation of another spectral range was discussed in literature [6,7].

In this work the experiments were carried out on the layers of As-S system. He-Ne lazer irradiation was introduced into the waveguide with the help of GaP prism. The track was influenced by Ar-ion laser with the wavelength 0.51 μm (Fig.3). This decreased mode intensity (I).

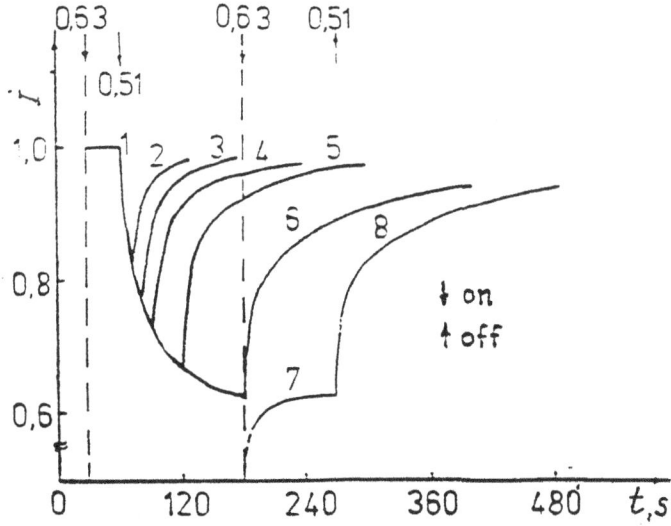

Fig.3. Light-operated waveguide in $As_{20}S_{80}$ thin-films: dynamicsof switch and recovery.

Photodarkening curve for the mode (rate of intensity change or modulation depth) depends both on external light intensity and on intensity of mode itself. Waveguide photodarkening, evoked by the external light, may be stored in the absence of waveguide light transmission. In the case when waveguide exposure is prior to mode excitement in it, the level of photodarkening which is reached and may be stored, is greater than at the simultaneous influence. It is explained by competition of recording-erasing procrsses on the level of constant optical memory. Modulation depth is defined by the signal obtained by the photodetector directly from the track.

The second type of intensity modulators is based on the dynamical constituent of the light sensitivity effect [2] under which spontaneous restoring of refractive index and mode intensity take place. This phenomena can be observed under external periodical irradiation influence on the place of prism coupling as restoring and violation of resonance coupling conditions. Initial track intensity restoring time is~10 s.

By their rate of functioning such photooptic modulators are worse than electrooptic ones. Still now it is impossible to evaluate the limits of their parameters.

References

1. Suhara T., Nishihara H. (1986) Integrated Optics Components and Devices Using Periodic Structures, IEEE J. Quantum Electron., vol.QE-22, pp.845-867.

2. Vlasov V. et al. O Mechanisme Photoinducirovannich Izmenenij v Chalcogenidnich Steklach na Osnove As-Se. Izvestija Vuzov, serija Physica, pp. 48-52.

3. Kikineshi A., Mishak A.,Sterr A. Holographic Recording with Operated Sensitivity and Stability in Chalcogenide Glass Layers. Proc. SPIE, v.1731, pp.173-180.

4. Tada K. et al. (1982). Sputtered Films of Sulphur-and/or Selenium-Based Chalcogenide Glasses for Optical Waveguides,Thin Solid Films, vol.96, pp.141-147.

5. Suhara T. et.al (1976). Waveguide Holograms: a New Approach to Hologram Integration, Opt. Commun., vvol.19, pp.353-358.

6. Vlasov V. at.al. (1989). Photoinducirovannije Izmenenija I Relaxacionnije Processi v Tonkich Plenkach As_xS_{100-x}. In: Proc. Conf. Non-Crystalline Semiconductors-89, Uzhgorod, vol.2, pp.201-203.

7. Tanaka K., Odajima A. (1981) Photo-optical Switching Devices by Amorphous As_2S_3 Wavegides, Appl. Phys. Lett., vol. 38. pp. 481-483.

MATERIAL COMPOSITION CHANGES DURING HIGH TEMPERATURE ANNEALING AND ELECTROCHEMICAL MIGRATION RELIABILITY[1]

GÁBOR HARSÁNYI[2]

Technical University of Budapest, Department of Electronics Technology, H-1521 Budapest, Hungary, Phone: 36-1-463-3634, Fax: 36-1-463-4118
LIANA PERNES, AND W. KINZY JONES
Florida International University, Mechanical Engineering Department, University Park Campus, Miami, FL 33199, USA, Phone: 1-305-348-2345, Fax: 1-305-348-1401

Abstract. *Metals can exhibit dendritic short-circuit growth caused by electrochemical migration in conductor-insulator structures, which may result in failures and reliability problems in microcircuits. The classical model of electrochemical migration has been well known for several decades. This process is a transport of metal ions between two metallization stripes under bias through a continuous aqueous electrolyte. Due to the electrodeposition at the cathode, dendrites and dendrite-like deposits are formed. Ultimately, such a deposit can lead to a short circuit in the device and can cause catastrophic failure. Recent investigations have demonstrated that not only metallic components, but also oxides from the isolating layers can take part in the formation of migrated shorts, after a chemical reduction process. Material design aspects need to clarify the correlation between material composition, processing, chemical bonding state, and electrochemical migration failure rate in isolating compounds: this is the scope of the present study.*

Keywords: electrochemical migration, dendrites, metallization failures, material interaction during firing, in situ X-ray diffraction analysis

1. Introduction

Recently, in connection with the production of high density interconnection systems in integrated circuits and multichip modules (MCMs), the claim to conductor-systems with very high resolution and high reliability has emerged. The possibilities of integration are determined not only by the technological bases but also by those physical and chemical processes that can cause resistive shorts between adjacent metallization stripes during the operation. One of these phenomena is the electrochemical migration. This can be defined as a transport of ions between two metallization stripes under bias through an aqueous electrolyte. Electrodeposition also occurs forming dendrites or dendrite-like deposits. Ultimately, such a deposit can lead to a short circuit in the device and can cause catastrophic failure. The conditions are: a film of polar liquid (usually water) to form an electrolyte, bias, and operating time [1].

Migrated resistive shorts occur randomly in practice and mainly under extreme conditions. A device can operate for many hundreds of hours under normal operating conditions, and then, after a short exposure to special environmental conditions, fail [2],[3].

The classical model of electrochemical migration has been well known for four decades [4]-[6], however, several anomalous phenomena have initiated to perform some revisions and to add

---

[1] This work has been supported by the Hungarian National Scientific Research Fund, OTKA, project No. F007365, as well as by the COBASE program, a grant of the National Research Council, USA, project No.P-1-4568

[2] At present, visiting research fellow in the frame of COBASE program at the Florida International University

G. Harman and P. Mach (eds.), Microelectronic Interconnections and Assembly, 217-223.
© 1998 *Kluwer Academic Publishers.*

supplementary models to the conventional one. A recent discovery in this field was that metal ions forming dendrites can originate not only from the metallic but even from the isolating compounds of the layers, assuming that these constituents can chemically be reduced [7]. In some cases, also the reduction site may run through the dielectric film resulting dendrite-like conductive filaments without any ion formation and migration processes [8]. This latter phenomenon can be called "virtual migration." The failure rates depend on the reducibility that means the chemical bonding state of the given compound. Material design aspects need to clarify the correlation between them.

In ceramic or glass-ceramic based electronic components, packages, and interconnection systems, numerous composite materials are applied containing various secondary constituents such as metal oxides for inorganic binder, dye and other purposes. Several types of these oxides can chemically easily be reduced, for example by hydrogen, thus, they show the ability for metallic dendrite formation. Till now, the materials having been found to take part in dendrite formation corresponding to this model are the oxides of copper, bismuth, lead and iron [9]. Other materials may also show this ability. Most dangerous are those oxides types that remain in the state of oxides after the heat treatment (sintering, firing) of the structures. Thus, at reactive type oxides, which take part in the physical-chemical reactions during firing, the migration lifetime is determined by the excess amount of metal-oxide. Lifetime data may be improved by composition and technology changes. At non-reactive type oxides, which do not take part in any reaction during firing, no improvement of lifetime data can be expected by changing the composition and the firing parameters.

The models of the various migration failure mechanisms and evidences for the "reduced oxide type" as well as "virtual" migration have been described in earlier studies [7-9].

The present work concentrates on the real chemical state of the critical compounds investigated by in-situ X-ray diffractometry during heat annealing. The objective of the paper is to study the chemical bonding state of the mentioned auxiliary oxide type materials which are inclined to chemical reduction and subsequent metallic dendritic growth during device operation. The interaction is of great interest with the functional dielectrics, such as $BaTiO_3$, which is used as ferroelectric compound in capacitor dielectrics.

2. Experimental

2.1. SAMPLE FABRICATION

Thick film capacitors were chosen as model systems for the experimental investigations. Samples were produced by traditional thick film technology using a non-migrating Au conductor composition. The films were printed onto 96% alumina substrates using a 200-mesh stainless steel screen for conductor layers, 160-mesh and 200-mesh screens for the two layers of the dielectrics, respectively. The films were dried at 150 °C for 15 minutes and fired in air in a conventional belt furnace, at a standard 850 °C profile, over a 60-min cycle time. The time at peak temperature was approximately 10 minutes. Two printed layer of the dielectrics were cofired separately from both conductor layers. The fired thickness was 12-15 mm for the conductor and 50-60 mm for the dielectric layers. The samples contained traditional square shaped, 3.32mm x 3.32mm (130 mil sq) capacitor structures.

Thick film dielectrics are compound materials containing not only a high permittivity ferroelectric powder, such as $BaTiO_3$, but various secondary constituents as inorganic binders and dyes, respectively. Therefore, special thick film dielectric materials were produced for the experimental purposes with controlled amounts of Bi_2O_3 and Fe_2O_3 additives to get a better understanding of the failure mechanisms and to separate the migration behaviour of the different oxide constituents. Originally, they have the following role in the pastes: Bi_2O_3 is a common reactive binder component in conductor and ferroelectric pastes. It provides a good adhesion due to the formation of different compounds [13], [14]. Fe_2O_3 is used as a dye in the dielectric pastes.

The components of the pastes used in the experiments were: $BaTiO_3$, the main constituent in capacitor dielectrics, a small amount of glass frit, Bi_2O_3 or Fe_2O_3 additives, respectively in various concentrations, and organic vehicle.

2.2. ACCELERATED LIFE TESTS

THB (Thermal Humidity Bias) testing was performed on the samples for up to 4000 hours which meant an accelerated life test at 95% relative humidity (RH), under 10 V dc bias at the temperature of 40 °C. The purpose of this was to estimate the mean time to failure data of the different samples. The empirical time dependency of the migration failure rates showed statistically a close correlation with lognormal distribution, according to the expectations [5], [11]. Main time to failure data were estimated using least square fitting using lognormal scale.

Figure 1 shows main time to failure data as a function of the metal-oxide content of the dielectrics. It is obvious that migration processes are strongly dependent from the Bi_2O_3 concentration, but less influenced by the Fe_2O_3 content.

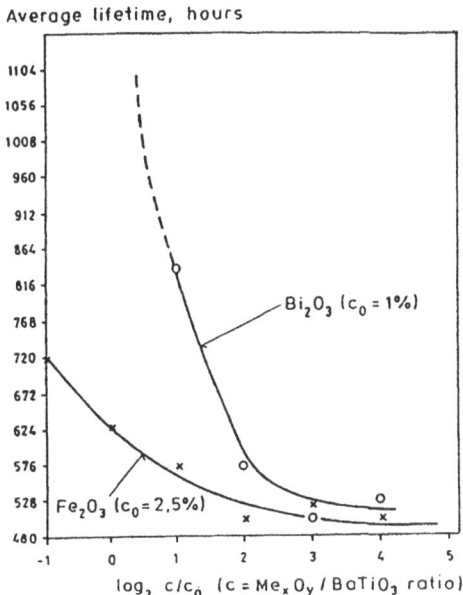

Figure 1. Thick-film capacitor lifetime values as a function of the composition of the dielectrics (THB test performed at 95% RH, 40 °C, 10 V dc)

A similar behavior was found when examining the influence of the fabrication parameters. Table 1 summarizes failure rate results after 1000 hours' THB test as functions of the firing profile: peak temperature and peak time. For Bi_2O_3, the migration behavior can be improved by increasing peak temperature and time. On the other hand, firing parameters can not influence the migration behavior of Fe_2O_3.

From these results, a general hypothesis can be drawn:

- At reactive type oxides which take part in the physical-chemical reactions during firing, migration lifetime is determined by the excess amount of metal-oxide (remaining in the form of oxide after firing). Lifetime data can be improved by composition and technology changes (e.g. Bi_2O_3).
- At non-reactive type oxides, which do not take part in any reaction during firing, lifetime data can not be improved significantly by changing the composition and the firing parameters. The use of this oxide-types in practical structures should be avoided (e.g. Fe_2O_3).

Table 1. Failure rates of thick-film capacitors after THB test with different dielectrics and firing parameters

Oxide type	Metal-oxide/ BaTiO₃ ratio, %	Failure rates, % (1000 hours, 40 °C, 95% RH, 10 V dc) Firing peak temperature (°C) / peak time (min)					
		800/5	800/10	850/5	850/10	900/5	900/10
Bi_2O_3	1	90±10	80±15	10±2	0	0	0
	0.5	50±10	15±5	0	0	0	0
Fe_2O_3	1.25	100	98±2	100	100	95±4	98±2
	0.6	80±15	90±10	80±10	80±15	90±10	90±10

2.3. IN SITU X-RAY DIFFRACTION ANALYSIS DURING FIRING

In order to strengthen the previous hypothesis and for a better understanding of the chemical processes, in situ X-ray diffraction spectra have been prepared during heat annealing. The application of real thick film compositions was not possible, because of the relatively low additive compositions. Thick film samples were also not suitable for the analysis because of the presence of the substrates. 1:1 mol-ratio Bi_2O_3/BaTiO₃ and Fe_2O_3/BaTiO₃ powder samples were prepared by mixing. The sample powder was pressed onto a heated Pt element for making X-ray diffraction (XD) spectra using a CuK_α source. The XD spectra (made by a Siemens D 500 diffractometer) are shown in Figure 2. The most important results can be summarized as follows:

1. In the case of Bi_2O_3/BaTiO₃ powder (see: Figure 2a), several strong chemical changes can be detected when heated up from 650 °C to 850 °C, the typical thick film firing peak: the peaks of several new compounds can be recognized such as: $Bi_{12}TiO_{20}$, and $BaBi_4Ti_4O_{15}$.
2. In the case of Fe_2O_3/BaTiO₃ powder (see Figure 2b): no reaction has been found under 950 °C.

These results prove our hypothesis described above. Accordingly, Bi_2O_3 reacts inside the dielectrics with the ferroelectric material and it can disappear when the temperature program is appropriate for completing the chemical reaction. The chemical reaction between Bi_2O_3 and BaTiO₃ was described already elsewhere [13], however, the postulated compound ($Ba_2Bi_4Ti_5O_{18}$) was not the same which has been found here. On the other hand, no reaction has been found between Fe_2O_3 and BaTiO₃ during a conventional thick film firing cycle. Thus, any modification of the firing parameters can not alter the migration behaviour of Fe_2O_3 inside the structure.

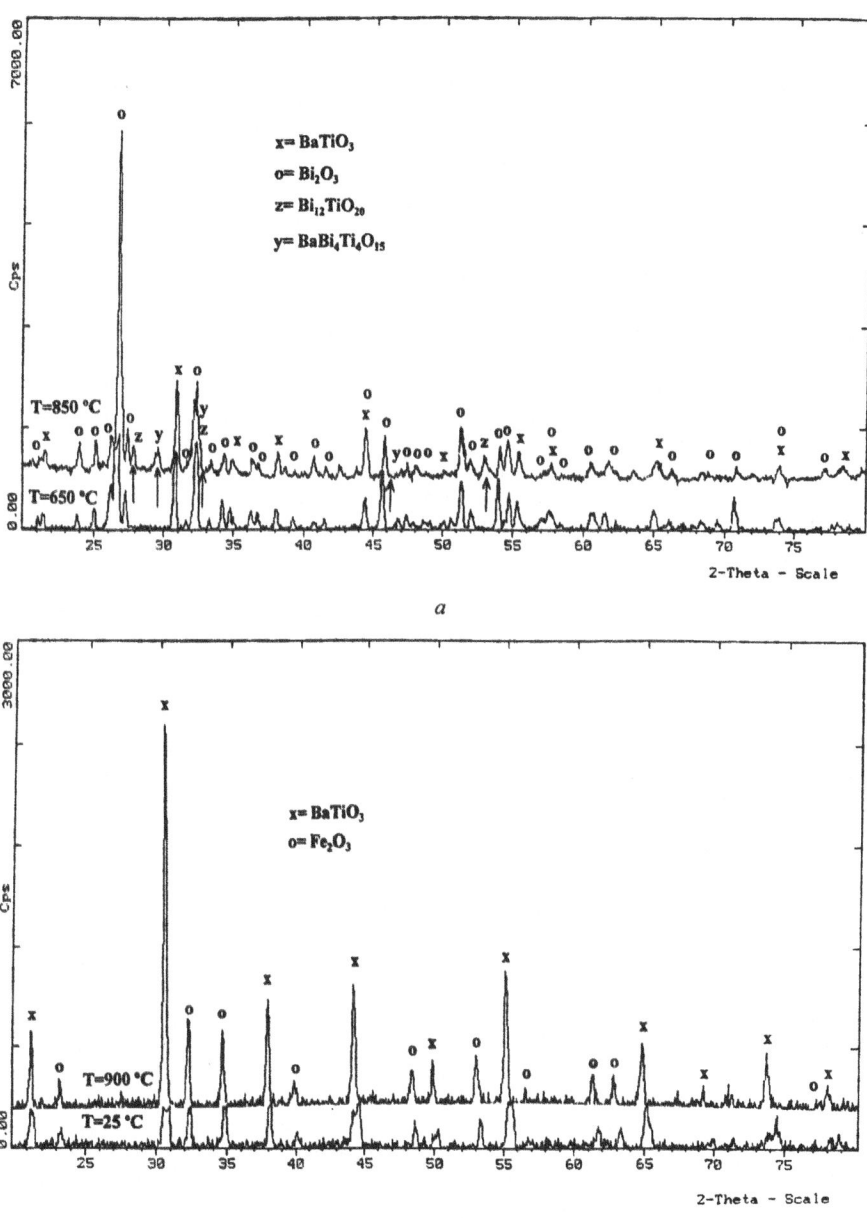

Figure 2. In situ X-ray diffraction analysis during heating (source CuK_α):
a) $Bi_2O_3/BaTiO_3$; b) $Fe_2O_3/BaTiO_3$

3. Conclusions

- Investigating short circuit failure processes caused by the electrochemical migration of reduced isolating compounds in thick film capacitors, a strong influence of the composition and processing parameters has been found.
- In situ XD spectra during heat annealing indicate that the determining factor is the availability and completeness of those chemical reactions in which the critical, easily reducible compound may participate.
- Thus, there are two types of metal-oxide constituents:
  Reactive type oxides, which take part in the physical-chemical reactions during firing. The migration lifetime is determined by the excess amount of metal-oxide (remaining in the form of oxide after firing). Lifetime data can be improved by composition and technology parameter changes (e.g. Bi_2O_3).
  Non-reactive type oxides do not take part in any reaction during firing. Lifetime data can not be improved by changing the composition and the firing parameters. The use of this oxide-types in the practical structures should be avoided (e.g. Fe_2O_3).
- A wide range of materials and processes should be investigated in the near future to understand their behaviour in this aspects to optimize compositions and processing parameters in order to improve the reliability of the structures where they are used.

4. References

[1] G. T. Kohman, H. W. Hermance, and G. H. Downes "Silver migration in electrical insulation," *Bell System Technical Journal* vol. 34, p. 1115, 1955.

[2] A. Shumka and R. R. Piety "Migrated-gold resistive shorts in microcircuits," in *Proc. Int. Reliability Physics Symp.*, p. 93-98, 1975.

[3] R. C. Benson, B. M. Romenesko, J. A. Weiner, B. H. Nall, and H. K. Charles, Jr. "Metal Electromigration induced by solder flux residue in hybrid microcircuits," *IEEE Trans. Comp., Hybrids, Manuf. Technol.*, vol. CHMT-10, p. 363, 1988.

[4] A. DerMarderosian, "The electrochemical migration of metals," *Proc. Int. Society of Hybrid Microelectronics*, p. 134, 1978.

[5] G. Ripka and G. Harsányi, "Electrochemical migration in thick-film ICs," *Electrocomp. Sci. Technol.*, vol. 11, p. 281, 1985.

[6] M. V. Coleman and A. E. Winster, "Silver migration in thick-film conductors and chip attachment resins," *Microelectronics Journal*, No. 4, p. 23, 1981.

[7] G. Harsányi, "Dendritic growth from dielectric constituents: a newly discovered failure mechanism in thick film ICs?" *Int. J. Microcircuits & Electronic Packaging* vol. 16. No. 3, p. 207, 1993.

[8] G. Harsányi, "New type short circuits at fritless thick film conductors - formed from reduced oxides," in *Proc. Intern. Symp. on Microelectronics, ISHM,* p. 140, 1992.

[9] G. Harsányi, "Electrochemical processes resulting in migrated short failures in microcircuits," *IEEE Trans. Components, Packaging, and Manufacturing Technology - Part A,* vol. 18.,No.3., p. 602, 1995.

[10] S. J. Krumbein, "Metallic electromigration phenomena," in *Electromigration and electronic device degradation* (Ed. by A. Christou), John Wiley & Sons, Inc., New York, p. 139, 1994.

[11] N. L. Sbar and R. P. Kozakiewicz, "New acceleration factors for temperature, humidity, bias testing," *IEEE Trans. Electron Devices*, vol. ED-26, No.1, p. 56, 1979.

[12] S. Peck and O. Trapp, *Accelerated Testing Handbook*, Technology Associates, Portola Valley, CA., 1989.

[13] L.C. Hoffman and T. Nakamaya, "Screen Printed Capacitor Dielectrics," *Microelectronics and Reliability*, 1968. p. 131.

[14] T. Ogawa, M. Ootani, T. Asai, M. Hasegawa, and O. Ito, "Effect of Inorganic Binders on the Properties of Thick Film Copper Conductor," *IEEE Trans. Components, Packaging, and Manufacturing Technology - Part A,* vol. 17.,No. 4, p. 625, 1994.

NONLINEARITY BETWEEN INTERCONNECTION
AND RESISTIVE LAYER

Pavel Mach
Czech Technical University in Prague, Faculty of Electrical Engineering,
Department of Electrotechnology
Technicka 2, 166 27 Prague 6, Czech Republic
E-mail: mach@feld.cvut.cz

The mechanism of electrical conductivity for an interconnection layer and a resistive layer is given by a composition of resistive and conductive ink. However, when the resistive film is contacted by the conductive film, a boundary area between the film materials of both the inks are mixed. Therefore different types of electrical conductivity can occur here. The current vs. voltage characteristics of these mechanisms will probably be nonlinear because different types of materials are in a contact here. Because the resulting C-V characteristics of the junction area is given by the sum of the partial mechanisms it will also be nonlinear. This nonlinearity was used for evaluation of properties of the contact area and for the investigation of changes caused by thermal ageing of inks. Nonlinearity measurement were realized by a modulating technique. The samples were investigated in the temperature range 373 K - 14 K. The low temperature measurements were carried out in a cryostat, measurement at the higher temperature was provided in an oven. Odd nonlinearity was measured by the use of a CLT meter, the measurement of even nonlinearity was realized by a lock-in amplifier. All measurements were carried out using Kelvin's four-point arrangement. It was found that nonlinearity of a junction area between interconnection and resistive layer is lower than nonlinearity of the resistive layer, depending weakly on temperature and substantially on ageing of the samples.

1. INTRODUCTION

Quality of the resistive and conductive layers is usually evaluated according to their time stability, thermal coefficient, noise, and ageing in higher temperature. This is necessary to consider that the resistivity of conductive and resistive layers is caused by a very complicated mechanism of conductivity, which is substantially influenced by the composition of the inks. However, the resistive films are contacted by the conductive ones and there are therefore areas where the materials of both layers are mixed. Properties of the mixed material could differ substantially from the properties of other layers because other types of mechanisms of conductivity could dominate here in comparison with the mechanisms of conductivity of the layers. It was assumed that the mechanism of conductivity of the junction area will be nonlinear and therefore the C-V characteristics of this area would be nonlinear, too.

The goal of this work was to find the level of nonlinearity of the junction area, compare it with the level of nonlinearity of both the resistive and conductive inks, and to investigate its

G. Harman and P. Mach (eds.), Microelectronic Interconnections and Assembly, 225-231.

dependence on the thermal ageing of the layers. The influence of temperature to the nonlinearity was also investigated.

2. GENERAL THEORY

The area of a contact between resistive and conductive layer is a mixture of different types of materials and therefore it was assumed that its C-V characteristics was nonlinear. When an electronic component with this type of C-V characteristics will be powered by a sinusoidal current; nonsinusoidal periodical voltage will occur across it.

Let us presume that it is possible to describe the C-V characteristics of the contact by a polynomial approximation and that the cubical polynome fits the contact characteristics with sufficient accuracy.

$$u\ (t)\ =\ \alpha + \beta\ i\ (t) + \gamma\ i^2\ (t) + \delta\ i^3\ (t) \tag{1}$$

When the contact will be powered by the sinusoidal current

$$i\ (t)\ =\ I_0 * \sin\ (\omega t) \tag{2}$$

the voltage across it will be given by the equation

$$u\ (t)\ =\ \alpha + \beta\ I_0\ \sin\ (\omega t) + \gamma\ I_0^2\ \sin^2\ (\omega t) + \delta\ I_0^3\ \sin^3\ (\omega t) \tag{3}$$

By the use of the basic properties of goniometrical functions it is possible to derive the following equation

$$u\ (t)\ =\ \alpha + \frac{1}{2}\ \gamma\ I_0^2 + (\ \beta\ I_0 + \frac{3}{4}\ \delta\ I_0^3\ )\ \sin\ (\omega t) - \frac{1}{2}\ \gamma\ I_0^2\ \cos\ (2\omega t) -$$
$$- \frac{1}{4}\ \delta I_0^3\ \sin\ (3\omega t) \tag{4}$$

The voltage u(t) can be decomposed in a DC component and three harmonics. The level of the DC voltage and the amplitudes of the harmonics depend on the amplitude of the sinusoidal current and on the parameters which are given by the nonlinearity of the C-V characteristics.

The situation presented was a simple example only. Generally, the C-V characteristics of the contact will be characterized by the continuous nonlinear function f. When the contact is powered by the ac current, the voltage across it will be given by the sum of terms of the Fourier series.

$$u\ (t)\ =\ c_0 + 2 \sum_{k=1}^{\infty}\ |\ c_k\ | * \cos\ (k\omega t + \phi_k) \tag{5}$$

Where

$$c_k = \frac{1}{T} \int_{-\frac{T}{2}}^{\frac{T}{2}} f(t) e^{-i k \omega \tau} d\tau \tag{6}$$

According to the eq. (5) the voltage u(t) is the sum of an absolute term c_0 and harmonic voltages with the frequency f, 2f, 3f etc.

It was shown how the higher harmonics across the contact originate when the contact is powered by the sinusoidal current. According to the type of C-V characteristics it is possible to decompose the voltage across the contact in the series of the odd, even, or both types of harmonic voltages. The amplitude of these voltages increases when the nonlinearity of the C-V characteristic increases and when the amplitude I_0 increases. The amplitudes of the second and third harmonics are usually the highest. Therefore these harmonics were measured and used for evaluation of nonlinearity.

3. EXPERIMENTAL TECHNIQUE

3.1. Samples Preparation

Thick film samples were prepared by an usual way. First, three contacts were printed on an alumina substrate. A resistive film was printed across these contacts . Topology of the sample is shown in the Fig. 1. Thickness of the substrate was 0.65 mm, contacts were made with AgPd conductive ink (type 9308, Du Pont), resistive film was made of ink TT 5041 (Tesla Lanskroun, Czech Republic). Because the contact nonlinearitiy was investigated, it was not necessary to adjust a nominal value after firing. The resistance of samples

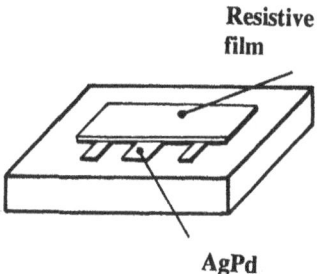

Resistive film

AgPd

Fig. 1 Topology of the sample

were near 11 kΩ. The influence of ageing to the contact nonlinearity was also investigated. Therefore some samples were aged in a firing oven. Artifical ageing in higher temperature (comparable with the temperature of firing) was used. The samples were terminated by 0.2 Cu wires contacted to the conductive layer by soldering (SnPb solder 60/40 , Kovohute Pribram, Czech Republic was used).

First, nonlinearity of AgPd layer and the contact AgPd layer - Cu soldered wire were tested. Special samples were prepared for this measurement. A conductive path was printed on an alumina substrate using an AgPd ink and the sample was fired a standard way. Then it was be contacted by three Cu wires by soldering. Nonlinearity of this arrangement was investigated using a low-resistance input part of equipment for measurement of nonlinearity. It was found that nonlinearity of this sample is substantially lower in comparison with nonlinearity of the contacts between conductive and resistive film. Then the samples were aged by the same way as the samples with the resistive film and the measurement of nonlinearity was repeated. The change of nonlinearity was very low. It was found that the nonlinearity of the conductive layer and a soldered contact between the Cu wire and this layer can be neglected.

3.2 Cryogenics

The samples were placed on a cooled chamber of the cryogenic equipment Cryogen RS 210 (Balzers). The thermal contact of the substrate and the cooled chamber was carried out using silicon grease. Every Cu lead was connected with a double-level thermal "anchor". This arrangement is necessary to avoid cooling losses caused by Cu leads (the maximum of acceptable cooling losses is 10 mW to achieve the minimum temperature). The arrangement was covered by a vacuum bell jar and pumped by the rotation and diffusion pumps to the pressure 5 mPa. The vacuum was used to reduce thermal losses. Temperature was measured by a semiconductor diode. For stabilization of temperature a stabilizer with the accuracy of 0.5 °C was used.

3.3 Measuring Technique

The principle of the nonlinearity measurement is presented in the Fig. 2. R_c represents contact resistance, i_ω is a source of the sinusoidal current, SV is a selective voltmeter which is tuned for the frequency of the investigated harmonic voltage. Instead of the SV, the use of a lock-in amplifier (LIA) is possible, too. Using this equipment the selectivity of the measurement increases, but the measurement is limited by the dynamic range of the lock-in amplifier.

The measurement can be realized by two ways:

* the sinusoidal current i_ω from a stable, low distortion generator is applied without a dc offset voltage (switch S is turned off) to the leads 1 and 2 and the harmonic voltage is measured between the leads 2 and 3. The amplitude of the current is changed from 0 to the $I_{0\ max}$. This measurement is used for the nonlinearity measurement very often. This way works equipment CLT meter (Radiometer, Copenhagen),

* the sinusoidal current i_ω with a low amplitude and the dc current are applied to the leads 1 and 2 together (switch S is turned on). The dc current is usually added from a slow sweep generator (0.01 - 0.0001 Hz). The amplitude of the dc current is slowly changed. The harmonic voltage is

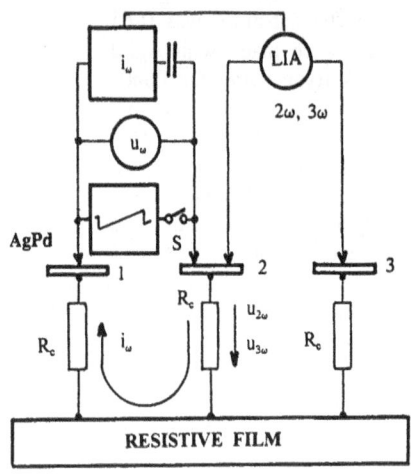

Fig. 2 Arrangement for measurement of nonlinearity of a contact

measured between the leads 2 and 3. The amplitude of the ac current is not changed and nonlinearity of the C-V characteristics is investigated by the change of the dc offset caused by the change of the sweep current. The block diagram presented in the Fig. 2 shows the arrangement which makes both these measurements possible. Because very low levels of voltages are measured the sample and the leads must be carefully screened, high attention has to be paid to grounding. It is necessary to avoid ground loops. The ground leads have to be short and has to be connected directly to a separate thick ground conductor.

4. EXPERIMENTAL RESULTS AND DISCUSSION

The results of the measurement of nonlinearity are usually processed the following way: measured data are plotted in log-log coordinates and the best linear approximation is found using regression analysis. Then a coefficient n is calculated like tg α, where α is an angle between regression line and the x axis. Nonlinearity is expressed by SHI (Second Harmonic Index) and THI (Third Harmonic Index). These indexes are defined by the equations

$$SHI = 20 * \log \left(\frac{u_{2\omega}}{u_\omega{}^n} \right) \tag{7}$$

$$THI = 20 * \log \left(\frac{u_{3\omega}}{u_\omega{}^n} \right) \tag{8}$$

The values of the SHI and THI describe nonlinearity of the contact and can be used for evaluation of the contact quality.

Results of the measurements of contacts nonlinearity between the conductive and resistive layers are presented in the Figs. 3-6. Fig. 3 shows the level of the second (U2) and third (U3) harmonics measured at the temperature 100 K for the sample which was not aged and the Fig. 4 shows the same characteristics for the sample after ageing. A dashed line in these figures was obtained by the measurement of nonlinearity of the resistive layer (U3T). The changes of U2 and U3 in the wide temperature range (14 K - 373 K) for the sample without ageing and after ageing are presented in the Fig. 5 and in the Fig. 6. The course of SHI and THI for both these samples are shown in the Fig. 7.

It was found that nonlinearity of the contact conductive layer- resistive layer was substantially lower in comparison with the nonlinearity of the resistive layer. An interesting discovery is that nonlinearity of the contact between these layers does not depend on the level of the first harmonic voltage (the characteristics is a straight line) in contrast to nonlinearity of the resistive layer which depends on the level of the voltage of the first harmonics. This characteristics can be explained by the very strong influence of the conductive layer to the property of a contact. This assumption was acknowledged by the measurement of nonlinearity of the samples which were thermally aged. After ageing nonlinearity of the resistive layer decreased and nonlinearity of contact decreased also. It is possible to assume that mutual diffusion of materials of both the conductive and resistive layers is caused by thermal ageing. Nonlinearity of such the contacts would be lower in comparison with the contact without ageing. This characteristics was actually found.

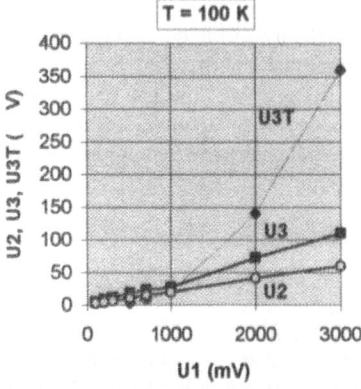

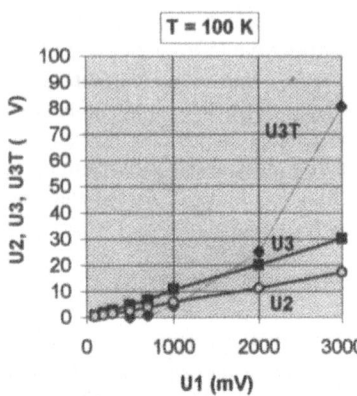

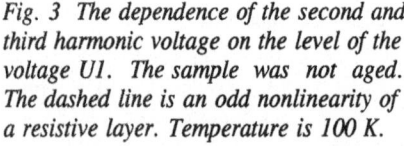

Fig. 3 The dependence of the second and third harmonic voltage on the level of the voltage U1. The sample was not aged. The dashed line is an odd nonlinearity of a resistive layer. Temperature is 100 K.

Fig. 4 The dependence of the second and third harmonic voltage on the level of the voltage U1. The sample was aged. The dashed line is an odd nonlinearity of a resistive layer. Temperature is 100 K.

As follows from our measurement, the properties of a contact area are substantially influenced by the properties of the conductive layer. Our hypothesis about high nonlinearity of the contact was not validated. It has been found that the dependence of the contact nonlinearity on the current flowing through the contact is weak only and substantially lower in comparison with the dependence of nonlinearity of the resistive layer. Contact nonlinearity increases with the decrease of the temperature weakly. This characteristics can be explained by the reduction of some mechanisms of conductivity at the low temperature.

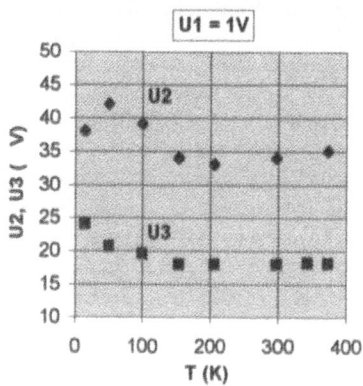

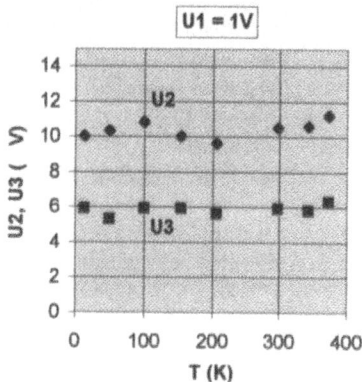

Fig. 5 Characteristics of the second and third harmonic voltages in the wide temperature range. The sample was not aged. Voltage u1 is 1 V.

Fig. 6 Characteristics of the second and third harmonic voltages in the wide temperature range. The sample was aged. Voltage u1 is 1 V.

5. CONCLUSIONS

The results of the presented experiments demonstrate not only the wide applicability of the nonlinearity measurements, but also the possibility of their utilization to specific applications. It should be added that this measurement can be used as a tool to study the conductivity mechanism of the contact between the resistive and conductive film. In the absence of a more general understanding, which would make quantitative prediction of contact effects possible, the measurement of resistance, completed with the measurement of nonlinearity, seems to be an useful method for evaluation of contact quality and stability.

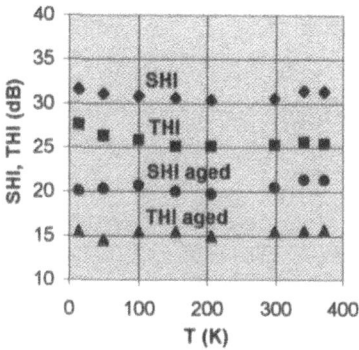

Fig. 7 Typical courses of the SHI and THI in the wide temperature range (Ul is 1 V).

6. ACKNOWLEDGMENTS

I would like to thank Dr. J. Urbancik from TU Kosice (Slovak Republic) and Prof. Dr. J. Urbanek from TU Prague for preparation of samples and Eng. P. Bak and Eng. A. Pospisilova for help with the experimental work.

7. REFERENCES

[1]   MACH, P. : Low resistance contacts on Y-Ba-Cu-O films, *CTU Workshop 94*, CTU Prague, Jan. 1994, pp.141-142

[2]   MACH, P. : Equipment for diagnostics of the contacts quality, *CTU Workshop 95*, CTU Prague, Jan. 1995, pp. 477-488

[3]   Ch. S. HSI, G. H. HAERTLING : A new method for measuring low resistivity contacts between silver and $YBa_2Cu_3O_{7-x}$ superconductor, *Rev. Sci. Instrum.*, Vol. 62, 5, May 1991, pp.1317-1320

This research has been conducted at the Department of Electrotechnology of the CTU Prague as a part of the research project supported by the Grant Agency of the Czech republic under the No. 202/94/0707

EDUCATION IN HYBRID MICROELECTRONICS AT THE UNIVERSITY OF BRNO

Ivan Szendiuch, President of ISHM Czech and Slovak Chapter, Associate Professor, Msc, PhD, Technical University of Brno,Faculty of Electrical Engineering and Computer Science, Department of Microelectronics, Údolní 53, 602 00 Brno, Czech Republic

Introduction

Hybrid integrated circuits (HIC s) are being teached at the Technical University of Brno since the time of establishing the Microelectronics Department in 1980. At this time I have turned my activities after twelve years of research work in the hybrid film circuits area to the education sector. Today the Department of Microelectronics with only 15 years of continuous development of teaching process is one of the largest of the 19 departments comprising the Faculty of Electrical Engineering and Computer Science.

Department of Microelectronics is involved in teaching, research activities and its applications in the following areas: microelectronics technology including hybrid integrated circuits, semiconductor devices, design of the microelectronic circuits, systems and electronic instruments, microelectronic sensors, optoelectronics, computer modeling and simulation of circuits, diagnostics and reliability. For the mentioned areas there are well equipped laboratories at the department. The laboratory of Hybrid integrated circuits was established during 1983 - 84 years in order to provide clasroom/laboratory instruction and research capabilities. Basicly the laboratory, designed for complet thick film process was later also extended for some surface mount technology steps. There were various types of research tasks which were worked out in the past time, as for example the application of thick films for microvawes, chip carrier reliability and thick film substrates for biosensors.

In the last ten years the general development in the microelectronics area was so extrem in all levels including technology, that the position of HIC s had to be understood different than in the previous years. At the begin the HIC s were mostly looked as separated components with very little specific characteristics, but today many of the technological parameters including the mechanical and electrical are collected to general construction so this area can be called hybrid microelectronics. As the word „hybrid" is generally not simple to defin because means a wide variety, here may be understand as the sum, not only of various techniques, but also of circuit construction forms with development trends. This is one of the reasons why we have developed this new methodical approach to the area of the „hybrids". Some principal steps and basic ideas concerning this new approach are described further in this paper..

The general education system and its activity in the Hybrid Microelectronics sector

The teaching process at the Technical University of Brno is divided in two basic steps as shown in figure 1. First step bring the undergraduate and the second bring the postgraduate, they doctorate study to be exact. The last two years there is the bachelor study, which is easier and more general then the undergraduate ones.

The area of hybrid microelectronics comprise generally of two parts. The first part is composed of two one semester courses, which are taken at the second level of the undergraduate study. And the second step is a additional one semester course taken taken at the postgraduate level with the focus on the actual problems and the latestlast news in the area evolving in literature, especielly in journals and proceedings.

The bachelor study is a little briefly compared to the undergraduate and postgraduate studies.The meaning of these lessons is to give clear overview and basic knowledges about hybrid microelectronics.

For main areas there are well equipped laboratories at the department. They are exploited in educational process and also in research and as well as for preparing diploma works. For a design of integrated circuits there is the design laboratory center with Hewlett-Packard workstations. Some of the activities are supported by Tempus projects and other by research programs often formed by manufacturing and production companies.

At the undergraduate level the manufacturing activities make up only a small part of one or more syllabuses. This is often convenient for this to be a part of the laboratory workshop. As compared to the postgraduate level, manufacturing and design activities may form the main part of the course.

G. Harman and P. Mach (eds.), Microelectronic Interconnections and Assembly, 233-236.
© 1998 *Kluwer Academic Publishers.*

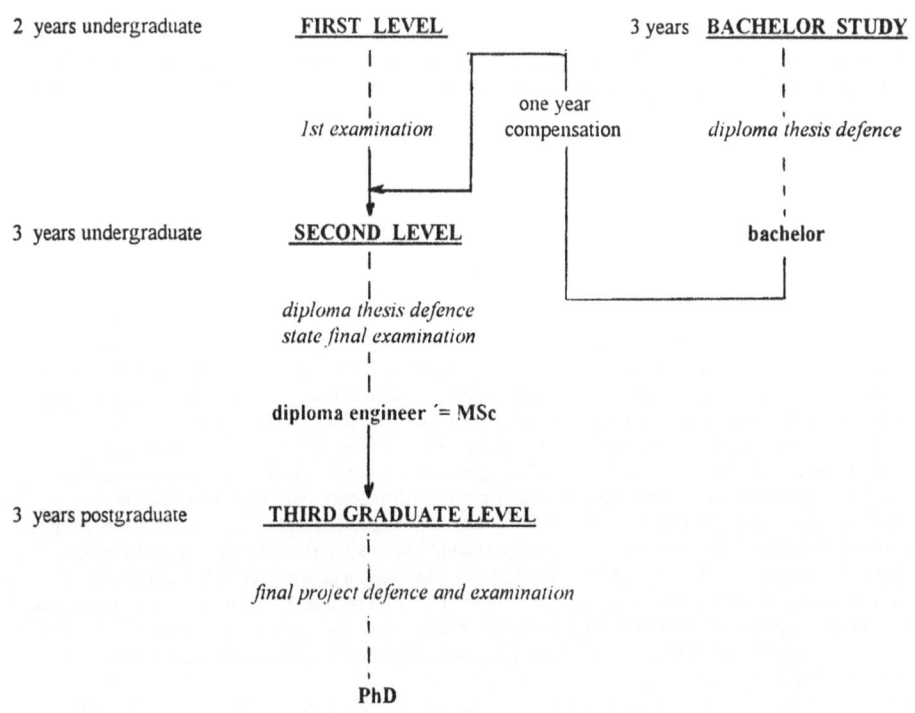

Fig. 1. Organization and system of education at the Faculty of Electrical Engineering

To ensure optimal extent of the education in the area of hybrid microelectronics most of the activities are aimed at the collaboration with industrial companies, like Tesla Lanškroun and Metra Blansko. In the course of time and particularly after political and economical changes in the year 1989 the attention was given more to the foreign companies and especially to the International Society for Hybrid Microelectronics. The reason was simple: to keep strong contact with topical development and to reach common aims under the identical conditions, where both of the parts wont take active interests in the education and training of people, which are studing and working in its industry. The annual Symposiums and Exhibitions all over the world give lot of useful information, and more help in establishing direct contact with experts working in the area of microelectronic.

Teaching of Hybrid Microelectronics

One of the most important sectors in today's microelectronics is the Microelectronic Technology. Generally meaning the science about the materials, processes and methods, which are applied in the production of electronic components, circuits and systems. Years ago we spoked only about hybrid integrated circuits, which have presented more or less as an independent area, but today it is necessary to take in something more complex, which may be known as „ Hybrid Microelectronics". What means the term „Hybrid Microelectronics" is explained in the figure 2.

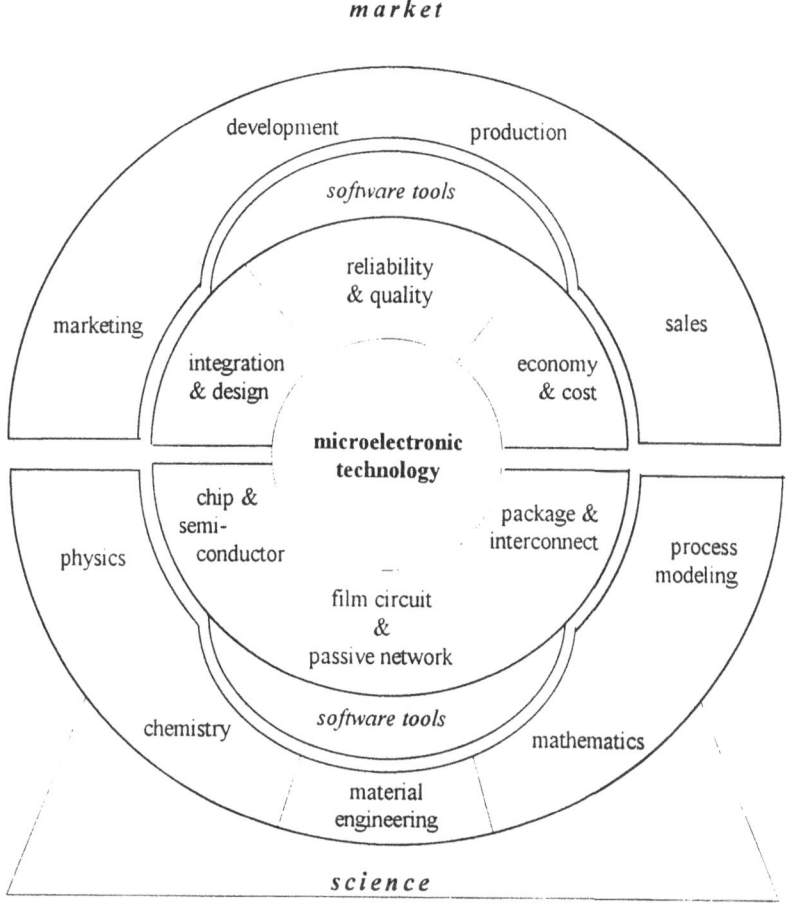

Fig. 2. What is the meaning of microelectronic technology

The substance of the hybrid microelectronics is formed by three parts, which are defineted by symbiose between chip&semiconductor devices, film technologies including theirs modifications, and interconnection techniques including packaging. Hybrid microelectronics may be understood as a part of microelectronics technology, associating with the complex of the superstructural technological processes which are being used in modern electronics production. Under this topic we must see practicly all modern technologies leading to final electronic production involving not only hybrid integrated circuits but also surface mount assembly, tape automated bonding, chip on board assemblies, ball grid arrays, multichip modules etc.

The organization of the Hybrid microelectronic courses in the Microelectronics department of the Technical University of Brno has arranged the second and third graduated level. The methodics are based on the logical syllabuses in a intuition to actual development in this area. There are generally three steps of courses ,, Hybrid microelectronics I, II and III", which offer following topic areas :

I. Hybrid Microelectronics I

1. Introduction to Microelectronics Technology
2. Basic materials used in today's microelectronics
3. Substrates and theirs characteristic properties
4. Monolitic integrated circuits fabrication and wafer processing
5. Thick film materials and their characteristics
6. Overview of thick film technology, thick film process
7. Thin films deposition techniques and materials
8. Overview of thin film technology
9. Design and trimming of passive film components
10. Role of polymers & Cofired ceramic technology

II. Hybrid Microelectronics II

1. Introduction to Hybrid Integrated Circuits
2. Bare and molded chips, semiconductors overview
3. Assembly techniques and microjoining development
4. Glues and soldering - flow and reflow, cleaning
5. Manufacturing and packaging
6. Non conventional thick and thin film applications
7. Introduction to surface mount technology
8. Basic steps in surface mount devices assembly, repair and rework
9. Design rules and assembly of surface mount devices, SMT solutions
10. Multi chip modules - technologies and alternatives

III. Hybrid Microelectronics III

1. Modeling and simulation of general technology process
2. Statistic process controll and reliability
3. Packaging and interconnection. future trends and economics
4. ISO, CE and certification of quality

Some courses are supplemented by instructional video tapes which help to present a detailed demonstration of the technological operations. There are video tapes made directly in the microelectronic production factories, as for example Delta elettronica or Vemer instrumenti elettronici.

The laboratory activities provides hand-on experience in thick film and surface mount technology. „Hybrid microelectronics I" is accompanied by realization work of thick film HIC from each student alone, starting by HIOCAD or HYDE design system and breaking up by tests and measurements. The second course is providing similar realization of surface mount circuits , where each student makes the complet surface mount process alone , in a simplified form. Some operations are followed through more detailed, especially from the point of view of the mathematical statistics and the evaluation of technological operations.

The cooperation with production companies is very important and necessary , as for example with Metra Blansko (Hybrid IC), with Jankovsky Comp. (SMT and Computer applications) or Krejci Engineering (biosensors and medical applications).

Conclusion

There is nothing to discuss about the fact that an extensive leaeniing experience such as this must be well organized for the creative and knowledgeable students, with both, theoretical and practical experience, well preparing them to accomodate the immediate industrial needs. Some of the passed students are working just succefull in industry, and more some of them manage to reach the leading positions.

Today, microelectronic technologies are dramatically and rapidly advancing. Each new technology creates a new product, and a new product creates a new technology. This situation requires a new approach. I hope that this paper may bring something in this resolution.

The obtaining of informations and good collaboration are the first steps which may only be supported by the ISHM activity.

THICK FILM INTERCONNECTIONS
FOR SENSOR APPLICATIONS

Darko Belavič, Srečko Maček*, Stojan Šoba, Marko Pavlin, Marko Hrovat*, Dubravka Ročak*

HIPOT Hybrid, Trubarjeva 7, 8310 Šentjernej, Slovenia
*Jožef Stefan Institute, University of Ljubljana, Jamova 39, 1001 Ljubljana, Slovenia

ABSTRACT - Signal processing for some sensor applications demands circuits with specific characteristics, such as miniaturization, high functional density, high number of input ports, and high signal speed. For some sensors operating in industrial environments protection from electromagnetic interference (EMI) is also required. Thick film technology is convenient way to realize such circuits, but some special technological processes must be used. In this paper some of these will be presented through some examples of thick film multilayer hybrids. First is the multipoint pressure measuring system for medical applications. It is designed for measuring 720 pressure points in a test mattress for hospital patients. The second example is thick film multilayer interconnections for sensor array readout with high function density, high signal speed, and low interlayer capacitances. The third example is of thick film hybrid circuits for proximity switches.

1. INTRODUCTION

Applications of sensors and sensor technology have been to rapid development within the last few years. Thick film technology, can be used in sensor applications for two purposes. One is to produce the sensor element itself and the second to produce an electronic circuit for sensor adjustment (nulling offset, calibrating sensitivity), themperature compensation and signal processing. This electronic circuit must be designed in most cases to be as small as possible. One reason for such miniaturization is the high number of sensors and input/output ports for connections within a limited area. A second reason for miniaturization is the limiting dimensions of the sensor itself.

Generally a transducer includes a sensor element and signal condition electronic circuit. The electrical signal from sensors is mainly very low level, and therefore very sensitive to any kind of interference. On the other hand, the first electronic stage has a large influence on the accuracy and long term stability of the transducer. For these two reasons the

G. Harman and P. Mach (eds.), Microelectronic Interconnections and Assembly, 237-247.

electronic circuit must be designed as electromagnetically compatible, precise, and with long term stability. This is especially important for precise applications in an industrial environment where electromagnetic interferences can be very strong. The transducers must be designed with emphasis on long term stability and on protection from electromagnetic interference (EMI).

2. SPECIAL REQUIREMENTS FOR INTERCONNECTIONS

Through many years experience in thick film interconnections for sensor applications we have met many different requirements. Some of those which occurred more frequently, are described in the following text.

2.1. Miniaturization

There are several multilayer techniques to increase the density of thick film hybrid circuits, e.g., low or high temperature cofired ceramics, screen printed multilayes on ceramic substrates etc. Our facilities enable the production of thick film multilayer technique on alumina substrate with printed thick film dielectric as isolator layers and printed thick film conductor layers for interconnections. Conductor width is down to 150 µm and distance down to 200 µm. Minimum vias dimensions, for interconnections through dielectric layers, are 250 µm square (exceptionally 150 µm). Thick film resistors are printed and fired on a dielectric layer or on alumina substrate. Discrete add-on components are surface mounted devices for soldering and dies for wire bonding. In some applications, the hybrid circuit can be of double sided construction, with through-hole connections.

One aspect of miniaturization is the use of small thick film resistors. In some of our applications we employed thick film resistors of dimensions down to 0.3 mm with commercial thick film pastes, whose quality for this size is not guaranteed by the supplier. Since this technological proceeding is not common, we evaluated the characteristics of untrimmed and laser trimmed resistors of small dimensions [1,2].

Test resistors were printed and fired on alumina substrates and on prefired multilayer dielectrics. Characteristics such are sheet resistivities, TCR, and the influence of repeated firings were measured. Shorter resistors have lower sheet resistivities. No significant difference in the TCR values could be observed between untrimmed and laser trimmed resistors, regardless of resistor geometry. For higher sheet resistivities the TCR of the smallest resistors are shifted to higher, more positive values, which is presumably due to the interaction between the conductor and resistor material. After refiring the sheet resistivities at first increase. Then, in most cases, the values tend to stabilize and then start to decrease after further refirings.

Test resistors were laser trimmed with different settings of the trimming parameters. While the standard deviation of the resistivity values of trimmed resistors increased with decreasing resistor dimensions, no significant correlation could be found between either the Q rate or laser power and the standard deviation.

To evaluate the influence of resistor geometry on stability, resistors were subjected to accelerated ageing to increase the rates of resistor change. Humidity, boiling water, thermal overstress (aged at 85°C, 150°C, and 200°C), and temperature cycling (between - 25°C and 150°C) were used as acceleration factors. The resistivity drift of resistors, aged 500 hours at 200oC, are shown in Fig.1. Using the Arrhenius equation for thermally accelerated ageing of resistors (ageing at different temperatures), the activation energy of the ageing mechanism(s) which causes the resistivity drift, can be calculated. Higher values of the activation energy indicate more stable resistors. The resistivity drifts after ageing in boiling water and after temperature cycling was higher for laser trimmed resistors, which could be attributed to residual stresses in the laser trimmed material. Similarly, activation energies, calculated from resistivity changes after ageing at different temperatures, were lower (meaning less stable resistors) for laser trimmed resistors.

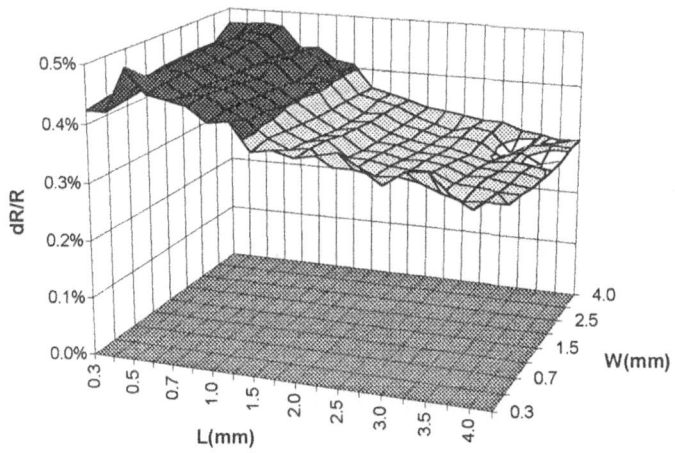

Fig.1
The resistivity drift of resistors after aged 500 hours at 200°C

Generally, in most cases resistivity drifts of small resistors are, to small extent, higher than those of larger ones, but, however, the results obtained indicate that is possible to produce stable small size resistors with commercial thick film resistor pastes.

2.2. Electromagnetic interference (EMI)

More and more electronic devices and equipment are designed for electromagnetic compatibility (EMC), in order to be immune from electromagnetic interference (EMI) and also not to be themselves generators of EMI.

The IEEE definition of electromagnetic compatibility [3] defines EMC as the ability of a device, unit or equipment or system to function satisfactorily in its electromagnetic environment, without introducing intolerable electromagnetic disturbances to anything in that environment.

Some essential requirements for EMC:
- The device or equipment should not interfere with radio or telecommunication equipment operation;
- The device or equipment itself must be immune from electromagnetic disturbances due to sources such as RF transmitters and other equipment emitting EM radiation;
- The device or equipment itself must be immune from electrostatic discharge in environmental and installation conditions

Some general design rules for EMI reduction in thick film hybrid circuits [4] are as follows:
- Do not mix analog and digital parts of a circuit
- Devices of one logic do not mix with devices of another logic if the logics have different functions or different clock speeds
- Sensitive signal lines for high speed logic (clock over 1 MHz) must be short, highly conductive, without branching (no connected lines), and without breaks (angles), or with only obtuse angles
- Ground and power lines must be formed in a "star" shape or one layer used for these two functions
- A block capacitor must be located as close as possible to IC with short connections to IC pins
- If possible a metal housing should be used as a shield

Some practical design rules for the distribution of interconnection layers in thick film multilayer circuits ([5]):

- The first layer (on the alumina substrate) was used for critical signal interconnection
- The second layer was used as ground reference plane for layers one and three
- The third layer was used for non-critical signal lines and clock lines
- The fourth layer was used for power supplies and as ground reference plane for layer five
- The fifth layer was used as a die attach surface and for I/O interconnections

2.3. Power line transients

High level impulse voltages can be coupled through the power line into the power supply of the transducer. Such power line transients may cause a noisy signal, or they may result in constant signal offsets in extreme cases.

Power line transients are usually caused by switching heavy inductive loads such as motors on and off. Thus, this type of interference can be minimized by making sure that no such activities take place while using the sensor, and that no transient producing loads be connected to the transducer. In practice, power line transients can be further reduced by using one of the many commercially available power line filters, or incorporating analogous filters in the transducer. In the last case the power line in the transducer must be designed with special rules (short as possible, "star" form, shield, …) and some new components (suppresser diodes, block capacitors, inductance, owerload resistor, …) must be added.

3. SOME ILLUSTRATIVE APPLICATIONS

3.1. Multipoint pressure measuring system

A multipoint pressure measuring system for medical application was designed to measure 720 pressure points in a test mattress for hospital patients. Based on these data, a specially designed mattress will be constructed. Due to the measuring speed (1000 measurements per second) data can not be obtained with one sensor, but with a group of sensors which were constantly connected to power supply and pressure. Signal switching from sensors to analog output was controlled by multiplexers and microprocessor. To meet all these demands the system was realized in three stages. The first stage (Fig.2) was a multilayer thick film circuit which integrated and interconnected eight silicon sensors, the second stage (Fig.3) was a PCB slot where fifteen hermetically encapsulated hybrid circuits were placed and interconnected (electrical and pressure). The third stage (Fig.4) was a complete system with six PCB slots, power supply unit, central control unit, personal computer, and mechanical construction of the pressure interconnections.

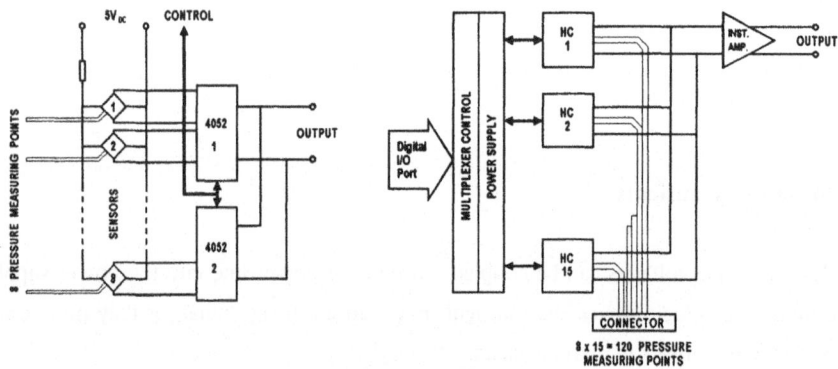

Fig.2 and Fig.3

Multipoint pressure measuring system   -   first and second stage of interconnection

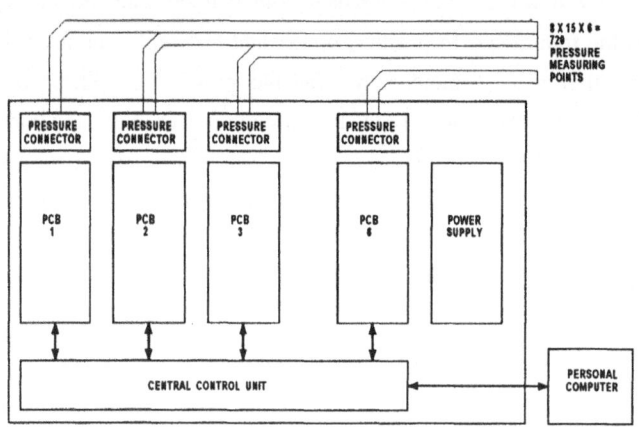

Fig.4

Multipoint pressure measuring system   -   third stage

This multipoint pressure measuring system could measure pressures of 10 mBar with an accuracy +/-10%. The 720 pressure measuring points were connected by plastic tubes (outer diameter 0.6 mm) and special connectors to the 720 pressure sensors. Each pressure sensor was a silicon piezoresistive sensor for gauge pressures of 0 to 100 mBar and with a sensitivity between 20 and 30μV/V/Bar. The bridge resistivity of the sensor was 380 Ohms and the voltage supply at the sensors 5V DC.

The first level of integration was the thick film multilayer circuit with eight silicon sensors, eight pressure inputs, one pressure reference input, components for calibration and temperature compensation, two 4052 multiplexers, and analog and digital I/O ports. The circuit is realized on an alumina substrate (dimensions 31.0 × 17.0 × 0.64 mm) with nine holes (1.0 mm diameter). Interconnections were realized in three layer thick film technology, chip and wire technology, and soldering SMD components. The hybrid circuit was encapsulated in a plastic housing with dimensions 33.0 × 20.0 × 10.0 mm (Fig.5). Eleven pins were used for analog and digital I/O ports and for voltage supply. For pressure supply, nine metal tubes of outer diameter 0.4 mm were used.

After calibration, temperature compensation, and multiplexing of the first stage, the output signal was between 50μV and 500μV. This signal was amplified and multiplexed on the second stage of integration. The second stage is two sided PCB slot with dimensions 180 × 100 mm (Fig.6). It integrated fifteen hermetically encapsulated hybrid circuits, a multiplexer and communications logic, instrumentation amplifiers, power supply regulators, an I/O connector, pressure interconnection with plastic tubes, and a special 120 pole pressure connector.

Fig.5
Thick film multilayer circuit with eight silicon sensors

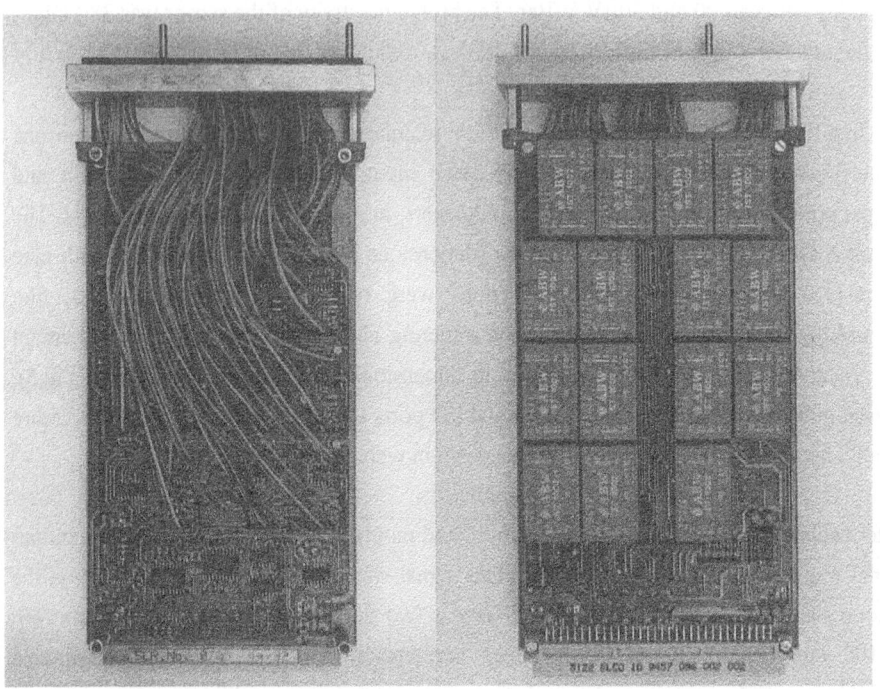

Fig.6
Rear and front side of PCB with fifteen hybrid circuits

3.2. Signal processing for diode array detector

Hybrid thick film multilayer interconnections for a sensor array readout with high function density, high signal speed and low interlayers capacitances were designed (Fig.7). The device serves as a mechanical support for detectors and amplifiers chips, and provides power control signals for electronics and to drive output signals. Such thick film circuits interconnect 512 inputs (from sensor array) to two ASIC dies with 512 low noise amplifiers with correlated sampling. Hold circuits are implemented for background and signal storage, with digital multiplexing stages with 512 channels for series readout via a flexible Kapton strip cable.

Interconnections are realized in four conductive layers with thick film gold and silver conductors. Minimum conductor line widths are 100 μm. Interconnections through dielectric layers are achieved with very small vias (150 μm square) and via gold fill material. Various conductors and dielectric materials have been tested, to allow a reliable low resistivity of less than 3 Ohms / 50 mm long line, a load capacitance less than 1.5 pF, a leakage current of less than 1 nA and line to back plane capacitance of less than 5 pF.

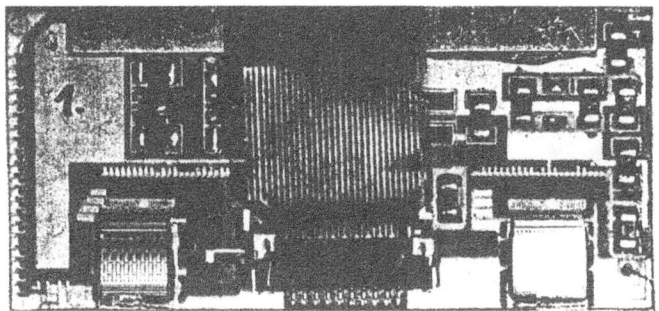

Fig.7
Hybrid thick film multilayer for diode array detector

Surface mounted components were soldered to a DP4596 glass bonded solderable Pt/Pd/Ag conductor. It has good solder acceptance and high adhesion on an alumina substrate and DP5704 dielectric with high leach resistance. Multilayered gold conductors and vias fills were made of DP5744 gold. Two interdigital conductor plane layers were incorporated to decrease cross-talk and increase read-out speed to 2 MHz, or 500μsec for next trigger digitalization.

Gold conductor material offers reliable bonding of VLSI chips to a multilayer circuit with 8 μm Al wire. The ceramic substrate is made of 99% aluminum oxide with a thickness of 250 μm to increase heat conductivity to the metal heat spread. A high density surface mounted 20 pin DC 0.5 mm flex connector is soldered to the top conductor layer, allowing interconnection via a Kapton flat gold conductor cable for digital and analog output electronics.

3.3. Proximity switch

Currently the majority of signal conditioned proximity switches employ thick film hybrid technology. We are producing several signal conditioned circuits for proximity switches in thick film hybrid technology (Fig.8). The technology for proximity switches has some specific requests, such as the restricted dimensions of the hybrid circuit which must fit into a tube with an inside diameter of 3.0 mm and a length of 25 mm, high component density, and electromagnetic compatibility. To meet these demands, we developed a technological process for two sided multilayer screen printing and component attachment (soldering and wire bonding), through hole (diameter 250µm) printing, and printing miniaturized resistors down to 0.3 mm size.

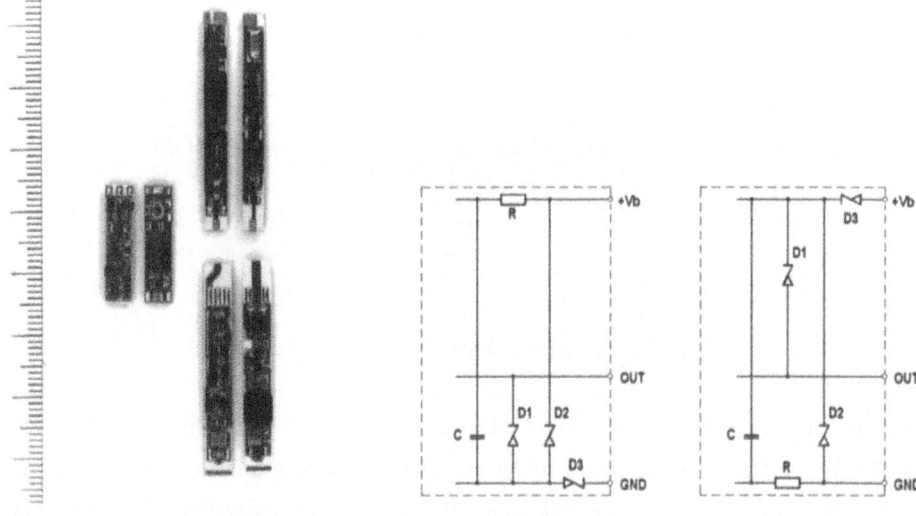

<table>
<tr><td align="center">Fig. 8
Hybrid circuits for proximity switches</td><td align="center">Fig.9
Added components for EMI protection</td></tr>
</table>

The proximity switches were designed to meet the generic European EMC directive, which lists certain criteria for tests to be carried out according to IEC standard 801. The proximity switches were designed to meet the following IEC specification:

- part 2 for electrostatic discharge to case at 8 kV (air discharge) and 4 kV (contact discharge)
- part 3 for radiated field from 27 MHz to 500 MHz at 10V/m
- part 4 for electrical fast transients applied to power ports, amplitude 2 kV

For EMI protection of proximity switches we used a special topological design (short lines, ground in "star" form, shield, ...) and added suppresser diodes, owerload resistors and block capacitors (Fig.9).

4. CONCLUSION

In this contribution we wished to demonstrate that thick film technology can be successfully used in sensor applications. Some realised examples of thick film multilayer hybrids were presented: a multipoint pressure measuring system for medical applications; thick film multilayer interconnections for diode array readout with high function density, high signal speed, and low interlayer capacitances; and thick film hybrid circuits for proximity switches. The ability to produce miniature, robust, and electromagnetic compatibility interconnections with flexible and low-cost production in small and medium quantity is a guarantee that thick film technology would play an important role in the "world" of sensors.

5. REFERENCES

1. D. Belavič, D. Ročak, S. Mojstrovič, M. Hrovat, M. Pavlin, Characteristics of small size thick film resistors, Proc. 23rd Int. Conf. Microelectronics MIEL-95 / 31st Symp. on Devices and Materials SD-95, Terme Čatež, 1995, 151-156
2. D. Belavič, D. Ročak, J. Fajfar Plut, M. Hrovat, M. Pavlin, An evaluation of stability of small size untrimmed and laser trimmed thick film resistors, Proc. 23rd Int. Conf. Microelectronics MIEL-95 / 31st Symp. on Devices and Materials SD-95, Terme Čatež, 1995, 157-162
3. International Standard IEC 801
4. L. Bosley, K. P. Slattery, D. Canestrari, Multichip modules: Analysis of electromagnetic interference effects by MCM implementation vs. single-chip package approach, Proc. Int. Conf. on Multichip Modules, Denver 1995, 120-125
5. R. Goyal, I/O modeling for EMI analysis of high speed digital design, Int. Symp. on Microelectronics ISHM-95, Los Angeles, 1995, 451-455

PTF (POLYMER THICK FILM) INTERCONNECTIONS IN TERM OF LONG TIME RELIABILITY

Ján URBANČÍK

Technical University of Kosice,

Department of Hybrid Microelectronics,

Park Komenskeho 2, Kosice,

Slovakia

Thick film technologies using polymer materials as a conductors, resistors and insulations in several layers were enhanced into electronic production as a routine technology at the present time. Together with expansion of the PTF (Polymer Thick Film) technology a lot of the interconnections based on the polymer materials arise everyday. Becase the failures of these interconnections are the important part of the complete electronic device features, the reliability and electrical facilities of interconnections are one of the essential attributes of there.

This paper deals with the reliability aspects of microelectronic interconnections based on PTF technology by using several types of rigid and flexible substrates. The aim of this contribution is to follow the influence of the long time processes in conductive layers to electrical properties of the PTF based interconnections from the quality and reliability aspects points of view.

I. Introduction

The Polymer based materials are characterized by macromolecular chain structure with a lot of mutual chemical binding. Either the disintegration of the macromolecular chain or the changes of the chemical structures followed by the chemical reactions of arised fragments are phenomena going along with aging process in PTF interconnection structures. Also, the ageing process may cause separation of the plasticators and it has next negative influance to their electrical and mechanical facilities. The depolymerization process usually causes the decrease of the mechanical resistance, stability, flexibility and adhesion of PTF layers. Further more the low range of working or storing temperatures may cause that PTF layer fragility increases. In term of electrical properties, decrease

G. Harman and P. Mach (eds.), Microelectronic Interconnections and Assembly, 249-255.
© 1998 *Kluwer Academic Publishers.*

of the electric strenght, increase of the dielectric loss and resistance changes are results of the ageing process and chemical reactions in polymer based layers and interconnections, too.

The speed of the chemical reactions in ageing or in the depolymerisation process mentioned above depends on the macromolecular structures and it is influanced by the temperature, the humidity, the light intensity and the presentation of chemical accelerators at the working or storing environment.

This paper discusses the polymer material properties and some measuring methods of the electrical parameters in term of the long time processes using accelerated life tests. It follows the influence of the material and substrate choice, interconnections starting quality, contamination of the active area and influance of the manufacturing, working or storing environment.

I. Accelerated life ageing process

Reliability tests have another denotation as operational unit tests which verified main functional properties and which are realized for relatively short time and in a limited environmental conditions. These tests must demonstrate that the tested device will be able to operate without the failure during definite period in concrete background and in required working conditions. The singularity of these tests is their higher demands to time, equipment and expenses, strict preparation and serious evaluation. Moreover the long time tests do not enable feedback to evaluate technological data during manufacture processes. On this accont these tests are substituted by accelerated tests in range from 1000 to 1500 hours. The failure evolution can be assumed from these results in term of long time interval (e.g. two years).

Physics of the failures is based on the process laws applied at the variable materials during vary conditions and energetic states. These laws are used for monitoring of the degradation process in materials or for specification of the technological failures during manufacturing process. The accelerated life aging process is based on increasing of degradation process speed which enables to estimate lifetime data and determines the properties with regard to long time processes. The basic accelerative value for physical and chemical reactions is the temperature.

The reactions which caused degradation processes depends according by the Arrhenius formula (1). This provides acceleration of the reaction speed if the temperature is increased.

$$r_A = C \cdot \exp\left(-\frac{\Delta E}{kT}\right) \tag{1}$$

where
r_A is the speed constant of the reaction
ΔE is the Activation Energy, consequently the minimun energy value necessary for initiate of the degradation process [eV]

C is The Constant
k is the Boltzman Constant, $k = 1,38.10^{-23}$ $[JK^{-1}]$
T is the Absolute Temperature at degradation process place

If the value of the Activation Energy for tested material is unknown, it can be calculated from the relation of two failure rates which were required from tests realized with two known different temperatures from the formula (2).

$$\frac{\lambda_1}{\lambda_2} = \exp\left[\frac{\Delta E}{k}\left(\frac{1}{T_2} - \frac{1}{T_1}\right)\right] \tag{2}$$

where
λ_1 is the failure rate at temperature T_1
λ_2 is the failure rate at temperature T_2
k is the Boltzman Constant, $k = 1,38.10^{-23}$ $[JK^{-1}]$
ΔE is the Activation Energy

II. Testing and the measurement methods

The diagnostics of the electrical interconnections is an integral part of device development and of the device quality and reliability assurance. The measurement methods are generally divided into two basic groups:

1, electrical methods, e.g. measurement of electrical parametres,

2, nonelectrical methods as thermal, ultrasonic, optical or holographic diagnostics and measurement of mechanical properties.

Another division knows destructive tests which represent especially mechanical methods as bond pull test or shear test and nondestructive tests, e.g. electrical measurements.

The life ageing and accelerated life ageing with the combination of the most appropriate choice of the methods and measurements listed above have separate position among test methods has. The combination which includes accelerated life ageing provides satisfactory results in term of the long time reliability.

Considering the type of the interconnections the measurement method was choiced from first group mentioned above. It was measured only the basic electrical parameters e.g. the interconnection resistance and their AV characteristics in range from 50 to 500 mA. On the ground of the very low values of the interconnection resistance it was selected measuring accurancy method which eliminates all of the interferences from background along with respect of the starting quality, choice of the matrix and the conductive materials in PTF, production technology, curing parametres, nonspecified contamination and chemical influences on the active area. The principe of this method, also called The Kelvin method, is based on the fact that through measured interconnection flows exact direct current and drop of the potential is measured concurrently by the Figure 1.

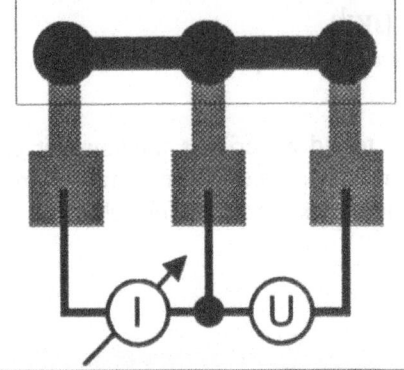
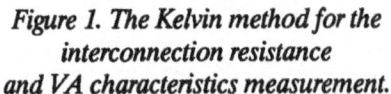
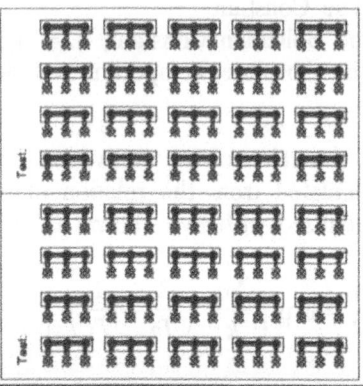

Figure 1. The Kelvin method for the
interconnection resistance
and VA characteristics measurement.

Figure 2. The complete test layout
on substrates.

III. Experimental part

The electrical resistance and VA characteristics were measured by the method described above. This method requires three contact points and for it we applied special direct connector with three gold plated contacts. All of the measurements were controlled by computer with separate interface and there were supported by software which enables display, evaluate and storage measured data on the hard or floppy disk. One of the advantages of this test method is the high speed of measurement and evaluation of measured results in real time. In addition, this method is nondestructive and can be used in regular circuits without any influance to the function.

Also, all of the bonds were subjected to the optical control before testing and interconnections which were damaged by handling were removed from another measurement.

The complete test layout of the tested substrates is drawn in the Figure 2. It consists of 40 basic subjects and all of them were subjected to measurement of the potential drop for more than 50 different values of constant current in range from 50 to 500 mA. The measurement results are represented by the AV characteristics and the average values of the interconnection resistance. Typical AV chart with calculated values of the resistance and its average value is drawn in Figure 3.

In compliance with ageing process by the Arrhenius rule (1) the substrates were exposed to effect of temperature 150 dgr. of Celsius at 1000 hours and the electrical measurements were realized in ageing intervals 0, 20, 50, 100, 200, 500 and 1000 hours. There were tested and compared two conductive polymer inks manufactured by DuPont (E5007, 18 mOhm per square at 25 um thickness) and by COATES (XZ 250, 30 mOhm per square at 25 um thickness). The well compatible polymer ink COATES 40-317 with resistivity $5,5.10^9$ Ohms at 500V and with breakdown voltage 800kV was used as a cover dielectric layer.   As substrates were used standard Al₂O₃ ceramics, flexible foils and standard Printed Circuits Boards with first conductive layer by Figure 4, etched from copper.

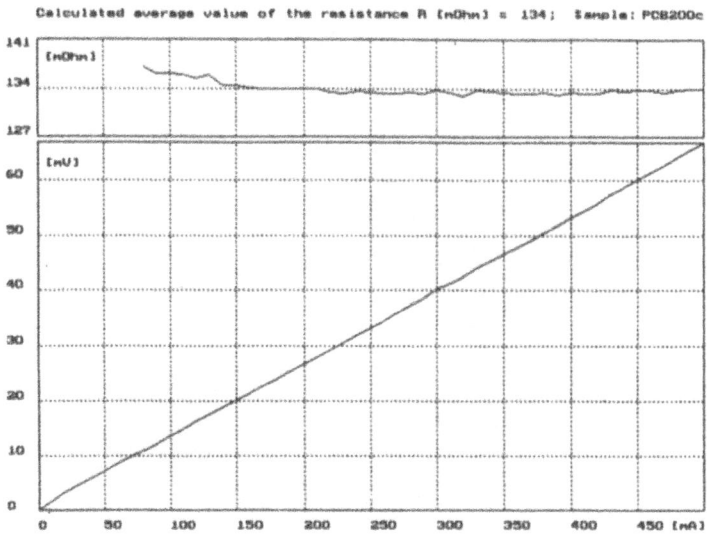

Figure 3. Typical measured AV chart
(Substrate: PCB, First conductive layer: etched copper,
Second conductive layer: COATES PTF, Ageing interval: 200 hours).

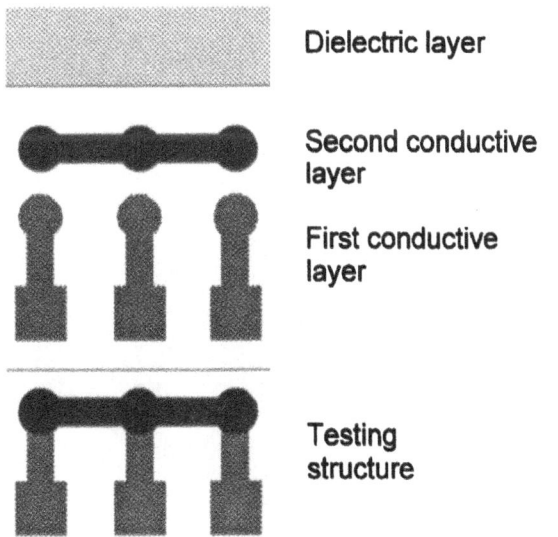

Dielectric layer

Second conductive layer

First conductive layer

Testing structure

Figure 4. The layout of the measured interconnections.

254

The measured results show Figures 5, 6 and 7. All of them represent expected values from more then 100 thousand measurements after data processing by methods of mathematical statistics. The most interesting resalt is the low value and the high stability of the interconnection resistance for the DuPont polymer conductive ink at PCB during all measurement intervals.

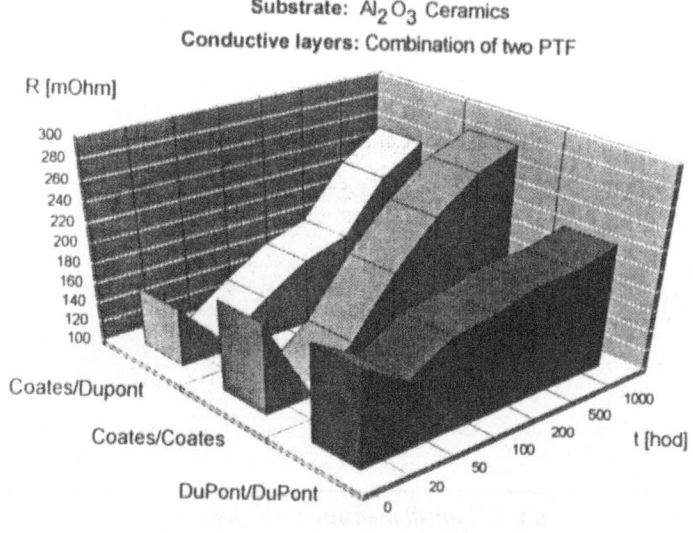

Figure 5. The continuance of the average interconnection resistance at standard ceramics.

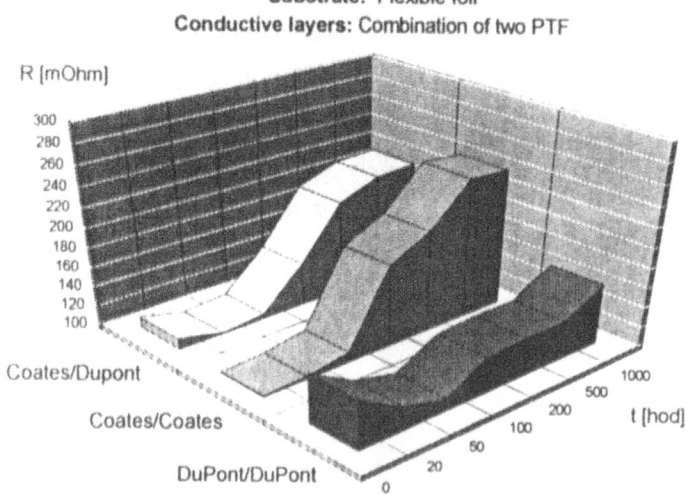

Figure 6. The continuance of the average interconnection resistance at flexible foil.

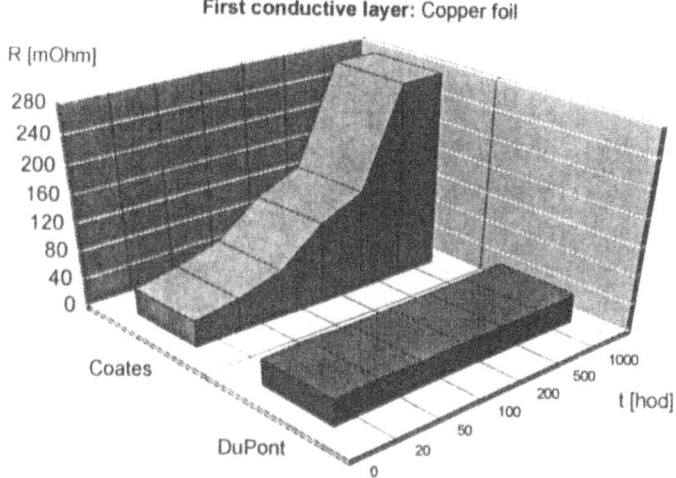

Substrate: Printed Circuits Board
First conductive layer: Copper foil

*Figure 7. The continuance of the average interconnection resistance
at standard Printed Circuit Board.*

Illustrated results reference about the fact, that described method of tests applied at various input configuration of conducting materials reports satisfactory results. Following these tests it can be evaluate the rate of materials compatibility and the influance of production technology changes in term of long-life durability and reliability. Test method also enables to choice the best combination from various type of substrates, conductive inks and adhesives for interconnection technology.

The Usage of the Polymer Inks in Interconnection Technology

František KOLESÁR, Miloš SOMORA

INTERMIKRA s.r.o., member of Science and Technology Park CASSOVIA TECHNOPOLIS
Vysokoškolská 4, 042 00 Košice, Slovak Republic

Abstract

Polymer inks are used in the interconnection applications mainly as a replacement of more expensive and environmentaly unfriend classic technologies. The paper deals with the selected problems of cross-overs realised on the base of polymeric inks onto the PCB's and the through-hole metalisation as well. The use of traditional two-layers or multilayers printed circuit boards with chemicaly plated through-holes, equipped with hole-mounted or surface mounted components, will not necessarily result in the best and cheapest product at the required quality level. It seems the usage of polymer inks is the good solution as an additional technology (for high temperature thick film, thin film or conventional PCB's) alone or in combination, a product can be designed to fit the given working, space, quality and price conditions optimally.

Introduction

The main advantages of the PTF are:

- low price

- fast production of prototypes and production volumes

- suitable for repairs / modifications of regular PCBs

- simple processes

- integrated resistors and potentiometer tracks can be made and trimed to required value

- PTF printed through holes can replace the more expensive plated TH process

- carbon printing can replace costly gold plating for keyboards, multilayer ceramic circuits

Specialties:

- membrane switch panels and front plates

- 3-D features on moulded PCBs.

The main disadvantages of PTF:

- satisfies only moderate environmental requirements (typical working temperature range from -40^0 $+85^0$ C)

- only moderate circuit complexity

- only low / moderate power aplications

- high conductor resistivity (>20x that of Cu foil, except when plated PTF conductors are used)

- variation in conductivity and other different properties from batch to batch of the materials, what cause increasing the tolerances in final circuit properties

- special design rules, not used in regular PCB design

- limited solderability

- limited availability (few subcontractors in Europe) [1]

G. Harman and P. Mach (eds.), Microelectronic Interconnections and Assembly, 257-261.
© 1998 *Kluwer Academic Publishers.*

258

The above list therefore cuts the market sections for application possibilities. The PTF advantages bring reasons for a relatively high popularity of PTFs, the disadvantages and restriction using of them respectively. The wel-known screen-printing technology combined with polymeric inks also aply to the chemical sensing field. Moreover the development of screen-printing inks for the chemical sensitive layers is quite simple and facilitates modification of the composition of the layer. As was mentioned, there are very few producers of the commercial polymeric inks. The production is by customers divided into the materials for

- contacts and cross-over conductors,

- resistor systems,

- membrane keyboards,

- touch panels,

- assembly operations,

- additive circuits by silver and copper pastes,

- conductive adhesives,

- air dry type conductive pastes on the flexible boards respectively.

The most important producers are Du Pont de Nemours, Asahi Chemical Research Laboratory, Acheson etc. [2]

Several own experiencies with PTF´s.

1.a Carbon PTF for non-critical conductor straps and contacts

In our cooperation with industry at development new model of telephone set was request to place keyboard onto a single sided PCB where other contact is to be used with conductive rubber . Moreover to avoid the usage of double sided PCB there are several wire straps where resistivity is not crucial.

Solution brought single sided PCB based on CEM-1 (35μ Cu foil) with standard process with epoxy UV cured green solder mask, in places for crossover reinforced by one additional layer of the same material. During development was followed necessity of that additional layer but reliability tests showed that the additional layer is necessary because of lower insulation resistivity or higher direct shortcut probability respectively. Copper keypads were directly covered by PTF carbon thermally cured in conventional furnace without silver PTF innerlayer (as some authors recomend to reduce contact resistance between Cu foil and carbon). During printing of keypads are also print conducting straps which replace wirestraps. The keyboard is covered with temporary topcoat - „peelable solder mask", because of hot air levelling and wave soldering. All necessary electrical and environmental tests fulfill awaiting results and confirmed possibility of introduction PTF materials in combination with conventional PCB. Despite of the obvious simplifications of the new board, the exact economy comparison with the conventional board wasn´t done.

Carbon PTF — Peelable solder mask — Cu foil

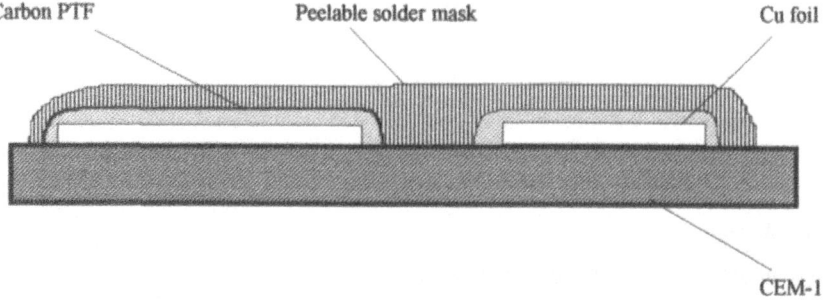

CEM-1

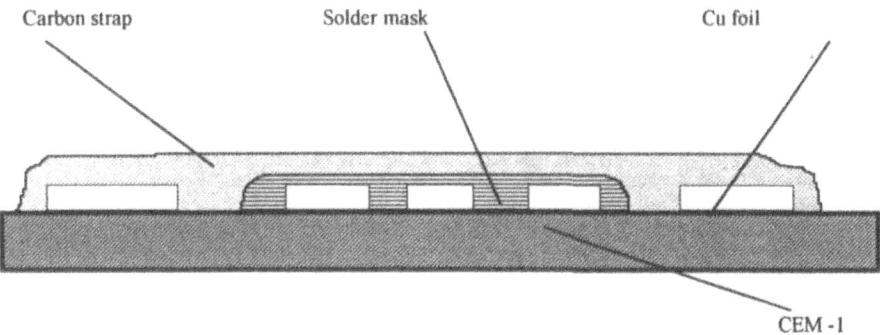

Carbon strap   Solder mask   Cu foil

CEM -1

1.b PTF printed through-hole contacts: silver- through hole (STH).

To use PTF silver in replacement conventional through hole plating must designer thoroughly take into account restrictions mentioned in introduction of this paper and special rules valid for design the layout:

- STH is suitable for electrical contact only. Silver filling in hole not allow to use the hole for component leads because wave soldering causes leaching and damages the conection inside the hole.

- The layer structure in CAD system must include two separate layers reserved for the top and bottom through - hole PTF silver lands.

- From the CAD system must be possible to get drill coordinates for the PTF STH separately from the coordinates of other holes, since the STH hole coordinates must be known for making vacuum plates.

- Solder from the wave soldering process must never come into contact with the PTF silver. Therefore the STH must be protected with solder mask coating. For maximum safety a dot may be added in the component notation layer on the wave solder side.

Despite until now PTF STH wasn't introduced in mass production in our country we have some experimental results that are in accordance with published sources, which we are going to offer our partners in innovation their production. In our experiments we concentrated only on way of metalisation of hole itself using vacuum and evaluation the resistivity of metalisation. Reliability and environmental tests will be subject of exploring at real application.

The PTF silver paste was printed on one side and then sucked approximately 75% through the holes with vacuum under strictly controlled conditions. Then the PCB was turned around and the process was repeated. This resulted in overlapping of silver paste from the two sides in the hole and good electrical contact from side to side.

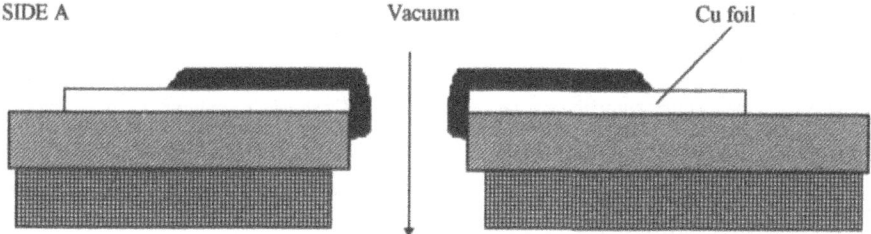

SIDE A   Vacuum   Cu foil

SIDE B   PTF silver

                                                                                    PCB

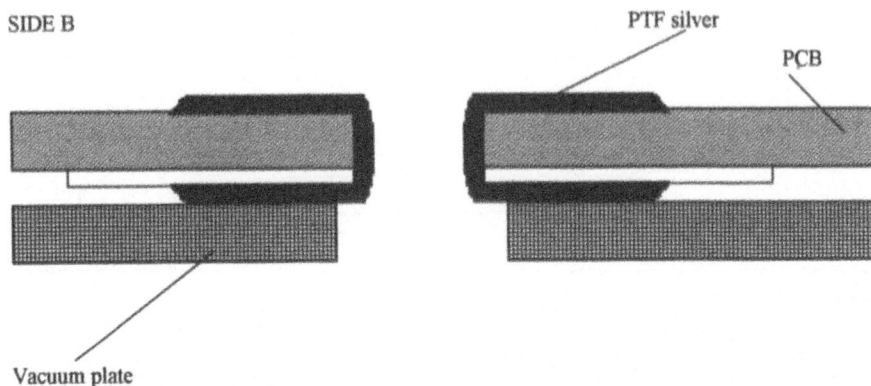

Vacuum plate

2. PTF for membrane applications on flexible substrates.

Former solution of membrane switch construction needs 5 different prints e.g. twice conductive layers twice dielectric layers and one carbon layer on flexible treated polyester substrate. Each layer requires separate curing process. The sequence of printing is as follows:

a) Print of first - conductive layer - which gives basic information about displacement of 16 touch key contact areas and about contacts for end conector. Here is given initial conductivity of membrane loop and in our case moves in range 10-25 ohms. It is cured in IR oven at 110^0C.

b) Print of second - carbon layer - which protects Ag conductive layer in areas which must stay conductive and to face environmental and mechanical stresses. It is cured at the same condition as conductive layer.

c) Print of third - dielectric layer - which ensures good insulation for crossover tracks only on the local places and is UV cured.

d) Print of fourth - conductive layer - which slightly modifies end conductivity and ensures logical properties of membrane. From this step is possible to check all electrical properties of product. Layer is cured as in item a).

e) Print of fifth - dielectric layer - which overlay all parts which are created by Ag conductive layer. It is cured by UV.

That solution offers full technical comfort which requires customer and what is given in his technical conditions for that product.

Technical requirements:

1. Resistance "on-loop" of membrane is max. 100 Ohm
2. Electrical withstand: 1.000 V, 50 Hz / 1 min.
3. Insulation resistance: min.20 MOhm (500 V DC)
4. Humidity warm Ca 4 in accordance with Slovak technical standards STN 03 8824
5. Withstanding against Sulphuric dioxide 2N STN 03 8130

Experience during experiments shows that the most critical operation is correctly positioning (registration) first conductive layer and carbon layer despite relatively free design of artwork . Analyse of this phenomenon showed that wrong positioning caused up to 2-3% rejects in production depending on concentration and skill of operator.

To avoid this problem, reduce rejection and save labour we decided change the technology with agreement of customer but without whatsoever changes of technical conditions. The main idea is blending polymer inks - conductive and carbon, set the correct value which fulfill all technical requirements as above.

Using blended ink means to have ink with reduced conductivity but with improved withstanding against ambient conditions, what allows to reduce one step during printing process. The question is which two layers we will print together. One can print first conductive together with carbon, and/or carbon with second conductive together. While second layer has less "squares" than first layer and in addition there is enough place to reduce number of squares (extending of width), we decided to use second possibility hoping to reduce technical properties of product as low as possible (conductivity of loop as high as possible, carbon on contact areas as much as possible).

Additionaly we have changed former polymer inks and replaced them with COATES series because of price. Results with COATES inks are comparable (probably lower electrical withstand of dielectric ink caused by too thick stencil and printed layer not bubble free and without pinholes). During process of changing technology we prepared several samples of blended inks which acknowledged results from data sheets of COATES (see figure of blended surface resistivity). For our purpose it was chosen with blended ratio 70% Ag / 30% C.

Due to new technology there was prepared other artwork where until now is not included possibility of extension conductors. Layout for first conductive layer differs from the former. It is almost the same, excluding 16 touch key areas which are omitted Sequence using blended ink is now as follows:

a) Print of first - conductive layer COATES XZ 250 - handling the same way as in former sequence

b) Print of second - dielectric layer COATES 40-317 - handling the same way as in former sequence

c) Print of third - blended ink 70% XZ 250/ 30% 26-8203 (carbon) - handling the same way as in former sequence

d) Print of fourth - dielectric layer - handling the same way as in former sequence

This way were prepared several samples which were tested at Tesla Stropkov laboratories. Electrical tests didn't show any problems. Climate tests are in progress therefore we cannot give more information now. In addition customer wants to make one test more "shelf life test" e.g. withstanding of blended layer against conductive rubber - withstanding at least 500.000 on-off cycles without any change of surface of blended layer. The test itself takes one month time.

Conclusion

To make serious conclusions now is impossible it is neccesary to wait rest of results from oficial tests and check this new technology in large production. Nevertheless we suppose improving total yields - increasing capacity of production, decreasing rejection, saving labour etc., what creates space for reduction price at adequate quality of product. It means not overpaid quality.

Literature

[1] Halbo L. - editor, Polymer Thick Film Technology, Special Issue of the ISHM - NORDIC, 1990

[2] Catalogue sheets of Du Pont Electronic Materials, ASAHI Chemical Research Laboratory, ACHESON, COATES etc.

Production of metallic patterns with the help of highresolution inorganic resists

Alexander V.Stronski

Institute of Semiconductor Physics, National Academy of Sciences, Ukraine
45 Prospect Nauki, Kiev, 252650, UKRAINE
Phone: 380(44)265-62-05, Fax: 380(44)265-83-42, E-mail: sekret@ spie-ukr.kiev.ua

Abstract

Properties of chalcogenide inorganic resists and their applications, in particular in microelectronics, are considered. The development of microelectronics suppose the utilization of technologies providing elements sizes with submicron sizes and less. Important place in the microlithography processes have lightsensitive media to which rigid demands are applied, that are conditioned by the necessity of the high quality of microrelief patterns. Chalcogenide inorganic resists in recent time were succesfully applied for various spheres of application , including microlithography.

In present work the peculiarities of physical, chemical processes that proceed in such resists under image formation are considered. Among them: the factors that are influencing the lightsensitive properties of inorganic resists, including the influence of inorganic resist formation condition, dependence from the parameters of layers, that are forming the inorganic resist, spectral and temperature dependencies of photostimulated interaction, properties of products of photostimulated interaction,including their chemical, optical, electrical properties. The results of the investigations of image formation in such systems, mechanisms of photostimulated interaction are discussed, as well as technologies of inorganic resist production and design forms of their realization. Results of their application in microlithography are presented.

The main resist parameters enable them to be well inserted into the modern production technological lines. They showed very high resolution capability: spatial frequency under HOE production - up to 4000 lines per mm, 0.2 μm lines under direct laser recording, nanolithography possibilities were shown, optical lithography: element sizes (layers of various metals:Cu,Cr,Mo,etc.) 0.4-0.5 μm as in positive,as well in negative lithography.The wide range of possible chalcogenide composition, and forms of realization, enables choosing the proper sensitivity on the given wavelength, if necessary. Sensitivity values ordinally are within 1-10 cm^2/J, but this values can be increased under the direct laser recording (up to 300 cm^2/J), under the temperature increase, different design and scheme realization. Such media are also sensitive to electron beams (10^{-3} - 10^{-4} C/cm^2) and ion beams. Contrast coefficient values are within 0.5- 10. Such media can be prepared with the help of vacuum evaporation or deposition from the solution. Postexposure treatment can be wet or dry.

As seen from this data on the resolution capability values, spectral sensitivity range and on range of contrast values the inorganic resists satisfy the demands applied at present time. The additional advantage of this resists that they are sufficiently higher than organic resists in mechanical and thermal strengths, and also in the absence of shrinkage under the treatment.

CONTENTS

G. Harman and P. Mach (eds.), Microelectronic Interconnections and Assembly, 263-293.
© 1998 Kluwer Academic Publishers.

Introduction

The investigations of the photoimaging properties of the thin films of chalcogenide vitreous semiconductors (ChVS) and structures on their base began with two primary publications [1,2] by the group from the Institute of Semiconductors (now Institute of Semiconductor Physics, NAS,Kiev, Ukraine) that showed the presense of two effects - photostimulated transformations in ChVS layers [1] and photostimulated diffusion of metal into the semiconductor [2] - 'photodoping effect' . This two effects enabled the creation of the two types of registering media - lightsensitive ChVS layers and ChVS-metal (or metal containing substance) systems.

Since the discovery of this effects the numerous publications and patents appeared. The interest to this effects is stimulated by the application possibilities in various spheres of activities. Most works were performed on the investigations of possibility of photolithography, electron and ion beam lithography applications (high resolution inorganic resist) [3-18], high density information recording [19-22], fabrication of diffraction gratings and holographic optical elements [23-27],holographic recording [28-30],offset lithographic printing [3,31], etc.

The essence of the effect of photostimulated diffusion of metal into ChVS is in ability of some ChVS, deposited on metal (or metal containing) layer to produce the visible image under the influence of irradiation. During the irradiation the interaction of the substances of ChVS and metal layers proceeds, the products of interaction are created. Image visualization take place due to sufficient difference of the optical properties of products and of the initial layers. A number of physical and chemical processes (contact and surface, diffusion,photostimulated changes in ChVS, interaction of components at the interface of metal and ChVS, etc.) are simultaneously undergoing, that is essential peculiarity of phenomena of photostimulated diffusion of metal into ChVS. In the following for the simplicity it will be mostly refered as photodiffusion (or photodoping).

As a result of irradiation of ChVS layers by light with wavelengths corresponding to the interband absorption, many of their properties, in particular, their optical characteristics, solubility, etc.,are changed. The change of solubility serves as a base of ChVS layers and ChVS-metal systems utilization as media for information recording and as inorganic resist.

The properties of the mentioned above two types of registering media and their application to the microfabrication technology are the subject of this article.

1. Design forms of realization

The design realization of the media,developed on the base of photodissolution effect, can be performed in various forms (Fig.1). The necessaary elements of such media are the ChVS layer and metal source (mostly silver or silver compounds). Most frequently used are the forms of I,II type with the metal layer[2,20,28,31-33] or with metal containing substances Ag_2X (X=S,Se) [3,34-37] , AgHal [38,39], that are used as metal source.The increase of sensitivity, contrast and thermal stability can be achieved with the utilization of III and V type of systems [20,40,41]. Schemes of IV type were succesfully used for the submicron lithography [42-46]. Lightsensitive ChVS layers are mainly realized in I,IV forms (but in this case the metal layer 2 is absent).

2.Kinetics of photostimulated interaction

Most investigations, performed on the ChVS-metal systems, have shown, that the photostimulated interaction kinetics consists of three parts: induction (with growing of process rate), the main part, and the third one when the strong slowing down of the process is undergoing [24,47-49]. The schematic illustration of the photodiffusion processes in ChVS-metal system is shown in Fig.2 [8]. During and as a result of exposure in the exposure sites metal (Ag) penetrates into ChVS layer forming the intermediate doped layer,that is growing as the sample is illuminated. The typical kinetic dependence for the most investigated As_2S_3-Ag system is shown in Fig.3 [49]. Under irradiation with electron,ion beams,or X-rays the type of kinetic curve is similiar (Fig.4-6) [3,29,50]. The durability of the induction part is inversely proportional to the intensity of the irradiation I [49], so that the exposure necessary for its completion is constant: $H = It_0$ = const. The durability of induction part can be influenced by the pre-exposure [51], conditions of systems formation [52,53], pre-exposure of semiconductor layer (before the metal deposition) [47]. Induction part in [47,48] was attributed to the

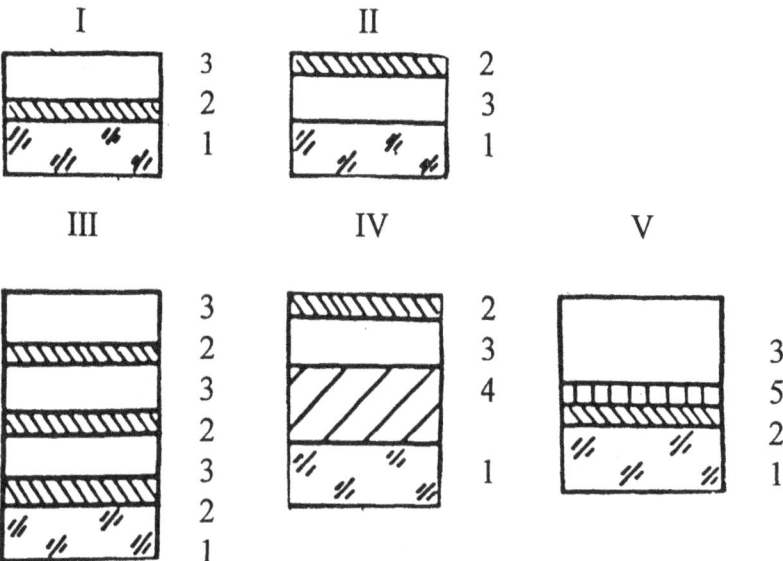

Fig. 1 Design forms (I-V) of lightsensitive materials on the base of photodiffusion effect : 1- substrate, 2- metal (metal containing substance), 3- ChVS layer, 4- polymer layer, 5 - barrier layer.

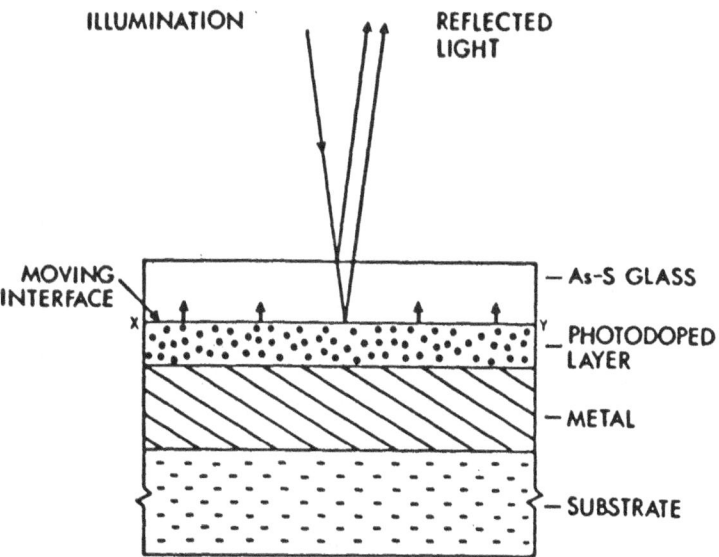

Fig. 2 Scheme of photodiffusion process. [8]

266

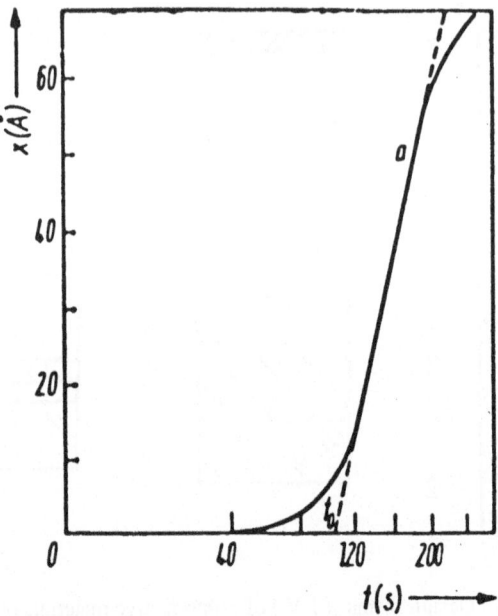

Fig. 3 Dependence of the quantity of silver (in terms of layer thickness change)
that have diffused into semiconductor versus the square root of the
illumination time for the As_2S_3-Ag system [49].

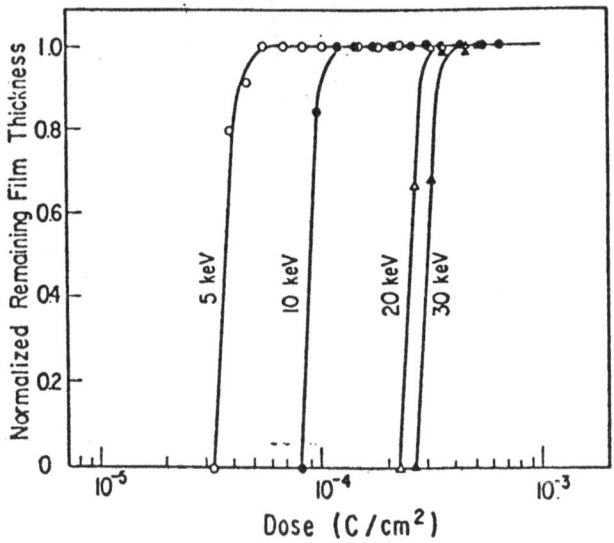

Fig. 4 Electron-beam exposure characteristics of Ag-$Se_{85}Ge_{15}$ system.
Remaining film thickness is normalized in the terms of the
initial 280 nm thickness [3].

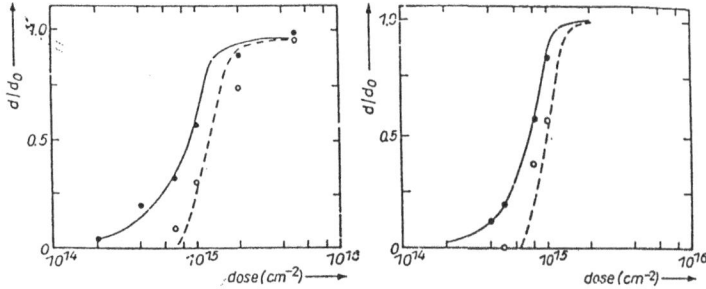

Fig. 5 Characteristic curves of the $Ag_2Se/GeSe_2$ resist system after irradiation
with 60 keV He+ ions and RIE etching in CF4 - (left) and SF6 (right) [50].

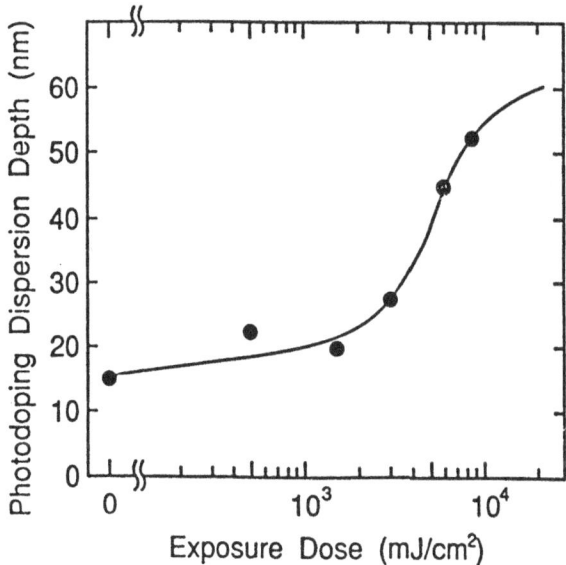

Fig. 6 X - ray exposure characteristics of photodoping dispersion depth into
an Ag-Se/Ge-Se layer [29].

process of defects accumulation under the influence of irradiation. From our point of view it is connected with the creation of the products layer of certain thickness [52]. The investigations of the growth of the products layer performed on the As_2S_3-Ag [54] (Fig.7), As_2S_3-Ag , As_2Se_3-Ag , $GeSe_2$-Ag [24 ,55] (Fig.8), Ag_2Se/Ge_xSe_{1-x} [56], Ag-$Ge_{25}Se_{75}$ [75], kinetics of metal expenditure in As_2S_3-Ag system and other systems [24,49,55] have shown the presence of the parabolic kinetics on the main part of photostimulated interaction (process is diffusion controlled).

On the other hand in [47] the constant rate of photostimulated interaction in As_2S_3-Ag system was obtained, that suppose the reaction on the As2S3-Ag interface as the controlling process. Different from parabolic dependence was obtained for Ag-(As-S) systems in [58]. The third stage which is characterized by the strong slowing down of the processes, is connected with exhaustion of the metal source or the ChVS saturation with metal.

In ChVS layers image formation proceeds in irradiated sites simulteniously through the whole depth of light penetration.

3. Sensitivity

3.1 Spectral distribution of sensitivity

In the first investigations of lightsensitivity of the ChVS-metal systems [60], it was shown, that the lightsensitivity range in general correspond to the region of intrinsic absorption of semiconductor layer, but is shifted to the side of lower frequencies in comparison to the ChVS absorption edge (Fig.9). The publications that followed, showed similiar results for various ChVS-Ag systems [24,28,54,61]. Under the exposure from the chalcogenide side the effective sensitivity is weakening and shift of the shortwave lightsensitivity edge due to optical absorption of chalcogenide was observed [54], that lead to the conclusion that photostimulated interaction is caused by the light absorbed near ChVS-metal interface. Comparison of the lightsensitivity spectral dependence of the As_2S_3-Ag system with the behaviour of As_2S_3 absorption have shown that their change with wavelength is not identical: if λ is changed from 460 nm up to 632.8 nm the absorption is changed in 4 orders and the rate of photostimulated interaction is increased only 15 times [47].

Spectral lightsensitivity dependence of ChVS layers is determined by the absorption of ChVS layers, that can be clearly seen on the example of As_2S_3 layers (Fig. 10) [16].

ChVS-metal systems and ChVS layers are sensitive to electron beams [3,14,18,35,62,63,64]. Sensitivity under the fixed energy depends from the thickness of semiconductor or metal layer, and under the fixed thicknesses of semiconductor and metal - from the electrons energy [63].

ChVS-metal systems are sensitive also to the ion beams [11,14,35,50,65]. Under exposure of such media by various ions (H^+,Ar^+,He^+) and the consequent treatment they behave as negative resists with the unique lithographic properties. But is necessary to mention that the mechanism of layers interaction in this case may be different.

X-ray sensitivity of ChVS based systems was investigated in [14,29,66,67]. It was determined that a high density resist, such as a Ag-Se/Ge-Se inorganic resist, is favorable for forming micropatterns using X-rays because of the short ranges of the electrons generated by the X-rays in the resist [66,67].

3.2 Factors, that are influencing the sensitivity values

The sensitivity of ChVS based systems and layers turn out to be dependent from the whole series of factors. Lightsensitivity of systems depends from the thicknesses of ChVS , metal , direction of exposure, utilized ChVS [68,3]. The best results were achieved with utilization of silver and copper (or their compounds) as a metal source. In ChVS-metal systems [54,58,69,70] and ChVS layers[71,72] (Fig.11) the sensitivity depends on the ChVS composition. For example, for the systems on the base of AsS_x the rate of interaction with metal increases with sulphur content [54,58,69]. The range of ChVS composition provided and the wide range of corresponding lightsensitivity spectra enables choosing the proper sensitivity on the given wavelength λ if

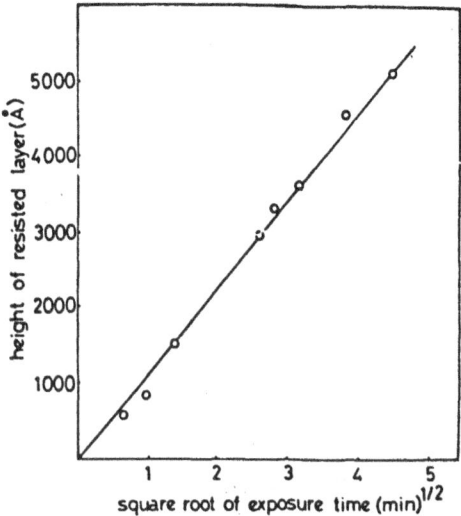

Fig. 7 Dependence of the products growth (Ag penetration depth)
in the As_2S_3-Ag system from the exposure square root [54].

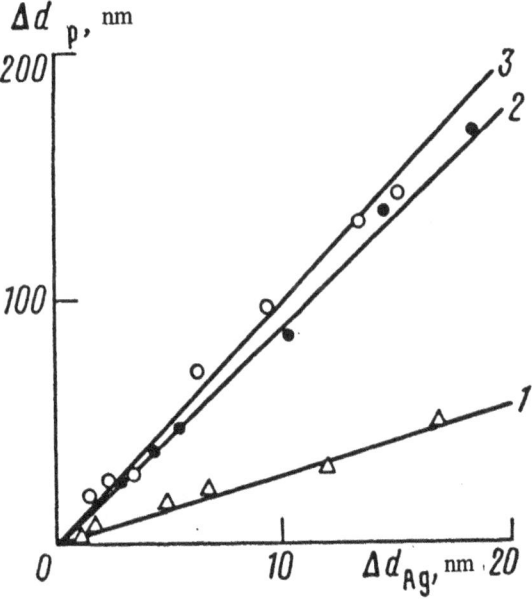

Fig. 8 Dependencies of the thickness of products layer from the
thickness of the used up metal layer (Ag) in the systems :
As_2S_3-Ag - 1; As_2Se_3-Ag - 2; $GeSe_2$-Ag - 3 [55].

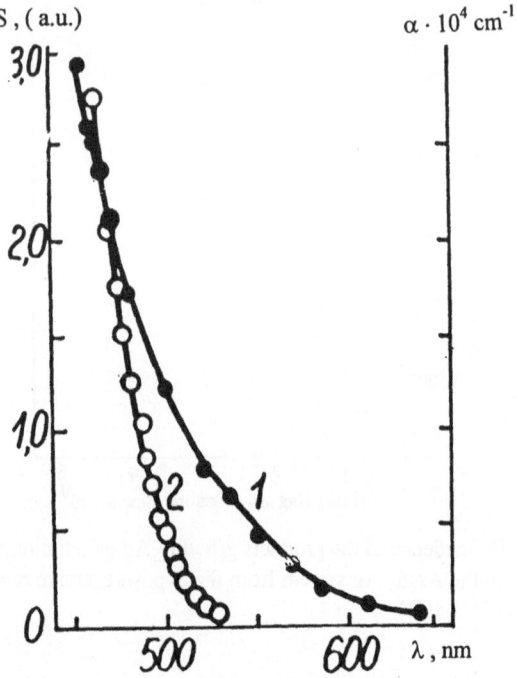

Fig. 9 Spectral dependencies: 1 - lightsensitivity of As_2S_3 - Ag system ,
2 - As_2S_3 absorption coefficient [60].

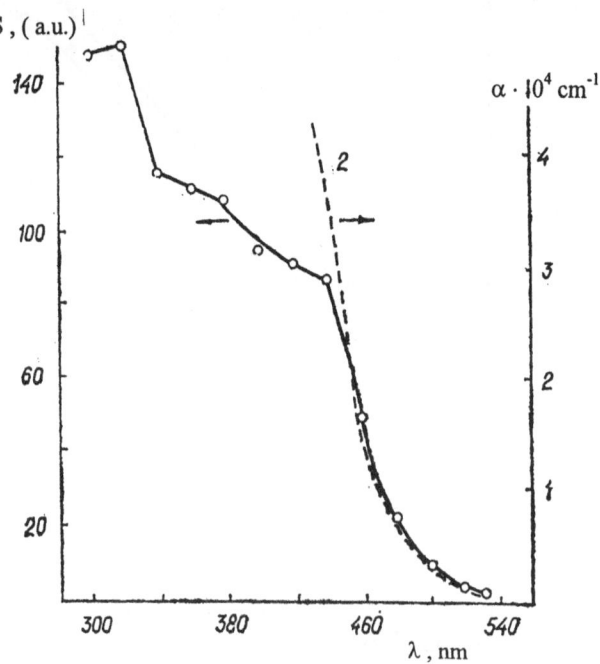

Fig. 10 Spectral dependencies : 1- lightsensitivity of As_2S_3 layers,
2 - As_2S_3 absorption coefficient [16].

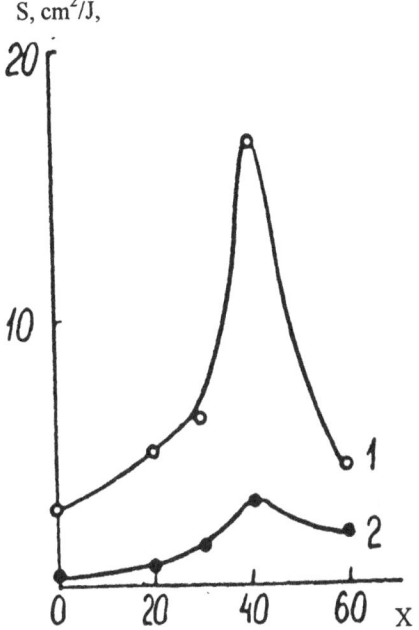

Fig . 11 Dependence of lightsensitivity S of $As_{40}S_{60-x}Se_x$ layers on composition (1- exposure on 436 nm wavelength, 2 - on 546 nm) [71].

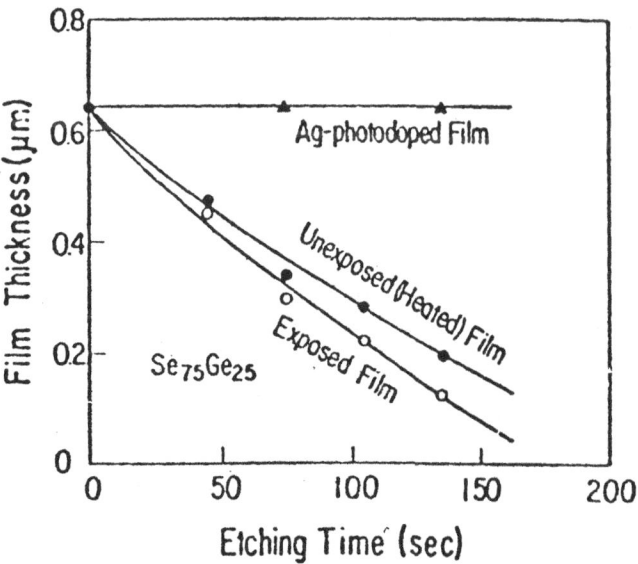

Fig. 12 Etching characteristics of $Se_{75}Ge_{25}$ glass films in aqueous solution of NH_3 [3].

necessary. For example, for the As_2Se_3 layers the lightsensitivity range covers all visible parts of the spectrum, which enables the use of argon and helium-neon lasers.

If ChVS is in crystalline, vitreous or amorphous state the rate of interaction differ considerably [73,74]. For the crystalline As_2S_3 it is 100 times smaller in comparison to the interaction of silver with As_2S_3 amorphous film under exposure. The exposure necessary for the photostimulated introduction of equal amount of Ag into crystalline, vitreous and amorphous As_2S_3 were 400; 3.5; 4 J/cm^2, correspondingly [74].
For the vacuum evaporated As_2S_3 films the smallest interaction rate was observed for the freshly prepared samples, higher for the annealed in the darkness in the nitrogen atmosphere and the highest for the samples preirradiated by Hg lamp [73]. There were no substantial changes in the effectivity of silver photodiffusion, connected with annealing or photodarkening effect, according to the investigations in [75-77], excluding Ge-S films, in which after thermotreatment there were evidences of presence of GeS and GeS_2 crystals and sesitivity was decreased in 100 times. Initial pre-exposure of As_2S_3 films lead with the increase of the polymerization degree to the growth of the photodiffusion rate and after polymerization saturation the effect of pre-exposure dissappeared [78].

Temperature dependencies of lightsensitivity of ChVS-Ag systems were investigated in [3,52,69,79,80-82]. It was shown that in certain temperature interval the lightsensitivity S can be described by the exponential dependence ($S=S_0 \exp(-E_a/kT)$, the activation energies E_a were determined, that were sufficiently lower of the values for the thermal diffusion of silver in ChVS in the darkness. At initial and main part of kinetics curve E_a differs [52], activation energy depends on aging, thermotreatment [79], intensity of actinic radiation [80]. The S values of systems can be sufficiently increased by using the thermal heating.

Systems lightsensitive properties depend on the time of aging [20,83,84], especially in chemically unstable systems [20,84]. Chemically unstable systems are stabilized by the introduction of barrier layers, that are slowing down the interaction in darkness [3,20,84,85]. Changes of systems lightsensitive properties mainly is attributed to the creation of intermediate layer of products of ChVS and metal interaction. Some part of metal layer interacts with ChVS already in the stage of systems formation [52,86,87]. Conditions, substrate temperature [52,86], order of deposition [86] determine the parameters and properties of the initial intermediate layers, and, correspondingly, the systems lightsensitive properties.

Lightsensitivity of systems [21,22,42,88] and ChVS layers [16,22] depends on intensity of photoactive irradiation. For As_2S_3-Ag system up to irradiation density 10^2 W/cm^2 there is no dependence from irradiation density P, then up to 10^5 W/cm^2 lightsensitivity is lowered, that is connected with the growth of probability of recombination processes, and begining from 10^5 W/cm^2 the growth of S begings, that is explained by the local heating of system during one impulse and from P > 10^6 W/cm^2 starts the region of thermal breakdown of system, connected with melting of the metallic layer [21]. With the help of local heating of $Ag_2Se/GeSe_2$ system it was possible to achieve sufficient increase of lightsensitivity - the recording was carried out using the single impulse with P ~ 5.2 mJ/cm^2 [42], for As_2S_3 layers the S values were ~ 7-10 cm^2/J [16,88]. In the latter case of laser lithography with sharp focused laser beam the sufficient increase of S was obtained due to interesting combination of effects [16,22,88]. The cross sections of light beam intensity on the substrates surfaces has Gaussian distribution:

$$I = I_0[\exp-(r/r_0)^2] \qquad (1)$$

where r_0 -beam diameter, I_0 is the intensity at the beam center and r is the distance from the center. Lightsensitivity S dependence on the temperature T is given by

$$S = S_0 \exp(-E_a/kT) \qquad (2)$$

where E_a is the activation energy. The distribution of exposure H in the line perpendicular to the beam trajectories on the sample surface,

$$H(z) = H_0 \exp (-z^2/r^2) \qquad (3)$$

determine the temperature T distribution under ChVS layer exposure ($T(z) = T_0 + \Delta T \exp (-z^2/r_0^2)$), where z is the distance from the beams center and ΔT is the temperature increase at the center of the beam.

The spreading of heat is not taken into account because the characteristic heat spreading distance (for the exposure time near some microseconds) is smaller than r_0. Thus the degree of photostructural transformations M in the ChVS layers is determined by the product of exposure H and lightsensitivity S

$$M = B \ H(z)S(z)$$ (4)

where B is coefficient. The estimates show that the dependence of M on z is substantially more abrupt and has smaller half-width than H(z) and, as a result, it is possible to obtain much narrower lines under the increased sensitivity (which is the unique case for the lightsensitive media). The estimated possible values of sensitivity S for ChVS layers may be within 300 - 3000 cm^2/J. Similiar effect is also present for the ChVS-Ag systems, but the sensitivity increase is smaller in comparison to the ChVS layers.

The lightsensitivity of the systems can be also increased by using various design forms: multilayer structures with the alternating ChVS and metal layers, introduction of system into complex system with multiple reflection [40,41].

4. Properties of products of photostimulated interaction (PPI) and exposed ChVS layers.

As a result of irradiating of the ChVS by light with wavelengths corresponding to the interband absorption many of their properties are changed, in particular, their optical characteristics (shift of absorption edge, changes of refractive index, induced dichroism), mechanical characteristics, etc. Most studied, however, are changes of their optical properties and solubility, because these changes serve as a basis for using ChVS as media for information recording and inorganic resist, because these represent the most practical interest.

The lightsensitivity of thin-film ChVS-Ag structures is determined by the photostimulated diffusion of metal into ChVS. During and as a result of exposure process of such structures in the exposure sites Ag penetrates into ChVS layer, forming the intermediate doped layer. The properties of such a layer are essentially different from those of initial ChVS. In particular, photodoped ChVS is highly resistive in relation to various alkaline etchants, which rather easily dissolve nondoped ChVS. This makes it possible to obtain relief images depicting the intensity of actinic irradiation.

4.1 Chemical properties

Easily dissolved in various alkaline etchants ChVS layers after photostimulated interaction with metal have good resistive properties to this etchants [3,6,17,18,24,35,42,72,88-92]. Selective etching properties of ChVS layers were investigated in detail also in many works [3,14,16,17,18,23,71,72,88,93-98].

With the increase of Ag concentration in the $GeSe_2$ the etching rate decreases [90]. The etching dependencies of photodoped and nondoped $Ge_{25}Se_{75}$ in aqeous solutions of NH_3 (Fig.12) are typical for most of ChVS-Ag systems [3]. It can be seen from the figure that after photodoping the created products are practically insoluvable for such concentrations of etching solutions that easily dissolve unexposed ChVS. Such drastic differences in etching rates (up to 500) provide possibility of obtaining relief images. ChVS-Ag systems have high contrast γ values (2.5-10), for example for As_2S_3-Ag and GeS_2-Ag systems γ values are 4.5 and 3.3, correspondingly [18,72], for $As_4Ge_{30}S_{66}$-Ag system - 3.7 [91].

ChVS layers also have good selective etching characteristics. The dependence of dissolution of As_2S_3 resistive layers from the time of etching is presented in Fig.13 (exposure with integral irradiation of Hg lamp, H = 2 J/cm^2) [88]. One of the important parameters, that are characterizing the acceptability of the lightsensitive layer as a photoresist is etching selectivity χ, that is the ratio of the dissolution rates of the nonexposed and exposed parts of layer. Such parameter (χ) can serve as a criteria of the quality of the ChVS lightsensitive resistive layer and used in comparison of parameters, in choice of optimal glass composition and etchants. In table 1 the parameters of As-Se and As_2S_3 layers in various etchants are prresented [96]. As seen from the table ,the etching character can be positive and negative. Quality of the treated surface when using most of described etchants , enables to obtain relief patterns with high resolution capability. The ChVS layers are very stable to chemical treatment with etching solutions widely applied in microelectronic fabrications.

274

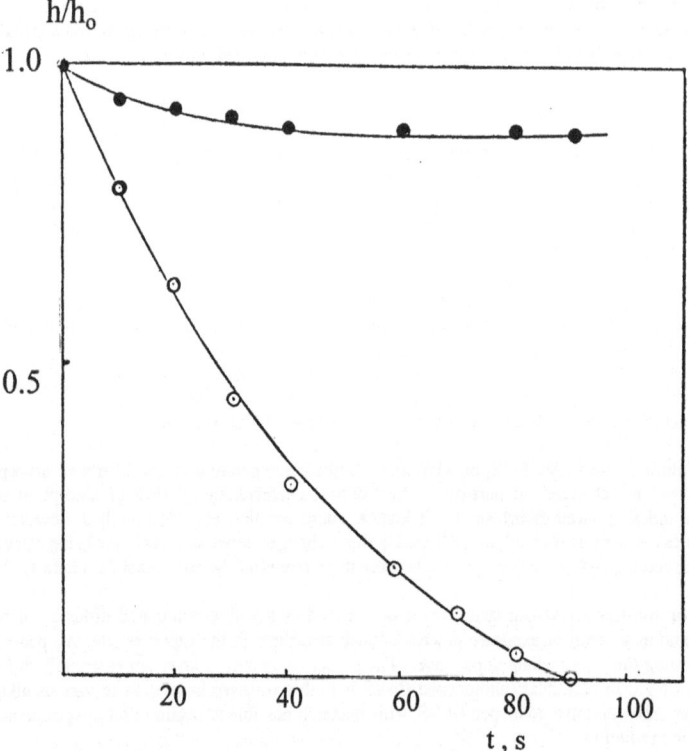

Fig . 13 Dependence of dissolution of $As_{40}S_{60}$ resistive from the time
etching (exposure with integral irradiation of Hg lamp, H= 2 J/cm^2) ,
o - exposed film , ● - nonexposed [88].

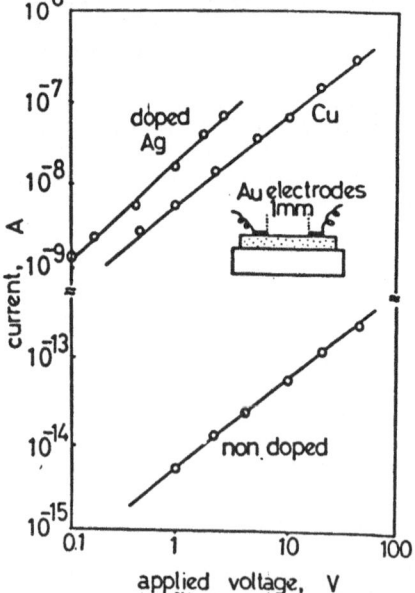

Fig. 14 I - V characteristics of doped and nondoped As16S80Te4
layers [89,103].

Table 1

Etchant	Temperature of dissolution, K	Dissolution rate of the exposed part, μ/min	Dissolution rate of the unexposed part, μ/min	χ	Character of dissolution
$As_{50}Se_{50}$					
KOH (10%)	295	0.15	0.064	2.3	Positive
$K_2Cr_2O_7$	295	0.012	0.03	2.5	Negative
$CH_2NH_2CH_2OH$	295	0.0071	0.036	5.0	Positive
$(C_2H_4OH)_2NH$	343	0.0429	0.086	2.0	Negative
$(CH_3)_2NC_2H4OH$	353	0.0396	0.0490	1.1	Negative
$(C_3H_7)_2NH$	363	0.086	0.014	1.3	Negative
$(C_4H_9)_2NH$	353	0.0206	0.0228	1.1	Negative
CH_3NH_2	363	0.00007	-	-	-
$(C_2H_5)_2NH$	323	0.0007	-	-	-
$As_{40}S_{60}$					
KOH (1M)	295	1.0	2.0	2.0	Negative
Na_2S (1M)	298	0.03	0.006	5.0	Positive
Na_2CO_3 (1M)	295	0.30	0.17	1.8	Positive
Na_3PO_4 (1M)	295	0.03	0.03	1.0	-
$C_6H_5NH_2$	363	0.0007	-	-	-
$(CH_3)_2NH$ (33%)	293	0.0067	0.0045	1.5	Positive
$(C_2H_5)_2NH(100\%)$	293	--	0.001	--	-

The effect of influence of different solvent as an environment for the chemical reaction of dissolution on the example of interaction between arsenic sulphide films and diethylamine was investigated in [23]. The dissolution rate was affected by the ionizing power of the solvent, also in general there is no simple correlation between the value of dissolution rate and any property of the solvent as far as the solution of the fresh films is concerned.

Impressive rate differences in „ dry " etching (300-500 times) were found between products of interaction (Ag photodoped films) and initial ChVS layers in ChVS - Ag systems [3,4,6,13,18,36,91,99,100] , high γ (3-5) values are kept unchanged. Plasma etching and exposure characteristics for Ag-Ge$_{25}$Se$_{75}$ systems were investigated [3] , the undoped films are etched in CF$_4$ very rapidly, with etch rate is about 55 nm/sec, after photodoping of silver etch rate decreases drastically to 0.15 nm/sec.

ChVS layers also can be applied in dry ething processes [12,14,99,101,102] , here the rates difference and γ values are sufficiently smaller than for the systems (for example γ - 1.25 for Bi-Ge-Se films [99], but nevertheless they can also be applied together with ChVS-Ag systems in microelectronics for microstructure pattern formation.

4.2 Electrical and photoelectrical properties of photodoped ChVS

As a result of metal photodiffusion drastically changed are also the basic physical properties of initial ChVS. The conductivity increases, conductivity activation energy is monotonically decreased with the growth of silver concentration [89,103]. I-V characteristics of doped and nondoped As$_{16}$S$_{80}$Te$_4$ layers are presented in Fig.14. During the investigations of the optical properties of the photodoped by Ag samples the photoerasable memory effect was discovered [89,103].If the voltage is applied to the photodoped sample the drastic decrease (more than 4 orders) of electrical resistance was observed. By means of light exposure it is possible to restore initial state characterized by the high resistance.

Investigations of the electrical properties, mostly performed on the layers of As$_2$S$_3$Ag$_{2.4}$ composition were performed in [104-107]. It was established, that conductivity value σ of the photodoped layer is 10 orders higher then the nondoped one, As$_2$S$_3$Ag$_{2.4}$ is mixed ion-electronic conductor. The electronic part of conductivity σ_e consists ~ 1/5 from the that for ions σ_i under the room temperature. Silver is moving in the doped As_2S_3 in the form of positively charged ions. Temperature dependencies of σ_e and σ_i were described by activation law with direct current activation energies - 0.53 and 0.44 eV, correspondingly. With the increase of Ag content the conductivity is also increasing, holes are main carriers up to 20at%, starting from 20 at% the ion conductivity appears, and above 29 at% of Ag σ_i is dominating.

4.3 Optical properties

The change of ChVS optical properties as a result of photodoping with metals was investigated already in the first investigations of this phenomenon [60]. It was discovered, that after photodoping the ChVS absorption edge shifts to the red side of the spectra. Such changes were confirmed by the investigations of various ChVS compositions, the E_g values were obtained under the change of Ag content [24,48,70,89,103,105,108-112]. Width of E_g is decreased with the Ag content increasing [89,103,105,111,112]. Also are changed the integral characteristics of absorption band, that determine the dispersion in the transparency region [111,112]. Optical constants of ChVS are also changed after exposure, but the change is small in comparison to the caused by the photodiffusion of silver, that can be clearly seen from the Fig.15 [24].

5. Resolution capability

Already in the first works the high resolution capability of new media was mentioned [2]. Very high resolution capability of the resists on the base of Ge-Se was shown in [3,37], in the experiments on ultra-fine Bi particles image transfer using Ge-Se based resists. The spatial resolution of Ag-photodissolution effect was found at least fine as 1 nm. This value can be basically regarded as intrinsic resolving power of the ChVS based resists, that approaches an order of atomic distances. Other works have also shown the nanoresolution possibilities of the ChVS based media and the perspectiveness of their application in nanolithography [15,113 -115]. In Fig.16 the examples of nanometer images are presented [15].

Possibilities in optical projection lithography were shown in [6,13,34,42,44,45,100], lines with 0.4-0.5 μm sizes were easily obtained (Fig.17). In electron beam lithography the elements with the sizes up to 30 nm were obtained [46,64,117]. In X-ray lithography 0.2 μm line/space pattern was obtained, as well 50 nm wide fine lines formed by utilizing Fresnel diffraction [67] . In laser lithography it was possible to obtain 0.2 μm lines [16, 22, 88]. Holographic gratings with spatial frequencies up to 3600 mm^{-1} were also obtained with the help of ChVS based inorganic resists [24,27].

6. Models of mechanisms of photostimulated interaction.

The photoinduced changes in ChVS layers have two components: reversible and nonreversible. In binary ChVS nonreversible changes are rather substantial and surpass the reversive component. The utilization of ChVS layers as an inorganic resists is connected with the structural changes in evaporated (or deposited from the solution) layers. For example, the structure of evaporated As_2S_3 film can be represented in the form of matrix, which consists of pyramidal units $AsS_{3/2}$ and containing considerable amounts of As_4S_4 and S_2 fragments, that is As-As and S-S „wrong"bonds. In the evaporated films, pores and hollowenesses are also present. Under illumination or annealing, polymerization of the molecular groups in the main matrix takes place, thus the number of homopolar bonds and hollownesses are diminished.

The model of photostimulated interaction in ChVS-metal systems must explain where the actinic absorption takes place, the kinetics of the process,profiles of metal in products of interaction, etc. Among the many works devoted to the mechanism of this phenomenon [9,61,68,69,70,89,118-124], the most comprehensive and update seems the model proposed in [107]. This photoelectrical model (considers As_2S_3-Ag system as a model one for photodiffusion processes) takes into account the processes undergoing in the photoactive interfaces [104] in trilayer As_2S_3 - $As_2S_3Ag_{2.4}$ - Ag structure. The doped layer provides channel for the diffusion flow of Ag from metal to the As_2S_3. The interface Ag- $As_2S_3Ag_{2.4}$ is a typical metal-solid electrolyte contact. The double charged layer is formed in this interface by means of Ag ion transition from the metallic layer. The extent of double layer is determined by the Debye length of screening and does not exceed 1 nm.The electric field of

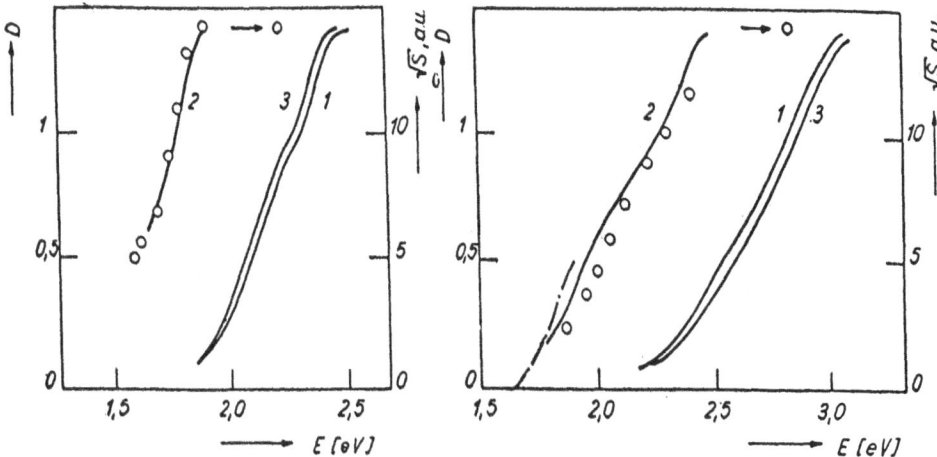

Fig . 15 Spectral dependencies of (left) As₂Se₃-Ag systems lightsensitivity S values - o;
optical density D - 1,3 of nonirradiated and irradiated As₂Se₃ layers,
correspondingly, and 2 - for D of As₂Se₃ - Ag systems products of
photostimulated interaction (PPI); and (right) the same for GeSe₂ - Ag systems
and GeSe₂ layers : o - GeSe₂ - Ag systems lightsensitivity S experimental
values; optical density D - 1, 3 of nonirradiated and irradiated GeSe₂ layers ,
correspondingly; 2 - D of GeSe₂-Ag systems PPI, -.-.- - optical density of PPI
during systems irradiation [24].

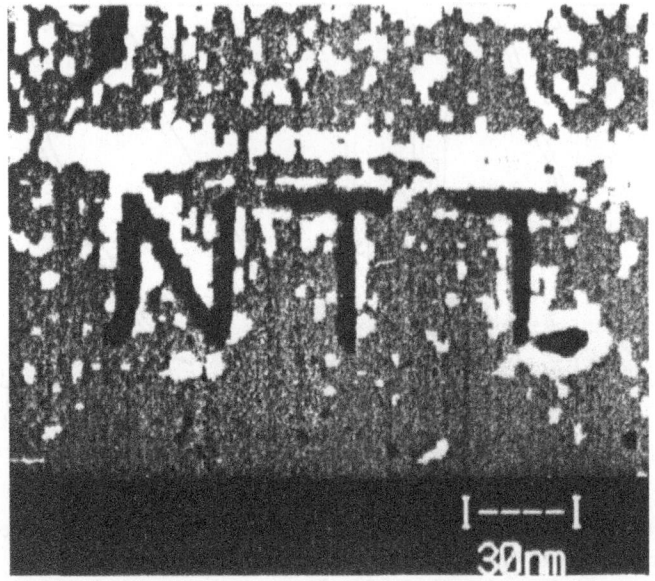

Figure 16aSTM image of nanometer-sized letters 'NTT' written on an Ag,Se surface. The frame is 176 nm by 149 nm. Each letter is about 30 nm wide and 30 nm high. The lines are about 3 nm in width and etched to a depth of about 1 nm. The latter were written with tip bias of +3.0 V. [15]

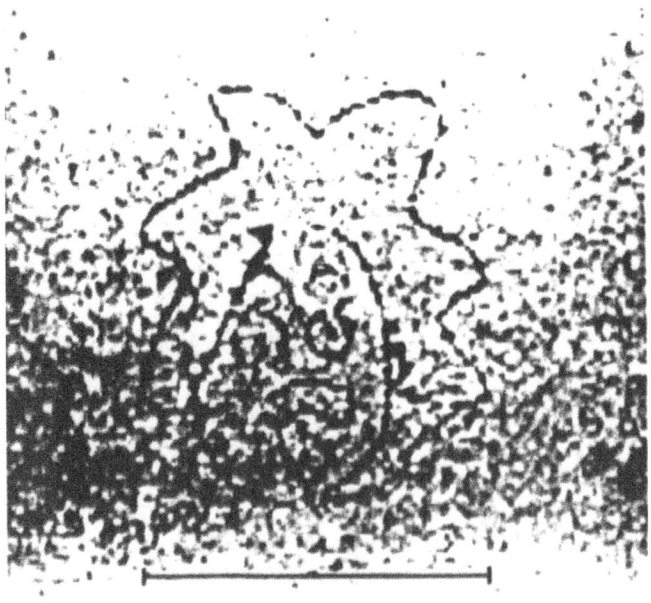

100 nm

Figure 16bSTM image of nanometer-sized figure including circles and curves. [15]

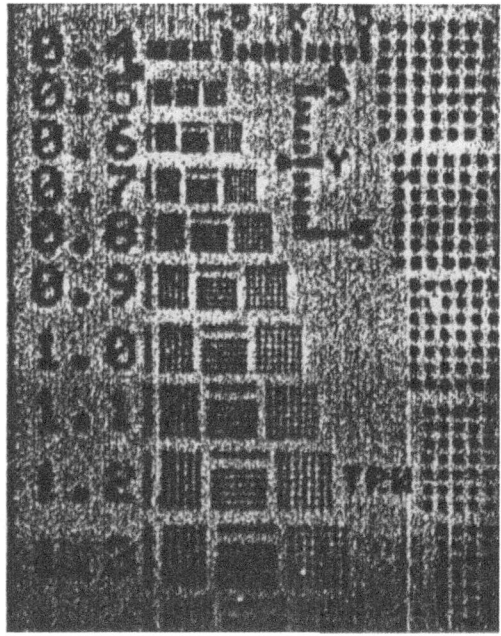

Fig. 17 Patterned spin/coated Ag/As$_{40}$S$_{60}$ resist. Linewidths of 0.5 μm were easily obtainable [13].

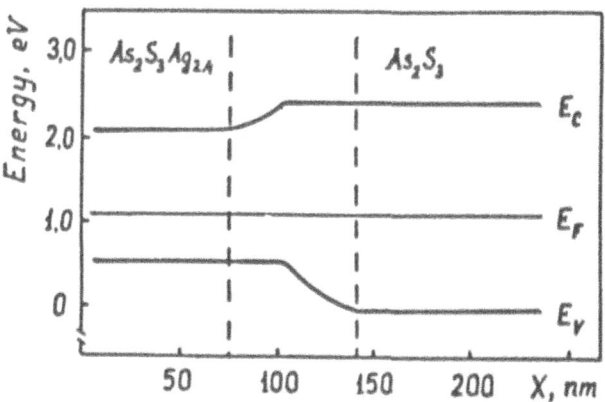

Fig. 18 Band diagram at the Ag - doped - undoped As$_2$S$_3$ interface, x is the distance in the direction perpendicular to the interface [107].

such layer hinders the Ag ion flow from metal to doped As_2S_3. Under illumination the photogenerated carriers decrease the double layer field. This causes a photo-emf and Ag flow through the interface.

At the doped-undoped As_2S_3 the processes are more complex. The band scheme of this heterojunction, that takes into account the Ag distribution in the diffusion front region (obtained with the help of secondary ion mass spectroscopy [105]) and also the dependencies of doped layer optical band and the hole conductivity activation energy on the Ag concentration is shown in Fig.18 [105]. It is seen from the figure that the diffusion front is a variable band structure. The actinic illumination creates free carriers, and photoexcited holes are shifted by the „quasi-electric"field of the variable band structure to the doped layer side. The drift of the photoexcited holes causes a photo-emf to be generated with the potential on the doped layer. At moderate illumination, it can be written in the form

$$\in \approx 1/e \, (\, \Delta\sigma/\sigma \,) \, \Delta E_g \qquad (5)$$

where σ is the conductivity , $\Delta\sigma$ is the photoconductivity and ΔE_g is the energy gap change in the region under consideration. The field connected with the photo-emf,

$$E \approx 1/e \, (\, \Delta\sigma/\sigma \,) \, \Delta E_g /d \qquad (6)$$

has a direction which is necessary to promote the enhancement of the Ag^+ ions flow (d is the width of the diffusion front). The numerical estimates have shown that this field can provide the observed Ag flow.
With the help of this model it is possible to describe all the phenomenological peculiarities of the photodiffusion phenomena in the As_2S_3-Ag structure (spectral and temperature dependencies of sensitivity, kinetics [123], etc.).

7. Application in microlithography

As already mentioned the most promising applications of ChVS based inorganic resists are in submicron and nanolithography. Examples of the results achieved in this direction are the subject of the concluding section.

7.1 Technology of fabrication and treatment.

Deposition of ChVS films and structures can be preformed by conventional film deposition technology, such as rf sputtering, vacuum evaporation, deposition from the solutions, melting, etc. The thicknesses of the layers are usually within dozens of nanometers up to several microns. As a metal source in ChVS - metal (metal containing substance) systems the vacuum evaporated metal films (most frequently Ag,Cu), silver halides, sulphides, selenides, etc. are used. The metal thicknesses (or its source) are usually within several nanometers up to hundreds nanometers. Most frequently under the resists fabrication the vacuum technologies of layers deposition is used, because in this case the high unifirmity of layers is provided, with easy control of layers parameters, besides that all the steps can be performed in vacuum, that provide the high quality of the systems and layers obtained [20,28,31,39,43,84,89,103]. In many cases it is necessary to avoid the presence of the initial thermally stimulated interaction of layers in systems, in this case the chemical deposition of silver is used [3,34,35,36,37,43,46,116]. Also the technology analogous to the utilized for the organic resists deposition can be used [125-126].

After exposure the image fixation is carried out with the help of selective etching, using the differences in dissolution rates of the photodoped ChVS or exposed ChVS layer in comparison to the nonexposed ones. Schematic illustration of the photolithographic process is shown in Fig.19 [3,122]. First a chalcogenide layer is sensitized by dipping. Heat treatment of the Se-Ge film is not necessary for the negative case. The next steps are removing of the nonreacted silver source and, finally, the undoped chalcogenide is dissolved in an alkaline solution. In such procedures a positive- or negative-type relief image is delineated. The underlying substrate layer (polysilicon, oxide, nitride or metal layer) is etched using the patterned ChVS layer as etching mask. Several variations in each step are possible, for example the use of dry-processing technique is very effective [3,6]. In detail the inorganic rests technology under LSI fabrication process is described in [6].

Some characteristics of ChVS layers and ChVS - Ag systems are listed. below. In Fig.20 (a-c) are presented the characteristic curves for the $As_{40}S_{60}$ -a, $As_{40}S_{40}Se_{20}$ -b and $As_{40}S_{20}Se_{40}$ -c, correspondingly [72]. Here h_o - ChVS initial thickness,h-thickness of ChVS layer that remained after exposure H and chemical treatment in the

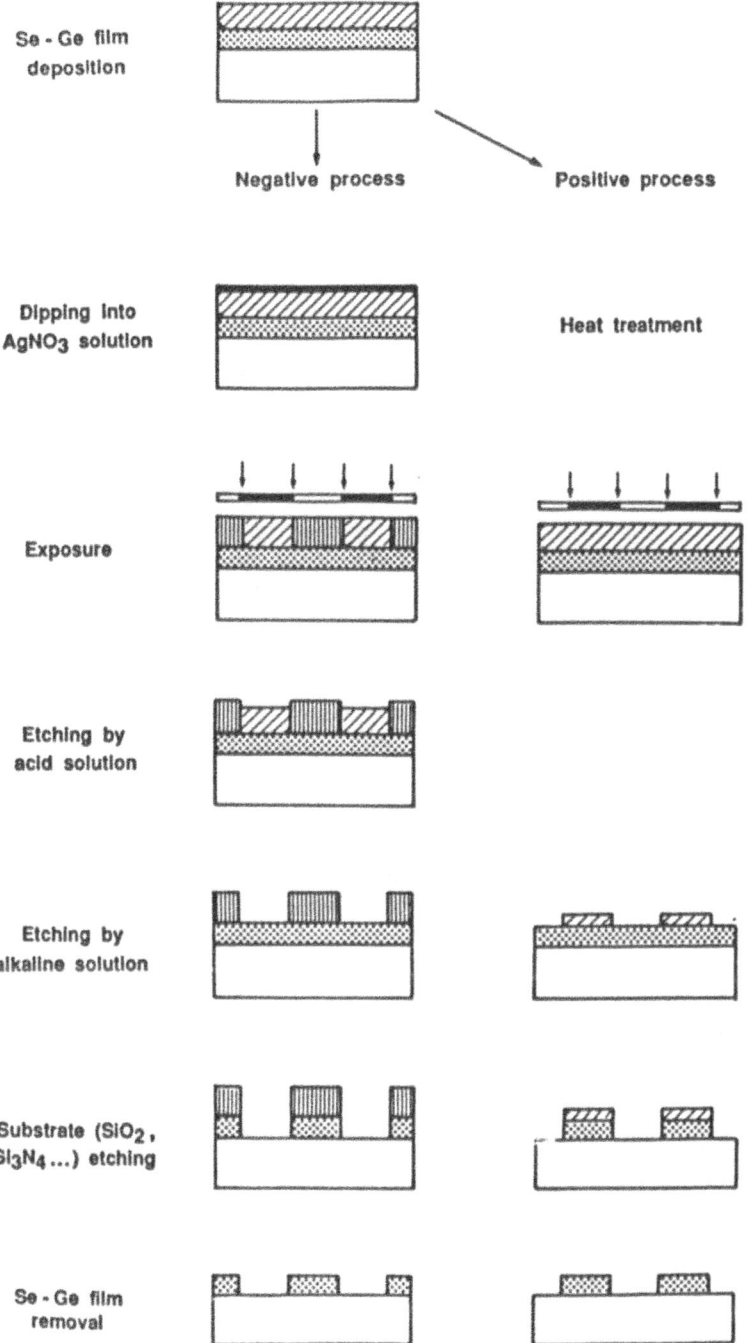

Fig. 19 Schematic illustration of the photolithographic process utilizing ChVS based inorganic resists [3, 122].

282

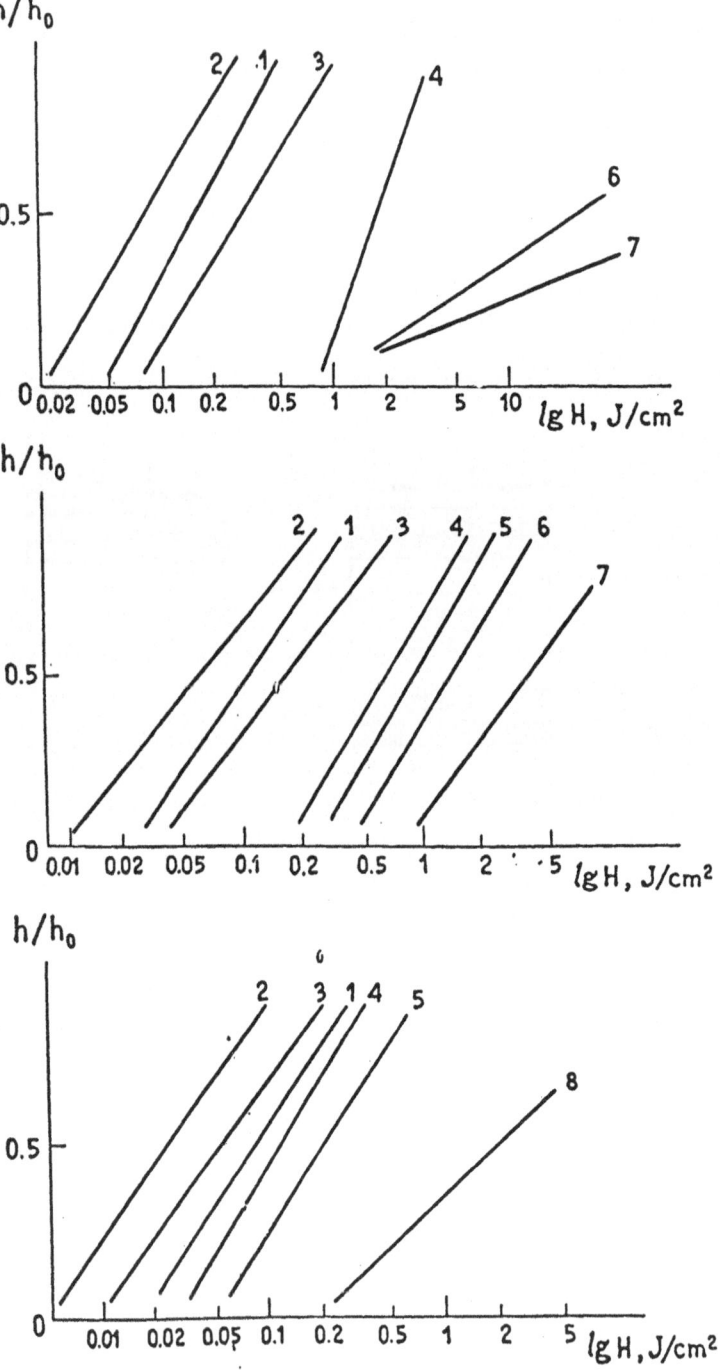

Fig. 20 Characteristics curves for the $As_{40}S_{60}$ - a, $As_{40}S_{40}Se_{20}$ - b and $As_{40}S_{20}Se_{40}$ - c, correspondingly [72].

amine based etching solution. Curves 1-8 correspond to exposure on 365,405,436,488,514,546,579 and 633 nm wavelengths, correspondingly. Sensitometric characteristics : lightsensitivity S_k (criteria $H_{0.5}$ was used - exposure value under which the half of the initial resist thickness remains) and contrast γ are presented in Table 2.

Table 2. Sensitivity $S_{0.5}$ (cm^2/J) characteristics for As-S-Se layers

Curves in Figures	1	2	3	4	5	6	7	8
wavelengths, nm $\Rightarrow$	365	405	436	488	514	546	579	633
Composition $\Downarrow$	$S_{0.5}$ (cm^2/J)							
$As_{40}S_{60}$	5.9	10.0	2.86	0.4		0.06		
$As_{40}S_{40}Se_{20}$	8.7	14.2	5.0	1.61	1.18	0.61	0.25	
$As_{40}S_{20}Se_{40}$	11.1	38.5	16.7	8.3	4.4	3.1	'2.9	0.44

The γ values in given spectral range for $As_{40}S_{60}$ consist 0.24-1.0; for $As_{40}S_{40}Se_{20}$- 0.6-1.0; for $As_{40}S_{20}Se_{40}$ - 0.4-0.9.

In Table 3 the sensitivity characteristics of As_2S_3 layers for the case of exposure by photons, electrons and ions are presented [14].

Table 3. Sensitivity of evaporated As_2S_3 layers

Photons (mJ/cm^2) (band 380 - 450 nm)	Electrons (C/cm^2) (30 kV)	X - rays (mJ/cm^2) (band 30-60 A)
500-600	$1 \cdot 10^4$	550
space/line 0.8 μm	space/line 0.5 μm	spot 25 mm^2

Usually sensitivity values of various ChVS-Ag systems are within 0.2-20 cm^2/J, but in special cases it can be increased (as already mentioned) up to ~ 200 cm^2/J [42]. Such systems are also characterized by high values of contrast γ (2.5-10). Characteristic curves for $As_{40}S_{60}$ -Ag and GeS_2 -Ag systems are presented in Fig. 21 [72]. Reactive ion etching for $Ag_2Se/GeSe_2$ as an ion-beam resist gives γ values for etching in SF_6 and 2.5 - 3 for CF_4 [50]. Sensitivity of Ag-Se/Ge-Se inorganic resist to the synchrotron radiation was 0.3-3 times that of PMMA resist, the γ values of 3.1 -3.5 obtained indicated high contrast [67]. A particular promising area for such resists is all-vacuum lithographic process [8,12,101].

Examples of obtaining of mask and metallic relief patterns are presented in fig. 22 - 25 for the cases of photolithography, laser lithography, electron beam and X -ray lithography. It can be seen that the high quality of patterning can be achieved with the help of ChVS -based resists.

Conclusion.

The chalcogenide based inorganic resists having many unique properties can be succesfully applied in submicron lithography and are very promising (according to the data of their resolution capability) in nanolithography.

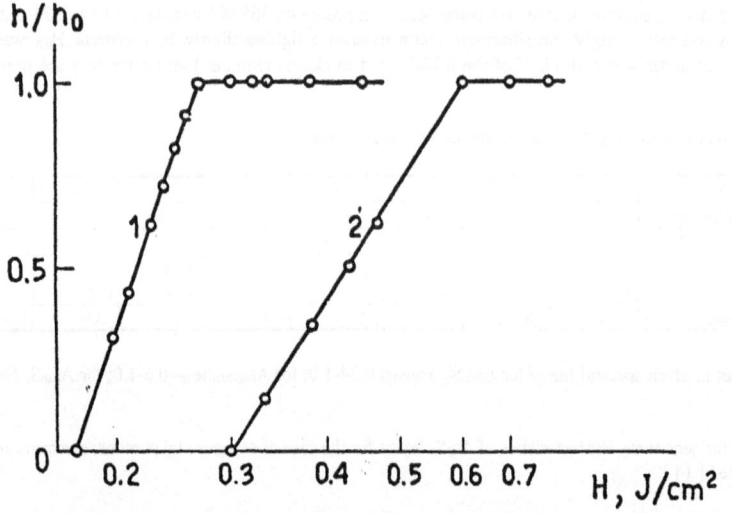

Fig. 21 Characteristic curves for GeS$_2$-Ag (1) and As$_2$S$_3$-Ag (2)
      systems [72]

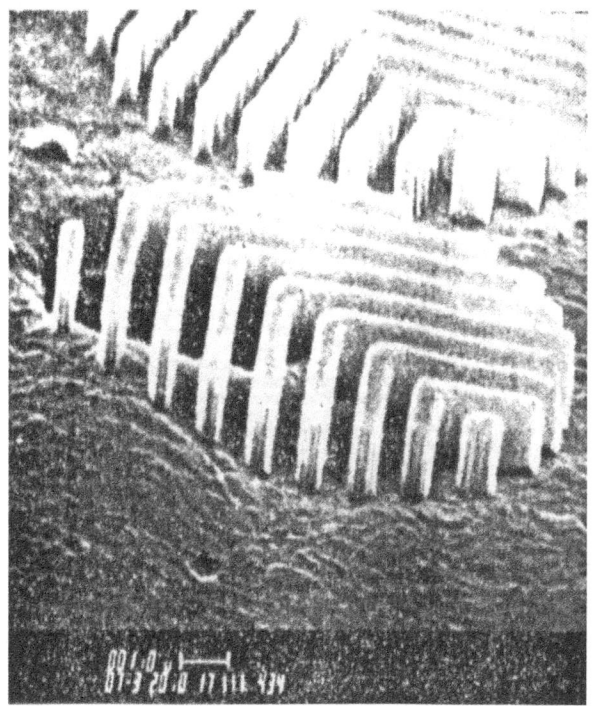

Fig. 22 a- line/spaces of 0.5 µm size on the aluminium relief obtained with photostepper, lens N = 0.35; λ = 405 nm. Two layer resist [127].

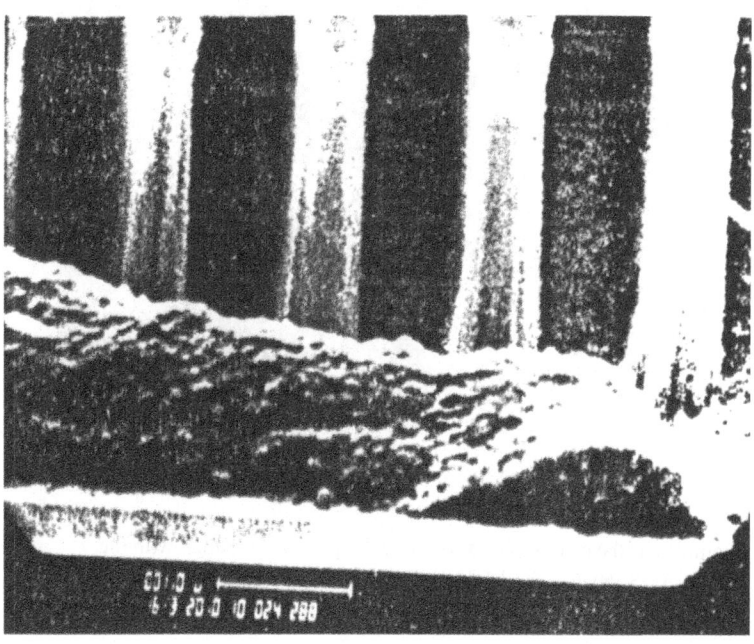

Fig. 22b bilevel 0.5 µm line and space pattern obtained by 5 × reduction photoexposure Ag/Ge$_{0.2}$Se$_{0.8}$ (200 nm) on hardened HPR-204 (2 µm) over Al device topography [34].

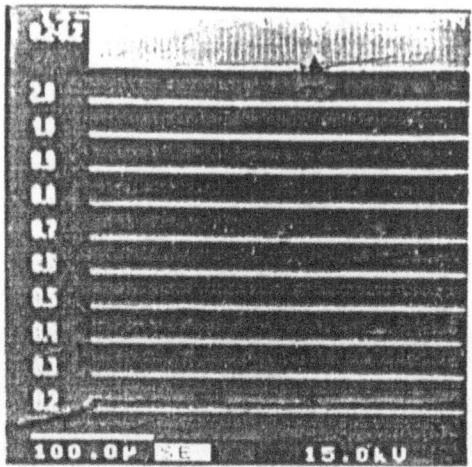

(a)

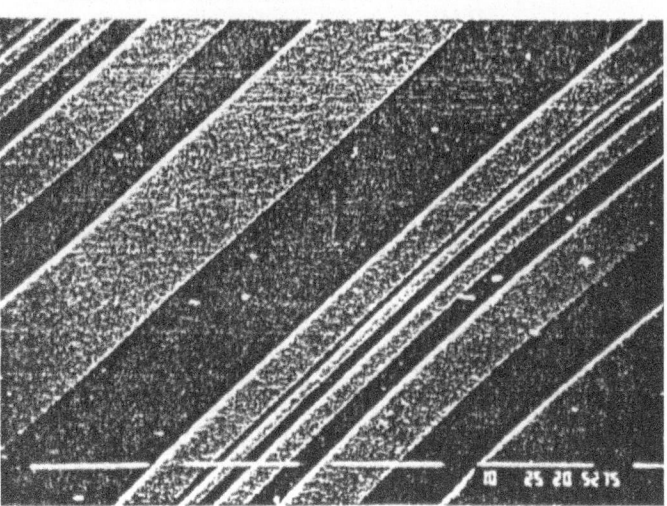

(b)

Fig. 23 Examples of the electron beam lithography patterns obtained with the help of As_2S_3 layers : a - [18] ; b - [14].

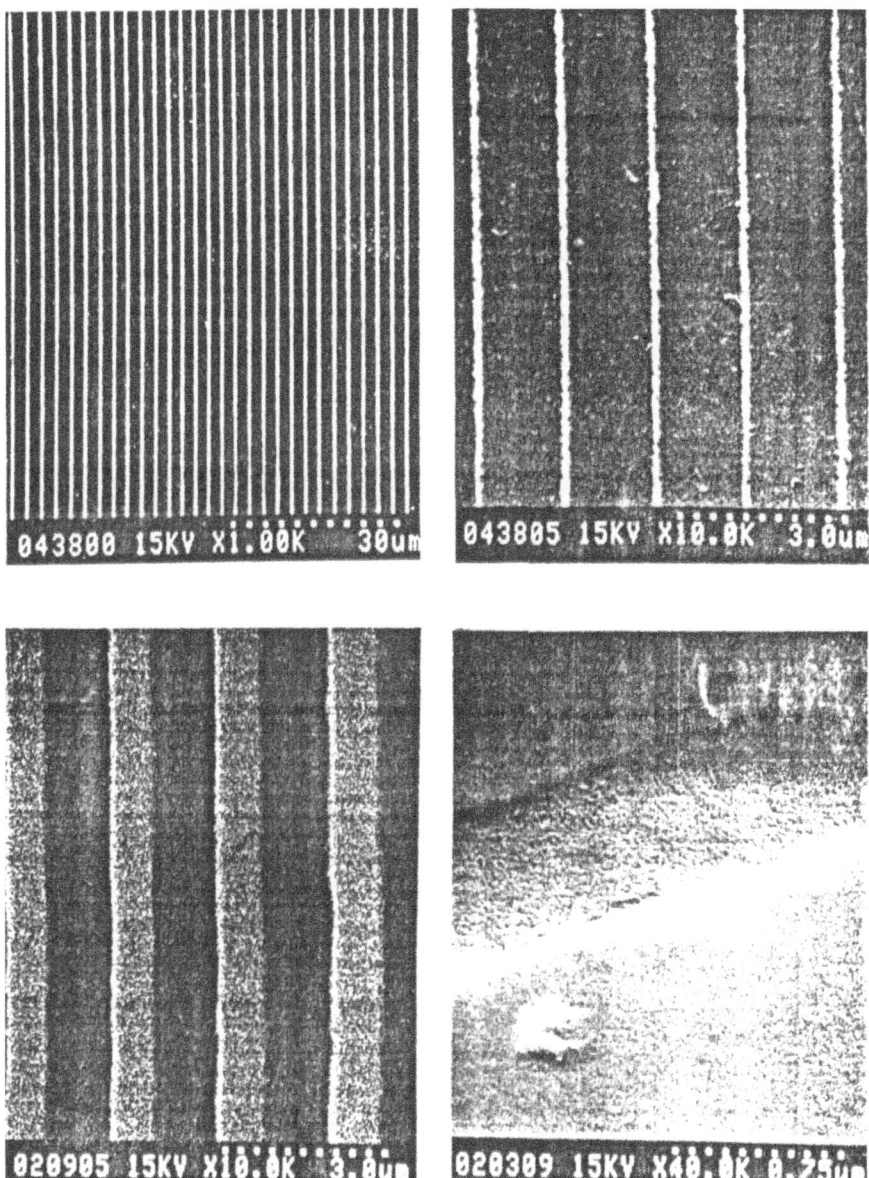

Fig. 24 Examples of laser lithography (Cr patterns) utilizing
As$_2$S$_3$ layers [18] .

288

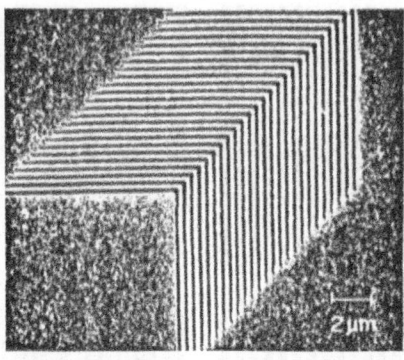

Fig. 25 X - ray lithography utilizing inorganic resist -
0.2 μm line/space pattern [67] .

REFERENCES

1. M.T.Kostishin, E.V.Michailovskaya, P.F.Romanenko and G.A.Sandul „About the photographic sensitivity of the thin semiconductor layers", J.Sci.Appl.Photogr.Sinematogr. 10(6), 450-451 , (1965).
2. M.T.Kostishin, E.V.Michailovskaya, P.F.Romanenko „On the effect of photographic sensitivity of the thin semiconductor layers deposited on metal substrates", Sov.Phys.Solid State 8(2), 571-572, (1966).
3. Y.Mizushima and A.Yoshikawa „Photoprocessing and lithographic applications", In: Amorph. Semicond. , Technolgies & Devices : Tokyo e.a. Amsterdam, 277-295, (1982).
4. M.S.Chang,T.W.Hou,T.W.Chen, K.D.Kolwitz and J.N.Zemel „Inorganic resist for dry processing and dopant applications „, J.Vac.Sci. and Technol., 16(6), 1973-1976, (1979).
5. Y.Mizushima and A.Yoshikawa „Inorganic photoresists"Electronic Ceramics (Japan), 12(63), 22-26, (1981).
6. Y.Utsugi, A.Yoshikawa and T.Kitayama „An inorganic resist technology and its applications to LSI fabrication processes „, Microelectronic Engineering, 2 (4), 281-298, (1984).
7.K.J.Polasko,C.C.Tsai,M.R.Cagan and R.F.W.Pease „Silver diffusion in $Ag_2Se/GeSe_2$ inorganic resist system", J.Vac.Sci.Technol.,B4(1), 418-421, (1986).
8. P.J.S. Ewen and A.E.Owen " Photo-induced changes in chalcogenide glasses and their applications", In: "High-performance Glasses", M.Carble and J.M.Parker - Editors,"Blackie",Glasgow&London, Chapman and Hall,NY,USA, 287-337,(1992)..
9. A.V.Kolobov and S.R.Elliott „Photodoping of amorphous chalcogenides by metals „, Advances in Physics, 40(5), 625-684, (1991).
10. R.Vadimsky „Three-dimensional photolithography with conformal GeSe resist", J.Vac.Sci.Technol.,B6(6), 2221-2221, (1988).
11. R.Klabes,A.Thomas and G.Kluge "Utilization of $Ag:GeSe_2$ films as inorganic positive resist", Phys.Stat.Sol., (a)110 (2), K810K84, (1988).
12. B.Shpanenberg, V.Orlinov, K.Popova,V.Obukhov, A.Baranov, A.Sharenko,E.Spasova, G.Danev "Two-layer resistive system for laser formation of microstructures", Bulg.J.Phys. 16(5), 495-502, (1989).
13. 13. R.E.Belford,E.Hajto and A.E.Owen „The selective removal of the negative high-resolution photoresist system Ag-As-S", Thin Solid Films, 173(1), 129 - 137, (1989).
14. P.Guttmann, G.Danev, E.Spasova and S.Babin „Behaviour of amorphous semiconductor As_2S_3 layers after photon,electron or X-ray exposures", In Physical Concepts of Materials for Novel Optoelectronic Device Applications, Manijeh Razeghi - Editor, SPIE Proc. 1361, 999-1010, (1990).
15. Y.Utsugi „Chemical modification for nanolithography using scanning tunneling microscopy", Nanotechnology, 3 , 161-163, (1992).
16. I.Z.Indutnyi, S.A.Kostioukevitch, V.I.Minko, A.V.Stronski and P.E.Shepeljavi „Laser lithography on the As_2S_3 layers „, Optoelectronics and Semiconductor technique, Kiev, „Naukova Dumka", (25), 52-59, (1993).
17. P.E.Shepeljavi „Phototechnical characteristics of inorganic resists on the base of chalcogenide glasses", In : Abstracts of I Int.Conf. on Material Sci.of chalcogenide and diamond-structure semiconductors, Chernivtsi, Ukraine, 2, 213, (1994).
18. P.E.Shepeljavi, A.V.Stronski and I.Z.Indutnyi"Fabrication and properties of vacuum inorganic resists,"
In : Vacuum technique and vacuum technologies, Kiev, Ukraine, Editors - V.F.Zelensky, V.T.Cherepin, V.M.Shugaev, V.B.Yuferov, Proc. of Ukrainian Vacuum Society, 1, 324-327, (1995).
19. I.Shimizu,H.Sakuma,H.Kokado and E.Inoue "Metal-chalcogenide systems as imaging materials", Photographic Science and Engineering, 16 (4), 291-295, (1972).
20. M.T.Kostishin, P.F.Romanenko, N.G.Khotynenko"New photographic materials: lightsensitive semiconductor-metal systems with barrier layer ", J.Sci.Appl.Photogr.Sinematogr. 25(1), 14-18, (1980).
21. M.T.Kostishin , S.A.Kostioukevitch ," Behaviour of As_2S_3 - Ag lightsensitive system under exposure with impulse laser irradiation of high power density", Ukrainian Journal of Physics, 26 (9) , 1561 - 1563 , (1981).
22. I.Z.Indutnyi, S.A.Kostioukevitch, V.I.Minko,P.E.Shepeljavi and A.V.Stronski „Application of inorganic resists in high density information storage technologies", In „Advanced Image and Video Communications and Storage Technologies", N.Ohta, H.U.Lemke and J-C.Lehureau- Editors, SPIE Proc., 456-467, (1995).
23. I.Y.Yusupov, M.D.Mikhailov, R.R.Herke, L.I.Gorai, S.B.Mamedov, O.A.Yakovuk „Investigation of the arsenic sulphide films for relief-phase holograms", In"Three-Dimensional Holography: Science, Culture, Education", T.H.Leong, V.B.Markov- Editors, SPIE Proc. 1238, 240-247, (1989).
24. I.Z.Indutnyi,M.T.Kostishin,P.F.Romanenko and A.V.Stronski „Recording of holographic diffraction gratings on lightsensitive semiconductor-metal systems", J.Inf.Rec.Mater. 19(3), 239-248, (1991).
25. P.J.S.Ewen,A.Zekak,C.W.Sliuger, G.Dale, D.A.Pain and A.E.Owen „Diffractive optical elements in chalcogenide glasses", J.Non-Cryst.Sol., 164-166(Pt.II), 1247-1250, (1993).
26. A.V.Stronski,P.F.Romanenko,I.I.Robur,I.Z.Indutnyi,P.E.Shepeljavi „Recording of holographic optical elements on As-S-Se layers", J.Inf.Rec.Mater, 20(6), 541-546, (1993).

290

27. I.Z.Indutnyi,A.V.Stronski,S.A,Kostioukevitch,P.F.Romanenko,P.E.Shepeljavi ,I.I.Robur „Holographic optical element fabrication using chalcogenide layers", Optical Engineering, 34(4), 1030-1039, (1995).

28. H.Sakuma,I.Shimizu,H.Kokado and E.Inoue „Certain properties of metal-photodoped chalcogenide glasses", Supplement to Oyo Buturi, 41, 76-84, (1972).

29. Y.Somemura,A.Yoshikawa and Utsugi „Characteristics of Ag-Se/GeSe as Recording medium for X-ray Holograms", Jpn.J.Appl.Phys., 31PtI(11), 3712-3715, (1992).

30. I.I.Robur,P.F.Romanenko,A.V.Stronski,L.I.Kostrova,P.E.Shepeljavi,S.A.Kostioukevitch " Chalcogenide layers as holographic media ", In " Optics as a Key to High Technology ", Gy.Akos, T.Lippeny, G.Lupkovics, A.Podmaniczky- Editors, SPIE Proc.1983. 593-594, (1993).

31. R.Stoycheva-Topalova „Chalcogenide glasses as sensitizing layer for offset printing plates", J.Photogr. Sci., 31(2), 45-50, (1983).

32. D.I.Blezkan,V.S.Gerasimenko,I.M.Grankin,et.al. „Technology of inorganic photolithography on the base of semiconductor-metal systems", Ukrainian Journal of Physics, 26(1), 14-18, (1981).

33. M.Chang and J.T.Chen „A dry-etched inorganic resist",Appl.Phys.Lett,33(10),892-895,(1978).

34.A.Yoshikawa,O.Ochi,H.Nagai and Y.Mizushima „A new inorganic electron resist of high contrast", Appl.Phys.Lett., 31(3), 161-163, (1977).

35. K.Balasubramanyam,J.Adesida,A.L.Ruoff and E.D.Wolf „Germanium selenide as a inorganic resist for ion beam microfabrication", Microelectronics Journal, 14(1), 35-42, (1983).

36. A.Yoshikawa, O.Ochi, and Y.Mizushima "Dry development of Ge-Se inorganic photoresist", Appl. Phys. Lett., 36(1), 107-109, (1980).

37. A.Yoshikawa,S.Hirota, O.Ochi, et.al. „Angstroms resolution in Se-Ge inorganic photoresist", Jap.J. Appl. Phys., 20(2), 181-183, (1981).

38. K.D.Kolwicz and M.S.Chang „Silver Halide-Chalcogenide Glass inorganic resists for X-ray lithography", J.Electrochem.Soc., 127 (1), 135-138, (1980).

39. B.Singh, S.P.Beamont, G.Stewart and C.D.W.Wilkinson „Silver chloride/arsenic trisulphide as an electron beam resist", In: "Inorganic Resist Systems"Electrochem.Soc.Proc. 1982, 82-9, Pennington,N.J.,USA, 121-126, (1982).

40. R.B.Shafizade. A.M.Kasumov „Interaction of layers in multilayer structure As₂S₃-Ag", Izvestia of AN of Azerbaidzan SSR (phys.,techn. and math. sci.), 4 (5), 77-80, (1983).

41. E.A.Lupasko, V.V.Mussil, A.P.Ovcharenko „On the dynamics of photoinduced processes in the Ag-As₂S₃-mirror structure", Quantum Electronics, Kiev, (45), 104-108, (1993).

42. K.J.Polasko, R.F.Marinero, and Cagan M.R. „Excimer laser exposure of Ag₂Se/GeSe₂: high contrast effects", J.Vac.Sci.Technol., B3(1), 319-322, (1985).

43. P.G.Huggert, K.Frick and H.W.Lehmann "Development of silver sensitized germanium selenide photoresist by reactive sputter etching in SF₆ ", Appl.Phys.Lett., 42(1), 592-594, (1983).

44. K.L.Tai, R.G.Vadimski, C.T.Kemerem, et.al. „Submicron optical lithography using an inorganic resist/polymer bilevel scheme", J.Vac.Sci.and Technol., 17(5), 1169-1175, (1980).

45. K.L.Tai, E.Ong,R.G.Vadimski, „Inorganic resist systems for VLSI microlithography", In:"Inorganic Resist Systems", Electrochem.Soc.Proc.,Pennington,N.J.,USA, 82-9, 9-35, (1982).

46. B.Singh,S.P.Beamont, A.Webb, et.al. „High resolution patterning with Ag₂S/As₂S₃ inorganic electron-beam resist and reactive ion etching", J.Vac.Sci.Technol. B1(4), 1174-1177, (1983).

47. D.Goldschmidt and P.S.Rudman "The kinetics of the photodissolution of Ag into amorphous As₂S₃ films", J.Non.-Cryst. Solids, 22(2), 229-243, (1976).

48. B.X.Kudoyarova, T.D.Dzhafarov, M.D.Mikhailov, „Influence of photodoping on the optical properties of vitreous arsenic sulphide", Fizika i techniksa poluprovondikov (USSR), 13(10), 2040-2042, (1979).

49. M.T.Kostishin, Yu.V.Ushenin "He-Ne laser radiation stimulated diffusion of Ag into vitreous As₂S₃", Phys.Stat.Sol., 66(1), K47-K50, (1981).

50. R.Klabes, A.Thomas, G.Kluge, W.Beyer, R.Grotzschel, and P.Suptitz "Ion-beam induced silver doping in Ag₂Se/GeSe₂ - restst system",Phys.Stat.Sol., A106(1), 57-65, (1988).

51. M.T.Kostishin, P.F.Romanenko, N.G.Khotynenko „On the influence of pre-exposure on the lightsensitivity of As₂S₃-Ag system", In: „Methods of recording on non-silver media", Kiev, „Vysshaya shkola", (8), 155-160, (1977).

52. M.T.Kostishin,A.V.Stronski, Yu.V.Ushenin „Temperature dependencies of initial and diffusion parts of photoprocess in arsenic chalcogenide-silver systems", J.Sci.Appl.Photogr.Sinematogr. 29(6), 468-470, (1984).

53. E.Inoue" Photodoping of Ag into chalcogenide glasses and its application for imagics", Tokyo Institute of Technology, Japan, October 1, 1-69, (1981).

54. E.Inoue, H.Kokado, I.Shimizu „Photodoping of metal into chalcogenide glasses", Suppl. to J.Jap.Soc. Appl. Phys., 43(3/4), 101-105, (1974).

55. M.T.Kostishin, A.V.Stronski, Yu.V.Ushenin"On the kinetics of photostimulated interaction in ChVS-metal systems", In: Pap. of USSR Conf. „Vitreous semiconductors", Leningrad, „Nauka", 314-315, (1985).

56. W.Leung, N.W.Cheung, and A.Neureuther „Studies of Ag photodoping in Ge$_x$Se$_{1-x}$ glass using microlithography techniques", Appl.Phys.Lett., 46(5), 481-483, (1985).

57. P.Suptitz, A.Fisher "Lateral diffusion of photodoped silver in amorphous Ge$_{25}$Se$_{75}$ layers", Phys.Stat.Sol., A82(1), 157-161, (1984).

58. T.Wagner, M.Frumar, and L.Benes „Photoenhanced dissolution and diffusion of Ag in As$_2$S$_x$ layers", J.Non.-Cryst.Solids, 90 (1-3), (1987).

59. T.Wagner, M.Vlcek, V.Smrcka, P.J.S.Ewen and A.E.Owen „Kinetics and reaction products of the photoinduced solid state chemical reaction between silver and amorphous As$_{33}$S$_{67}$ layers", J.Non.-Cryst.Solids, 164-166(PtII), 1255-1258, (1993).

60. M.T.Kostishin, E.V.Mikhailovskaya,P.F.Romanenko „On the spectral distribution of photographic sensitivity of the layers of orpiment „ , J.Appl. Spectroscopy (USSR), VII(3), 410-413, (1967).

61.H.Kokado,I.Shimizu and E.Inou „Discusion on the mechanism of the photodoping „, J.Non-Cryst.Solids, 20(1), 131-139, (1976).

62. K.L.Chopra,L.K.Malhotra,K.Solomon, et.al. „Photon, electron and ion beam induced physical and optical densification in chalcogenide films", Bulletin of Material Science, 6(6), 1013-1018, (1984).

63. M.T.Kostishin, V.P.Zakharchuk "Reaction of some „semiconductor-metal systems on the electron beams of mean energies", J.Sci.Appl.Photogr.Sinematogr., 18(5), 347-349, (1973).

64. G.H.Berstein, W.P.liu,Y.N.Khawaja, M.N.Kozicki and D.K.Ferry, „High resolution electron beam lithography with negative and inorganic resists", J.Vac.Sci.Technol., B6(6), 2298-2307, (1988).

65. G.Kluge,R.Klabes,A.Thomas, G.Grotschel, and P.Suptitz „High energy ion-bean induced Ag diffusion within Ag:GeSe$_2$ " Phys.Stat.Sol., A110 (1), K9-K11, (1987).

66. Y.Utsugi and M.Kakushi „X-ray monitoring using a new soft x-ray plate: Ag/chalcogenide film", Rev.Sci.Instrum., 60(7), 2295-2298, (1989).

67. K.Saito,Y.Utsugi, and A.Yoshikawa „X-ray lithography with Ag-Se/Ge-Se inorganic resist using synchrotron radiation", J.Appl.Phys., 63 (2), 565-567, (1988).

68. M.T.Kostishin,P.F.Romanenko „Dependence of lightsensitivity of semiconductor-metal system from the thickness of semiconductor layer", Ukrainian Journal of Physics, 13(8), 1345-1349, (1968).

69. M.T.Kostishin,P.F.Romanenko,V.M.Sharyi, et.al. „Temperature dependencies of rate of photochemical transformations for As-S-Ag systems", Izvestia AN USSR, „Inorganic Materials", VI(6), 1073-1076, (1970).

70. A.Yamaguchi,I.Shimizu and E.Inoue „Diffusion of silver associated with photodoping into amorphous chalcogenides", J.Non.-Cryst.Solids, 47(3), 341-354, (1982).

71. A.V.Stronski,P.E.Shepeljavi, S.A.Kostioukevitch,P.F.Romanenko, I.I.Robur,I.Z.Indutnyi „Fabrication of Fresnel lenses and other optical elements with the help of inorganic resists", In „Nanofabrication Technologies and Device integration", W Karthe-Editor, SPIE Proc. 2213, 114-120, (1994).

72. P.E.Shepeljavi, S.A.Kostioukevitch, I.Z.Indutnyi A.V.Stronski, „Fabrication of periodical structures with the help of chalcogenide inorganic resists", In „Integrated Optics and Microstructure II", M.Tabib-Azar, D.L.Polla and Ka-Kha Wong- Editors, SPIE Proc. 2291,188-192, (1994).

73. T.Imura,K.Kubota, A.Hiraki and K.Tanaka „Photodoping of Ag into Single Crystal As$_2$S$_3$ " , J.Phys.Soc.Jap., 52(7), 2459-2461, (1983).

74. M.T.Kostishin,V.V.Kolomietz, O.P.Kasyarum, „Influence of semiconductor ordering on the lightsensitivity of semiconductor-metal systems", Ukrainian Journal of Physics, 30(6), 916-917, (1985).

75. T.Yaii and S.Kurita „Photodoping sensitivity of Ag into amorphous As$_2$S$_3$ films", J.Appl.Phys., 16(2), 389-390, (1977).

76. K.Chatani,I.Shimizu,H.Kokado and E.Inoue "Influence of the localized structural transformations of As$_2$S$_3$ on the photodoping rate of Ag", Jap.J.Appl.Phys., 16(2),389-390, (1977).

77. M.Kasai and T.Hajimoto „The photodoping sensitivity in As- and Ge-S films", J.Appl.Phys., 47(8), 3494-3596, (1976).

78. M.T.Kostishin,O.P.Kasyarum,A.A.Kudryavtsev "Influence of initial light exposure of the system As$_2$S$_3$-Ag on the character of photostimulated diffusion of Ag into As$_2$S$_3$", Ukrainian Journal of Physics, 29(1), 142-144, (1984).

79. M.T.Kostishin,P.F.Romanenko "On the influence of thermal treatment and aging on the temperature dependence of the rate of photochemical transformations in As$_2$S$_3$-Ag system", Ukrainian Journal of Physics, 17(2), 230-233, (1972).

80. M.T.Kostishin, V.I.Minko „Dependence of lightsensitivity of photochemical transformations of As$_2$S$_3$-Ag systems from the intensity of irradiation", Ukrainian Journal of Physics, 22(9), 1560-1562, (1977).

81. A.V.Kolobov, V.M.Ljubin „Photodissolution of zinc into vitreous As$_2$S$_3$ ", Sov.Phys.Solid.State, 26(8), 2522-2524, (1984).

82. W.Leung,N.Cheung and A.R.Neureuther „Photoinduced diffusion of Ag in Ge$_x$Se$_{1-x}$ glass", Appl.Phys.Lett., 46(6), 543-545, (1985).

292

83. K.Kolev,M.Radoeva " Aging of photomaterial for the fabrication of flexible printing schemes", News on chemistry of Bulgarian Acad.Sci., 12(3), 450-456, (1979).

84. E.Inoue,H.Yasujima and H.Kokado " Imaging in ternary chalcogenide sensors by photo- and thermal doping", Photogr.Sci. and Eng., 21(3), 142-145, (1977).

85. M.T.Kostishin,P.F.Romanenko,N.G.Khotynenko „Influence of barrier layer on the chemical and photochemical transformations in As$_2$Se$_3$-Ag system", In: Collection "Fundamental base of optical memory and media", Kiev, „Vysshaya shkola", (10), 55-61, (1979).

86. M.T.Kostishin,S.A.Kostioukevitch, P.E.Shepeljavi „Influence of semiconductor deposition temperature on the behaviour of lightsensitive As$_2$S$_3$-Ag system", Ukrainian Journal of Physics, 29(4), 1417-1419, (1984).

87. S.A.Kostioukevitch, V.L.Gromashevski, N.V.Sopinski,I.Z.Indutnyi, A.V.Stronski „Lightsensitive acoustic absorption in LiNbO3- (As$_2$S$_3$-Ag) layer structure", Journal of Technical Physics, 54(6), 1231-1233, (1984).

88. S.A.Kostioukevitch, P.E.Shepeljavi, A.V.Stronski, I.Z.Indutnyi „Investigations and modelling of physical processes in high density information recording with the help of inorganic resists", In Holography and Correlation Optics", Oleg V.Angelski -Editor, SPIE Proc. 2647, 166-173, (1995).

89. E.Inoue, H.Kokado, and I.Shimizu „Photodoping of metal into chalcogenide glasses", In „Non-Silver Photogr. Processes, Proc.Symp.,Oxford, 1973, London etc., 1975, 71-96.

90. M.Marcus, A.Wagner „Optical monitoring of development kinetics of GeSe$_2$", In: „Inorganic Resist Systems", Editors M.Doane and A.Heller, Pennington,N.J.,USA, Electrochem.Soc.Proc., 82-9, 295-302, (1982).

91. I.Z.Indutnyi,S.A.Kostioukevitch, V.I.Minko, P.E.Shepeljavi „Spectral and resistive properties of lightsensitive structure As$_4$Ge$_{30}$Se$_{66}$-Ag", In: Pap. of All-USSR Conf. „Structural transformations and relaxation phenomena in noncrystalline solid states", Ukraine, Lvov, ZNTI, 86, (1990).

92. G.Kluge, A.Thomas, R.Klabes and P.Suptitz „Applicability of silver implanted amorphous GeSe$_2$ as inorganic negative resist", Phys.Stat.Sol., A113 (2), K171-175, (1989).

93. M.Frumar, M.Cvrkal,M.Vlcek and T.Wagner „The photostructural changes and reactivity of chalcogenide layers", J.Non.-Cryst.Solids, 164-166(Pt.II), 1243-1246, (1993).

94. S.Mamedov „On the macromolecular mechanism of dissolution of As$_2$S$_3$ films in organic solutions", Thin Solid Films, 226(2), 215-218, (1993).

95. J.Dikova, N.Starbov, K.Starbova „The mechanism of photoinduced transformations in amorphous As$_2$S$_3$ films", J.Non-Cryst.Solids, 167, 50-58, (1994).

96. K.I.Pinzenik, N.P.Frolova, I.I.Turyanitsa „Photoiduced changes in layers of chalcogenide vitreous semiconductors and the parameters of selective etching", Advances in Scientific Photography, 26, 50-52, (1990).

97. R.R.Gerke,S.B.Mamedov,M.D.Mikhailov, I.Yu.Yusupov, O.A.Yakovuk „Properties of holographic gratings on the arsenic sulphide films", Ibid., 52-55, (1990).

98. N.A.Alimbarashvili, L.I.Bekicheva,G.G.Dekanozoshvili,I.A.Eguliashvili, K.L.Mosulishvili, E.M.Shekhter "Formation of phase-relief holographic gratings on the As$_{40}$S$_{37}$Se$_{23}$ layers", Ibid., 55-57, (1990).

99. P.K.Gupta and K.L.Chopra „Plasma processed obliquely deposited Bi-Ge-Se and Ag/Bi-Ge-Se films as resist materials", Appl.Phys. A46(2), 103-106, (1988).

100.I.Z.Indutnyi,S.A.Kostioukevitch, P.F.Romanenko,A.V.Stronski and P.E.Shepeljavi "Possibilities of inorganic resists on the base of chalcogenide semiconductor layers ", In: Pap. of the All-USSR Conf. on Photoresists, Zvenogorod, 89, (1990).

101. A.V.Baranov,V.E.Obukhov,A.I.Sharendo, B.Shpanenberg „The two-layer resistive As$_2$S$_3$-polyimide system for submicron lithography ", Ibid., 90, (1990).

102. V.P.Khan,I.V.Fedotova, V.A.Kogai, A.S.Alashkin "Structural transformations and plasma-chemical etching of photosensitive chalcogenide glasses ", News of Acad.Sci. of USSR, "Inorganic materials", 27(2), 392-397, (1991).

103.I.Shimizu,H.Sakuma,H.Kokado and E.Inoue "Metal-chalcogenide systems as imaging materials", Photogr.Sci.and Eng., 16(4), 291-295, (1972).

104.V.A.Danko,I.Z.Indutnyi,A.A.Kudryavtsev, and V.I.Minko "Photodoping in the As$_2$S$_3$-Ag thin-film structure", Phys.Stat.Sol. A124(1),235-242, (1991).

105. .V.A.Danko,I.Z.Indutnyi,A.A.Kudryavtsev, V.I.Minko and A.I.Stetsun "Electrical and optical properties of thin As$_2$S$_3$ layers photodoped by silver",Ukrainian Journal of Physics, 36(6), 937-943, (1991).

106. V.A.Danko,I.Z.Indutnyi,V.Yu.Kulikovsky, and V.I.Minko " Influence of annealing on the characteristics of conductivity and diffusion masstransfer in As$_2$S$_3$ layers photodoped by silver", Physics and Chemistry of Glasses", 17(1), 148-153, (1992).

107. I.Z.Indutnyi "Photostimulated diffusion in thin-film light-sensitive semiconductor-metal structures", J.Sci.Appl.Photogr.Sinematogr., 39(6), 65-77, (1994).

108. M.J.Mitkova and P.Fallman "Some properties of silver photodiffused chalcogenide glasses", Phys.Stat.Sol. A75(2), K145-K148, (1983).

109. E.Marquez,J.B.Ramirez-Malo, J.Fernandez-Pena, P.Villares,R.Jimenez-Garay, P.J.S.Ewen and A.E.Owen "On the influence of Ag-photodoped on the optical properties of As-S glass films", J.Non.-Cryst.Sol. 164-166(PtII), 1223-1226, (1993).

110. P.E.Aspnes, J.Philips and K.L.Tai "Optical spectra and electron structure of crystalline and glassy Ge(S,Se)$_2$ ", Phys.Rew.,B22(2), 816-822, (1981).

111. I.Z.Indutnyi, A.P.Stetsun "Spectra of the interband optical transitions of As$_2$S$_3$ layers, photodoped by silver" , Optics and Spectroscopy, 71(1), 83-87, (1991).

112. I.Z.Indutnyi, A.P.Stetsun,V.I.Zimenko,V.G.Kravetz " Optical properties of the photodoped by silver GeS$_2$ layers", Ibid., 75(6), 1262-1266, (1993).

113. Y.Utsugi " Nanometre-scale chemical modification using a scanning tunneling microscope", Nature, 347(6295), 747-749, (1990).

114. Y.Utsugi " Scanning tunneling spectroscopy of nanofeatures on silver-selenide surface", Jpn.J.Appl.Phys., 32(PtI,N6B), 2969-2972, (1993).

115. H.Kado and T.Tonda " Nanometer-scale recording on chalcogenide films with an atomic force microscope", Appl.Phys.Lett., 66(22), 2961-2962, (1995).

116. K.L.Tai,W.R.Sinclair, R.G.Vadimski and J.Moran " Bilevel high resolution photolithographic technique for use with wafers with stepped an/or reflecting sufraces", J.Vac. Sci. Technol., 16(6), 1977-1979, (1979).

117. B.Singh, S.P.Beamont, A.Webb,et.al. " High resolution patterning with Ag$_2$S/As$_2$S$_3$ inorganic electron-beam resist and reactive ion etching", J.Vac.Sci.Technol., B1(4), 1174-1177, (1983).

118. A.Matsuda and M.Kikuchi " Observations of fundamental processes in Ag photodoping of amorphous As$_2$S$_3$ films", Suppl.J.Jap.Soc.Appl.Phys., 42, 239-248, (1973).

119. R.Ishikawa and M.Kikuchi " Photovoltaic study on the photoenhanced diffusion of Ag in amorphous films of Ge$_2$S$_3$", J.Non.-Cryst. Solids, 35&36 (1), 1061-1066, (1980).

120. T.Suzuki, Y.Hirose and H.Hirose " Modelling of photodoping mechanism in Ag/As$_2$S$_3$ system through ESCA analysis", In: " Inorganic Resist Systems ", Electrochem.Soc.Proc., Pennington, N.J., USA, 82-9, 255-264, (1982).

121. S.T.Lakshmikumar " A new model for photodissolution of silver in amorphous chalcogenides", J.Non.-Cryst.Sol.,88(2/3), 196-205, (1986).

122. S.A.Lis, J.M.Lavine " Ag photodoping of amorphous chalcogenides ", Appl.Phys.Lett., 42(8), 675-677, (1983).

123. I.Z.Indutnyi,A.V.Stronski,S.A.Kostioukevitch, P.F.Romanenko,P.E.Shepeljavi,I.I.Robur "Photostimulated processes in chalcogenide vitreous semiconductors and production of holographic optical elements on their base", In: Pap. of Int. Workshop on Advanced Technol. of Multicomponent Solid Films and Structures","Patent", Uzhgorod, Ukraine, 50-51, (1994).

124. N.Yoshida and K.Tanaka "Photoinduced Ag migration in Ag-As-S glasses", J.Appl.Phys., 78(3), 1745-1750, (1995).

125. K.S.Harshavardan, K.N.Krishna and K.J.Rao "Investigations of surface activity and photoinduced diffusion of metal in solution deposited amorphous films of As$_2$S$_3$", J.Mater.Sci., 20(9), 3253-3259, (1985).

126. J.Lauks, G.C.Chern and K.Y.Toh " Novel inorganic resist systems", In: " Inorganic Resist Systems", Electrochem.Soc.Proc. ,Pennington, N.J.,USA, 82-9, 93-99, (1982).

127. "Plasma technology in VLSI fabrication", N.G.Einsprich and D.M.Brown Editors, Moscow, "Mir", 136, (1987).

Subject Index

X.

Y.
Yields   4.2

Z.

The manufacturer's authorised representative in the EU is Springer
Nature Customer Service Centre GmbH, Europaplatz 3, 69115 Heidelberg,
Germany. If you have any concerns regarding our products, please
contact ProductSafety@springernature.com

Printed and bound by CPI Group (UK) Ltd, Croydon, CR0 4YY
23/04/2026
02095622-0001